ANALYSIS OF WATER RESOURCE SYSTEMS

by
LADISLAV VOTRUBA
Faculty of Civil Engineering of the Technical University,
Prague, Czechoslovakia

Co-authors:
Zdeněk Kos
Karel Nacházel
Adolf Patera
Faculty of Civil Engineering of the Technical University,
Prague, Czechoslovakia
Václav Zeman
Water Research Institute, Prague, Czechoslovakia

ELSEVIER

Amsterdam – Oxford – New York – Tokyo 1988

Published in co-edition with SNTL – Publishers of Technical Literature, Prague

Distribution of this book is being handled by the following publishers

for the U.S.A. and Canada
ELSEVIER SCIENCE PUBLISHING COMPANY, INC.
52 Vanderbilt Avenue, New York, N. Y. 10017

for the East European Countries, Chinese People's Republic, Cuba, Korean People's Democratic Republic, Mongolia and Vietnam
SNTL – Publishers of Technical Literature
Spálená 51, 113 02 Praha 1, Czechoslovakia

for all remaining areas
ELSEVIER SCIENCE PUBLISHERS
Sara Burgerhartstraat 25
P.O. Box 211
1000 AE Amsterdam, The Netherlands

Library of Congress Cataloging-in-Publication Data

Votruba, Ladislav.
Analysis of water resource systems.
(Developments in water science; 32)
Translation of: Vodohospodářské soustavy.
Includes bibliography and index.
1. Water-supply engineering – Data processing.
2. System theory. I. Title. II. Series.
TD353.V6813 1988 628.1 87-22150

ISBN 0-444-98944-7 (Vol. 32)
ISBN 0-444-41699-2 (Series)

With 99 illustrations and 61 tables

Printed in Czechoslovakia

DEVELOPMENTS IN WATER SCIENCE, 32

OTHER TITLES IN THIS SERIES

1 G. BUGLIARELLO AND F. GUNTER
COMPUTER SYSTEMS AND WATER RESOURCES

2 H. L. GOLTERMAN
PHYSIOLOGICAL LIMNOLOGY

3 Y. Y. HAIMES, W. A. HALL AND H. T. FREEDMAN
MULTIOBJECTIVE OPTIMIZATION IN WATER RESOURCES SYSTEMS:
THE SURROGATE WORTH TRADE-OFF-METHOD

4 J. J. FRIED
GROUNDWATER POLLUTION

5 N. RAJARATNAM
TURBULENT JETS

6 D. STEPHENSON
PIPELINE DESIGN FOR WATER ENGINEERS

7 V. HÁLEK AND J. ŠVEC
GROUNDWATER HYDRAULICS

8 J. BALEK
HYDROLOGY AND WATER RESOURCES IN TROPICAL AFRICA

9 T. A. McMAHON AND R. G. MEIN
RESERVOIR CAPACITY AND YIELD

10 G. KOVÁCS
SEEPAGE HYDRAULICS

11 W. H. GRAF AND W. C. MORTIMER (EDITORS)
HYDRODYNAMICS OF LAKES: PROCEEDINGS OF A SYMPOSIUM
12 – 13 OCTOBER 1978, LAUSANNE, SWITZERLAND

12 W. BACK AND D. A. STEPHENSON (EDITORS)
CONTEMPORARY HYDROGEOLOGY: THE GEORGE BURKE MAXEY MEMORIAL VOLUME

13 M. A. MARIÑO AND J. N. LUTHIN
SEEPAGE AND GROUNDWATER

14 D. STEPHENSON
STORMWATER HYDROLOGY AND DRAINAGE

15 D. STEPHENSON
PIPELINE DESIGN FOR WATER ENGINEERS
(completely revised edition of Vol. 6 in this series)

16 W. BACK AND R. LETOLLE (EDITORS)
SYMPOSIUM ON GEOCHEMISTRY OF GROUNDWATER

17 A. H. EL-SHAARAWI (EDITOR) IN COLLABORATION WITH S. R. ESTERBY
TIME SERIES METHODS IN HYDROSCIENCES

18 J. BALEK
HYDROLOGY AND WATER RESOURCES IN TROPICAL REGIONS

19 D. STEPHENSON
PIPEFLOW ANALYSIS

20 I. ZAVOIANU
MORPHOMETRY OF DRAINAGE BASINS

21 M. M. A. SHAHIN
HYDROLOGY OF THE NILE BASIN

22 H. C. RIGGS
STREAMFLOW CHARACTERISTICS

23 M. NEGULESCU
MUNICIPAL WASTE WATER TREATMENT

24 L. G. EVERETT
GROUNDWATER MONITORING HANDBOOK FOR COAL
AND OIL SHALE DEVELOPMENT

25 W. KINZELBACH
GROUNDWATER MODELLING

26 D. STEPHENSON AND M. E. MEADOWS
KINEMATIC HYDROLOGY AND MODELLING

27 A. M. EL-SHAARAWI AND R. E. KWIATKOWSKI (EDITORS)
STATISTICAL ASPECTS OF WATER QUALITY MONITORING

28 M. JERMAR
WATER RESOURCES AND WATER MANAGEMENT

29 G. W. ANNANDALE
RESERVOIR SEDIMENTATION

30 D. CLARK
MICROCOMPUTER PROGRAMS IN GROUNDWATER

31 R. H. FRENCH
HYDRAULIC PROCESSES ON ALLUVIAL FANS

32 L. VOTRUBA, Z. KOS, K. NACHAZEL, A. PATERA AND V. ZEMAN
ANALYSIS OF WATER RESOURCE SYSTEMS

CONTENTS

PREFACE

The process of integration, in the broadest sense, has been a characteristic feature of the development of politics, economics and specific economic and technological fields in the last thirty years. The greater the necessity for relationships between functionally interdependent elements, and the more easily these relationships can be realized, the sooner have systems been formed from these elements. The large-scale systems developed in economics, energetics, transportation, and, since the sixties, water management, require a qualitatively new approach and treatment.

The modern engineer, now and in the future, has to supplement individual reasoning and creative work with a scientific approach to engineering tasks. Basically, the research-scientist investigates and explains existing natural phenomena and the relationships between them; the engineer creates something which is new and which will become useful. Their activities mutually influence each other. In earlier times engineering was more closely related to art, since engineering and art have the element of creation in common; now engineering is closer to science.

In accordance with social needs, the research scientist is trained to demonstrate theoretically the possibility of constructing a technical creations as a new, not yet existing, reality and to bring it into existence. The classification of creative work in applied research and development as scientific activity can be adopted without hesitation. The relationship science–development–implementation conceives of science as an integral part of production.

If issues of water resource systems are to be treated at the contemporary level of scientific knowledge, the appropriate, related scientific fields, such as systems analysis, the probability theory, operations research etc. need to be understood.

The aim of this book, which deals with water management, is to present the basic facts on complex water resource systems using the systems approach. Since no final interpretation of the basic terms and concepts is available, they have been defined in relation to another application of systems science, namely economics.

The fundamentals of the new, related scientific disciplines, the results of which are used in complex water resource systems, have been included in the book to facilitate reading and reduce the number of necessary reference books.The references included in this book are intended for further study. The examples serve to show the application of the general theory. Knowledge of the elements of theory of probability, mathematical statistics, computer programming, hydrology and water management has been assumed.

The book concentrates mainly on the problems of effective water supply and water resource conservation in water resource systems. The issue of water quality in watercourses and the special issues of waterworks, sewerage, navigation and hydroelectric power generation have not been analysed in detail apart from the aspect of water resource management.

More attention should be paid to the subsystem of flood control, its relationship to the environment and its technical and economic relationships to water supply-demand integration. A reference for more detailed investigation is given.

In the branches of science which deal with systems many issues and questions have to be answered. This is also true of those disciplines where the systems approach and the application of systems sciences were first introduced, e.g. automatic control and economics; it naturally applies to the problems of water management, as proved, for example, by five international symposia – "Water Resource Systems" held in Czechoslovakia, first in Karlovy Vary, 1972, and the last in Znojmo, 1987.

The book sets out to offer the reader a set of principles and methods for dealing with water resource systems on a scientific basis. It is hoped that, along with the basic facts, it will provide an impetus to the further development of promising progressive methods of dealing with the problems of water management by means of a systems approach.

The theme of this book won a place in a competition held by the Czech Society of Engineers and the Publishers of Technical Literature. The authors are indebted to these institutes for the publication of this book.

The authors wish to express their special thanks to the following institutions: Water Resources Development and Construction, Technical University of Prague, the Faculty of Civil Engineering; Hydroproject, Prague, the Committee for Water Management of the Czechoslovak Academy of Sciences, and to individual members of these institutions for their helpful suggestions. The authors wish to acknowledge their gratitude to many colleagues, reviewers and students who have contributed to the completion of this book.

The Authors

INTRODUCTION

The characteristic nature of water management requires a synergetic line of reasoning. The effects of human activities on water resources have great consequences for water users. The natural and social relationships are mutually dependent and are co-existent in place and time. The behaviour of resources and demands is frequently stochastic and non-stationary.

In addition to economic objectives, "intangibles" are increasingly involved in the evaluation of alternative procedures. The recent development of this trend is common to other world problems and intangible objectives are now given the same weight as economic effectiveness, or are afforded an even higher priority, which can influence the direction of optimization methods. However, caution is necessary, as the methods of optimization have become very sophisticated, and the results may be misleading or arbitrary.

The extensive world systems, water resource systems included, require new methods of treatment common to a number of scientific disciplines.

Probability and systems approaches are involved in the striking contemporary tendencies in scientific investigation and knowledge. It can be taken for granted that these tendencies will persist in the future development of scientific methods of cognition.

The natural, technical, and social sciences such as physics, astronomy, cybernetics, biology, economics, psychology, etc. include *probability conceptions*. The process of development and implementation of probability conceptions and methods has continued for about 300 years from the time of Pascal and Fermat. However, probability has long been considered a measure of lack of knowledge and not an objective property of phenomena and processes.

The penetration of probability and statistical methods of investigation into different areas of knowledge proceeded hand-in-hand with their development and perfection. The probability theories and ideas have been used by such recently developed branches of mathematics as information theory, game theory, operations research, the theory of reliability, etc. Max Born declared probability to be a fundamental physical concept.

Social and population statistics use probability and statistical methods for the investigation of the quantitative laws of human society. Statistical laws are, without hesitation, considered natural laws.

For cybernetics, which investigates the relationships of complex dynamic systems, the concept of probability is fundamental to the understanding of its principles and associated ideas (information, organization, control, etc.). The deterministic approach is not adequate here and can be misleading; the correct alternative is the probability approach.

The *systems approach* assumes the holistic investigation of objects as a whole but also their separation into subsystems and elements, which facilitates the investigation of the structure, organization and functional behaviour of an object.

The object investigated is defined as a system; it is treated as a *whole* and its parts are not considered independently. Moreover, with respect to the behaviour of the object, the systems approach is interested in the total activity of the system even if change is considered only in a single or in a few parts.

As in other areas of scientific research, ancient Greek natural philosophy raised the questions inherent in the systems approach to the investigation of the world. The followers of Pythagoras, with their principle of the harmonic wholeness of the universe as one of the fundamental results of human reasoning, contributed to our understanding of the world (Volkov, 1975). According to van der Waerden, the new and characteristic feature of Greek mathematics was the systems approach, using a sequence of proofs of theorems. Dialectics, inherent in the systems approach, is one of the most important gifts of the Greek philosophers to mankind. They comprehended the world as the reflection of an infinite labyrinth of interrelationships and effects, where nothing remained what it was, where it was or how it was, but where all was in movement, changing, being created, rising and declining.

The systems approach can be defined as a comprehensive method of investigation of phenomena and processes, including their internal and external relationships. It thus meets the basic condition of the dialectical method which postulates that consideration of interrelationships is a necessary condition of correct cognition. Similarly, therefore, water resource systems cannot be identified and investigated without considering the internal relationships between their elements and their relationship to the system's environment, even if the separate elements and environment are known very well.

The systems approach can be used as a methodological tool in every branch of science; it is not the basis of independent scientific research with its own subject and method of investigation.

It is characteristic of the new scientific fields mentioned that the systems approach is closely related to the probability ideas concerning the structure and behaviour of complex dynamic systems. The principle of the structure and behaviour of complex systems having a probability character, it is evident that in their investigation both the methods mentioned have to be applied *simultaneously*, viz. the probability and the systems approaches, although the probability concepts were developed independently of the systems ideas. These relationships involve the need to analyse the concept of probability and the nature of probability relationships on the basis of the systems ideas.

In spite of the fundamental importance of the probability approach, there are, even now, some tasks where the deterministic approach is adequate and, on the contrary, some tasks where the use of the probability theory alone is inadequate. In problems of water resource systems there exist, together with the probability phenomena, certain (or almost certain) phenomena that can be expressed only qualitatively, or that are only partly recognised, if at all. According to the nature of these phenomena, probability or deterministic methods, but also heuristic and other methods or combinations of these, are applied in the investigation and optimization of the structure and behaviour of water resource systems.

However, the systems approach can never be disregarded in water resource treatment as it is the fundamental and most general standpoint. This view does not eliminate simplification of systems intended as a methodological instrument in their investigation.

New ways of planning, designing and operating extensive systems based on these ideas date from approximately 1960. The possibility of their practical application was related to the development of digital-computer hardware and software. The combination of the new scientific approaches and the use of computers constituted one of the great qualitative methodological achievements of our age. However, complete implementation requires more work to be done. As in every new, rapidly developing sphere, there is much uncertainty concerning the possibilities and manner of its application and further development – the new science has been established, is expanding, and the limits of its development cannot be foreseen.

1 SYSTEMS SCIENCE AND ITS DISCIPLINES

1.1 THE CONCEPT OF SYSTEMS SCIENCE

Systems science is a new scientific discipline the subject of which is systems, and the methodology of which is the systems approach to problems. These systems are sets of interrelated elements, which may be defined with reference to natural or social subjects, may be physical or informational, concrete or abstract, existing or planned, static or dynamic, etc.

The methods of systems science are used to define systems, to distinguish them from the environment, represent them, analyse and optimize their structure and behaviour. To these ends, the methodology of new scientific fields and modern technical media are applied. Therefore, systems science is interdisciplinary and consists of the systems theory and its application.

The subjects and the methods of the various systems sciences (cybernetics, operations research, etc.) are typically so characteristic that they are often classified as individual fields of science.

The classification and terminology of systems science disciplines has been developed. Taking the aspects of water resource systems (WRS) as a technical and economic entity, we may consider the following classification:

– general systems theory,
– cybernetics,
– systems engineering,
– operations research,
– systems analysis.

The first and second fields may be classified as parts of the systems theory, the others as applications. There is no precise boundary between the subjects of these fields. Each is further categorized and uses a number of methods that have been developed into individual fields of research.

1.2 CONTENTS OF SYSTEMS SCIENCE

1.2.1 General Systems Theory

The general systems theory deals with the most general scientific principles of the existence, description and behaviour of systems. After the initial work of von Bertalanffy, 1934, 1950, 1956, it was further developed by Boulding, 1956; Ackoff, 1963;

Churchman, 1963; Mesarovic, 1963, 1964; Rapoport, 1963, 1966; and Bennis, 1962. The general systems theory includes the following subjects:

– the general systems terminology and metalanguages for the description of systems,
– the conditions of the system's existence, and its definition,
– the principle of isomorphism between systems,
– the goals and behaviour of systems,
– a comparison of methodology in natural and social sciences,
– the systems approach in physics, biology, linguistics, sociology.

Reseachers approach the theory of systems so differently that distinct schools have evolved. Von Bertalanffy's approach is quite general and philosophical, it is not confined to the formalization of systems and it links the general systems theory with its applications.

On the other hand, Mesarovic, 1963, uses a formalized approach to the systems theory, and he concentrates on abstract systems (conceptual systems) that can serve as models for the whole group of concrete (real) systems, mainly technical ones, with regard to the method of formalization.

A formalized approach to the systems theory was used by Churchman, 1963, who concentrated on axiomatic construction of the general systems theory and applied it to systems optimization.

In contrast to von Bertalanffy, Ackoff, 1963, attached little importance to the investigation of isomorphisms between laws in different fields as a method of phenomena comprehension and identification. He also rejected the investigation of problems by the application of aspects of several fields as an assumption of the subsequent generalization. On the contrary, he suggested that the problem be treated as a whole and that the simpler properties be derived from the complex ones by abstraction.

The application of the ideas of Bennis, 1962, on the organization and operation of economic systems in WRS was a stimulating achievement. Bennis suggested the *multi-purpose criteria* as an index of system quality, since the system often has several purposes (e.g. the multi-purpose WRS). The systems that achieve the objectives to a required degree are considered sound systems. This method of evaluation of complex systems is promising if the required degree of achievement of the individual objectives can be defined.

According to Bennis, the *most essential feature* of a system is its ability to learn and to adapt to changes in itself and in its environment. A system which lacks this property cannot work well in variable (dynamic) conditions. Therefore, a sound system (WRS included) can monitor itself and its environment, its subsystems are governed by the objectives of the whole system, and it can adapt to internal or external changes.

Philosophical aspects are integral elements of the general systems theory. Philosophy, which uses the generalisation of the results of concrete scientific inquiry as the main source of its development, inevitably deals with concepts such as "system", "structure", etc. Different sciences (biology, mathematics, logic, etc.) express these concepts in different terms, and philosophy tries to formulate their common essence without favouring one approach or excluding others.

From a philosophical point of view, there are two categories of systems distinguished by their different gnostic essences (Kravets, 1970), viz., real (material) and conceptual (ideal) systems. Real (material) systems are perceived or inferred from observation and exist independently of an observer. Examples of these systems are the atom, a living being, a city, a WRS, i.e. not only natural objects but also systems created by man which exist, however, not only in his mind but in reality. Conceptual (ideal) systems are symbolic structures and exist only in the mind; examples are mathematical models, axiomatic theories, the simulation water supply model of WRS, etc.

In principle, there are profound interrelationships between these two categories of systems, deduced from the fact that conceptual systems are derived, generalized reflections of material systems in the shape of models, theories, concepts and the ideas of man.

Material and the corresponding conceptual (abstract) systems are conceptually different, but their functional relationships must be the same. This relationship (analogy) is called isomorphism and the corresponding conceptual system is an isomorphic reflection of the material (real) system.

Even if the ideal reflection is not attained in the behaviour of the two categories of systems (the relationship is then called homomorphic) the differences must not exceed a certain degree, so that in the transition of the results from the conceptual system to the real one the limits of the required accuracy are not exceeded.

1.2.2 Cybernetics

Cybernetics was defined by Wiener, 1948, in the title of his book as the science of control and communication in the animal and the machine. A further definition of the concept was given by Kolmogorov, 1956. He considers cybernetics to be a science which deals with systems of any nature that are able to input, store and process information and use it for control. Cybernetics may be defined as the science dealing with behaviour and control of complex systems. As a science, cybernetics has its own subject of investigation, theory and methods.

Dynamic systems are the *subject* of investigation of cybernetics. They are studied from the standpoint of exchange of information between them and their environment, and their structure is studied from the standpoint of the exchange of information between their elements (Klir and Valach, 1965). Since man and the machine may both be elements of the systems and information may be communicated in both directions, i.e. between people, between man and the machine and between machines, the subject of cybernetics comprises the process of information transfer and processing in automatic control systems.

Wiener's statement that the processes of communication and control in living organisms and in automatic control systems are *analogous* (not identical) is of primary significance. Therefore, both living organisms (people) and machines can contribute to the process of increasing information or to the reduction of entropy. Cybernetics is not interested in the exchange of material or the exchange of energy but only in the exchange of information both between the system and its environment and between the elements of the same system; the point is the *information approach* to the system.

The many-sided analogy between living organisms and machines in communication and the utilization of information is very important for theoretical applications, as it can be used for the improvement and perfection of technological systems and mechanisms.

Other terms that are defined include adaptivity and learning ability of the system. The living organism is able to improve itself using experience and knowledge gained. By analogy, cybernetics has raised the question: could a machine facility with a similar ability be constructed, which would mean a qualitative step in the advance of automation? Machine selfreproduction has been investigated and automata which would be able to produce more complex automata are under consideration.

The *methods of cybernetics* are related to the subject of cybernetics; they comprise:

- the theory of communication,
- information systems,
- the theory of adaptive control and learning systems,
- the general systems theory,
- analogy,
- modelling.

In systems, not the elements but their relationships are investigated by cybernetics; the question is *what the system performs*, and not *what the system is*. The response of the system to its stimuli, i.e. its behaviour, is what matters, and in this respect it has much in common with the general systems theory.

According to Ashby, 1956, cybernetics deals with all forms of behaviour, if these forms are regular, defined and reproducible. The advantages of cybernetics manifest themselves when it is applied to the behaviour of systems which are distinguished by their great complexity.

Ashby paid particular attention to those systems that are not fully accessible for direct observation, and he introduced the term "black box", derived from electronics, into cybernetics and into scientific methods generally. In the black box investigation the input and output states are the basic data. The way the inputs of the system (black box) are transformed into system output is of no interest.

Hýbl and Škabrada, 1968, considered the future utilization of cybernetics in economic and social systems. They state that in economic and social systems, man as a component is a cybernetic system with purposeful behaviour. If such a system is to function well, the different goals (interests) must be coordinated with the objectives of the whole system (see Bennis's ideas). Together with economics, scientific analysis of human behaviour (psychology, sociology) can help to do this.

Although cybernetics has developed mainly in technical systems, its achievements, general terms and methodology are also used in other complex systems.

1.2.3 Systems Engineering

Systems engineering is an applied systems discipline concerned with technical systems. Views differ both in regard to the subject of systems engineering and the methods used.

With respect to the application of systems engineering in complex WRS, the following definition seems to be adequate.

Systems engineering is an applied systems discipline concerned with the design, construction and operation of large-scale and complex technical and economic systems, using the systems approach and suitable methods of systems science to attain the optimal functioning of the system as a whole. A number of authors have put forward different definitions.

Hall, 1962, described the aims of system engineering from the standpoint of a firm: to provide information for the management for economic development, to formulate long-term objectives as a framework of the individual projects, by balancing the programs to control development in order to attain optimal use of manpower and resources, to subordinate minor projects to the overall objectives, to predict future situations and prepare the prerequisites for them, to be informed about new inventions and to secure their utilization, and to specify all operations, in great detail and with accuracy with respect to each stage of the process.

Goode and Machol, 1957, defined systems engineering as a set of problems related to the design of complex, highly automated, technical, large-scale systems.

Habr, 1970, wrote about systems engineering as an application of the systems approach in technical fields. In systems engineering the main stress is placed on the development of the concepts, and the design of a system and its implementation. Therefore, the term "system design" is encountered as an equivalent to "systems engineering". In systems engineering both the terms "optimum design" and "optimum conception" are used. They mean a design of the system such that the requirements concerning its function (objective function) are satisfied automatically or with a minimum of human intervention.

Dráb, 1973, wrote that the set of means, facilities, procedures and methods that make it possible to satisfy the accepted criteria in research, development, design, installation, production, verfication and operation constitute systems engineering.

Beneš, 1974, defined and explained systems engineering as a technique of large-scale systems concerned with design, issues of technical conception, projection, construction and operation of large-scale systems consisting mainly of non-living elements. It deals with the choice of subsystems and parts of large-scale systems so as to maximize their contribution to the required properties of the large-scale system as a whole; it is concerned with the creative design of a large-scale system, and the technical means to implement this design. In the design of large-scale systems, systems engineering is interested in technical, organizational and economic aspects.

Although comprehensive automation is one of the very important indices of technical development, in the evaluation of systems we shall direct our attention to completeness and comprehensiveness rather than to full automation. The comprehensive approach, especially, involves relatively high costs of the projects and their realization. Systems engineering, then, is used for the technical management of large projects.

The systems engineer has to be acquainted with the problems of complex systems, he needs to know how to manage a team in order to attain the objectives at the minimum cost. For this aim he has to think precisely and logically, to acquire and communicate information, to choose adequate methods for the solution of a problem, to utilize optimally the capabilities of the team and to gain its trust.

Although he may not be an expert in all the disciplines, he should have specialized in one of them (e.g. economics, water management, computer science, etc.). His most essential qualification is the ability to conceive the system as a whole in all its conditions, with the subordinate functions of the subsystems being combined to attain the overall objective most effectively ("systems thinking").

The range of methods and subjects involved in systems engineering is apparent from the book by Machol *et al.*, 1965, which deals with:

– Systems theory: the theory of information, game theory, decision theory, the simplex method, linear programming, dynamic programming, the theory of queues, Markov processes, the theory of feedback control, adaptive systems, etc.

– Systems techniques: information processing, simulation, testing of systems, economics, management of design, etc.

– Selected parts of mathematics: probability theory, Laplace's and other transformations, numerical analysis.

Sometimes all the theories and methods used in systems engineering are termed *systems analysis*, i.e. the application of the theory of systems to methods of analysis of the structure and behaviour of systems. Without specifying them taxonomically, we consider that in the design of systems, systems engineering uses all the theories and techniques that are available and can lead to the attainment of the required goal (Systems Engineering Department, 1973). Therefore, it is the methodological framework of the approach to complex problems that facilitates the effective use of all the special modern methods.

Although the application of systems engineering has made remarkable progress in technical systems, in economic systems it is only now being introduced.

Reliability is an important factor and it should be included in the economic evaluation of technical systems. Apart from general literature (e.g. Chorafas, 1960; Calabro, 1962; Roberts, 1964; Kapur and Lamberson, 1977), specialized references are growing in number, e.g. for electronics and machinery; in water management the number of references dealing with systems engineering is relatively low (Mirtskhulava, 1974; McBean *et al.*, 1979) but this gap is rapidly being filled by publications appearing all over the world.

1.2.4 Operations Research

Operations research or operations analysis is an applied systems discipline that is used for the comprehensive treatment of concrete, technical, organizational and economic problems of complex systems. It was first used in Great Britain in 1937 as a methodological basis for air-force operations (hence the name "operations"). During World War II the operations research groups dealt with problems of supply and transportation, etc. In civil contexts, operations research has been applied to a greater degree since 1950, in connection with the development of computers.

The subject of operations research has changed from the comprehensive, systems approach to the individual development of specialized methods. Therefore its subject is well described by the definition given by Walter *et al.*, 1973: "Operations research is a set of different methods, particularly mathematical that are used for scientifically approved results, for the solution of complex economic and technical problems, especially those comprising a number of alternatives".

Morse and Kimbal, 1951, defined operations research as a special method for obtaining a good basis for decision-making in the management of operations.

Vepřek, 1970, likened the principle of operations research to the work of a research team where a group of scientists, expert in different fields, deal with the complex problem of an object that is treated as a system, using knowledge from different branches of science for its investigation.

The relationship between operations and systems analysis can be formulated by saying that operations analysis is one of the methods of systems analysis. Hitch, 1955, regarded systems analysis as an extension of operations analysis.

Most of the methods of operations research (analysis) use models preferably mathematical ones. Interdisciplinary cooperation is necessary for the development of models of complex phenomena. The models of operations research often use computers, as the treatment of many alternatives with optimization is tedious.

The following methods are used in operations research: linear programming, graph theory, simulation models, game theory, the theory of queues, inventory control models, etc. As operations research has an interdisciplinary character, the methods of technology, economics, biology, mathematics, etc. are applied.

Operations research comprises a set of different, relatively independent methods from different branches of science and it is not used for the definition and formulation of the problem. For this purpose, i.e. for the definition of the system and for forming the assumptions underlying successful treatment, the systems approach must be used (e.g. systems analysis); and for the solution of the defined problem some of the methods of operations research can be applied.

In operations research the subject is not important, it is the method used for its treatment that matters; some methods are used for a single task and other methods are preferred for repetitive application of the model.

1.2.5 Systems Analysis

The concept 'systems analysis' comprises a set of principles and methods for the analysis of the behaviour and structure of complex systems.

First, engineering systems were analysed, then economic ones. The development, methods and possibilities of systems analysis were thoroughly investigated and described in a remarkable book by Habr and Vepřek, 1972, 1986.

The concept used at the beginning of systems analysis was well defined by Hitch, 1955: "Systems analysis is a technique of investigation of military problems in a broader context and over a long-term period". Hoag, 1956, also dealt with the application of systems analysis to military research. Both use operations research as the starting-point of systems analysis.

A comparison with operations research is included in the definition by Vlček, 1968: "Systems analysis deals with relations and dependencies between tasks, the elements of decision-making and control processes that can eventually be solved and realized by the methods of operations analysis." In 1969 Vlček specified the subject of systems analysis in tasks that provide some properties and functions of the systems.

Habr and Vepřek, 1972, 1986, summarized the views of some authors into the formulation: "Systems analysis is a set of logical and formalized principles that can be used for the effective combination of partial resources with corresponding knowledge for an effective attainment of the given goals." They classify the approaches of systems analysis in dealing with problems into three categories:

- systems analysis of real systems,
- systems analysis of information systems,
- systems analysis of control systems.

Systems analysis of real systems is suitable for the design of engineering systems, including the WRS. The analysis of several alternatives is used for the determination of an optimal design. Such a concept of systems analysis is very close to systems engineering. The military organizations (e.g. RAND in the U.S.A.) and the manufacturers of computers focus their attention on this concept. Vito-Don, 1967, applied systems analysis to problems of cost-benefit analysis.

Systems analysis is also applied in the design *of information systems*, assuming that the requirements to be met by them are specified by the management system. A well designed management information system should eliminate incomplete, redundant and irrelevant information. It should be suitable for long-term and short-term planning, for operations control and analysis of the behaviour of the system and it should provide an idea of the state of the object; this idea should be given in time, space and quality so that the decision-makers and managers can utilize is easily.

Systems analysis of control systems is applied in economic and technical subjects. Hare, 1967, defined three stages of systems analysis:

– the definition of the system,
– the analysis of the structure and behaviour of the system,
– the design of the system or its improvement and implementation.

In 1974, the Institute of Management answered the question of what applied systems analysis is in this way: "Applied systems analysis is not only a technique or a group of techniques such as the theory of probability or mathematical programming. It is a framework that is to aid the decision-makers in choosing an adequate (in some cases the best) way of action. Applied systems analysis is used for concentration and intensification in dealing with large-scale and complex problems. Applied systems analysis is a general term including fields like operations research, decision theory, cost-benefit and efficiency analysis; planning, programming and scheduling, design analysis; many aspects of cybernetics, the theory of information, artificial intelligence in management and information systems; dynamic modelling, the theory of behaviour, decision-making; the theory of organization."

Applied systems analysis is included directly in the name of IIASA (International Institute for Applied Systems Analysis). The range of the fields and problems contained in applied systems analysis is apparent from the research programme of this institute, which includes:

– resources and environment: ecological problems, problems of water resource research, problems of food and agriculture, etc.,

– human settlement and services: problems of population, health, education, communication, etc.,

– management and technology: man-made artifacts, institutes, economic systems, technologies, etc.,

– systems and decision sciences: mathematical and computational problems in the analysis of large systems, etc.

Among the fields included in the framework of systems analysis, *decision theory*[1]) provides methods showing how to choose in decision situations from a set of feasible decisions the scientifically approved optimum decision, and thus it is very important for WRS.

In decision theory, the methods of optimization, probability theory, mathematical statistics, utility theory, realiability theory and theory of games are applied in the first place.

Although mathematical methods are applied in the selection of the best decision or optimum alternative, a strong subjective component remains, and in many cases decision-making is not only a science but an art.

In a decision situation one or more participants are involved who are influenced by the consequences of the decision. According to the nature of these consequences, the decision situation may be either competitive (conflict) or bargaining (non-conflict). The *non-conflict situation* can arise if only one participant and a single consequence of a certain decision are involved. The mathematical method for this basically simple case of decision-making is often mathematical programming (linear or non-linear programming).

For WRS the second case is more interesting, i.e. when the decision is influenced by various participants and the *situation is competitive (conflict)*.

A survey of types of decision situations was given by Maňas, 1974, in a book on game and decision theory.[2]) The theory of games is a promising technique for the treatment of problems in WRS.

In the decision process the following steps are applied:

1) The *goal* is determined (attainment of certain parameters, risk or reliability, the requirement of environmental quality, etc.)

2) A list of all the feasible *alternative solutions* is constructed. A creative imagination is necessary for the definition of the feasible alternatives, and professional knowledge is required to omit the evidently non-effective alternatives from the list.

3) A list of *factors* (technical, economical, intangible, etc.) that influence the decision is made out. The danger of omitting an important factor is greater than omitting a feasible alternative.

4) The *number of alternatives is reduced* by an analysis of the important factors.

---

[1]) As analysis of real systems is sometimes assigned to systems engineering, decision analysis is sometimes included in systems analysis. The first books on decision analysis as a specific scientific field appeared in the fifties: Luce and Raiffa, 1957; Chernoff and Moses, 1959; Weiss, 1961; Dyckmann, McAdams *et al.*, 1969; Lee, 1971, etc.

[2]) Important initial references to the theory of games are Blackwell and Girshick, 1954; Karlin, 1959; McKinsey, 1952; von Neumann and Morgenstern, 1944. The book by Maňas, 1974, has 357 references and one of them, "Contributions to the Theory of Games"(Princeton University Press, 1959), includes more than 1000 references up to 1958.

This stage may be influenced by subjective views as estimates have often to be used (the art of decision-making).

5) The *number of factors is reduced* in relation to the remaining alternatives.

Using the results of these five steps we continue the analysis (if the results are satisfactory) or we try to find the reasons for the unsatisfactory results (oversimplification, or, on the other hand, an over-complicated model).

The *reliability* with which the system functions is its most important characteristic. A numerical value for this is obtained by the mathematical methods of probability theory, and mathematical statistics. Direct application of their outputs, however, can result in a situation that is far from reality. A scientific analysis of the phenomena is necessary and in complicated cases a heuristic approach is recommended.

The reliability, however, retains, its probability character, and it can be defined as the probability that the system (or its element) will, under given conditions, satisfy the requirements in a certain time period.

Most mistakes in the definition of reliability pertain to the idea of the long-term stability (constant value) of reliability, although this changes with time (it is time dependent) for various reasons; e.g. due to worn-out parts in some devices, or due to changes in water demand in water systems, etc.

An important index in the computation of reliability is the risk of failure of the system (or its elements). It can be expressed by a probable number of failures in a time unit (occurrence-based reliability), the duration of failures (time-based reliability) or the volume of deficits in production (quantity-based reliability).

The more complicated cases of reliability determination are derived from the coordinated functioning of several elements in the system with different reliabilities and with different relationships between them. The greater the number of elements in the system, the operation of which is necessary if it is to function correctly, the lower the total reliability. Therefore, the problem increasing system reliability is often dealt with in theory. The tools for achieving this increase may be:

- the installation of slack facilities or their parts ("redundancy"),
- simplicity of the structure,
- the use of standard and reliable parts,
- reduction in design load.

In decision-making generally, two questions may be raised:

a) How would an average person decide in a given situation? (e.g. what response in consumption can be expected from the inhabitants after the installation of a new water supply, what response in water deficits can be expected, etc.?)

b) What is the best decision in the given situation?

In WRS view b) is often applied which should lead to a logically correct decision. Of course, this approach to decision-making cannot eliminate the impact of subjective views and the intentions of the participants. The theory of games also includes these relatively frequent cases.

1.3 DEVELOPMENT OF SYSTEMS DISCIPLINES

In many scientific disciplines, the modern views were formulated 2500 years ago by the Greek natural philosophers on the basis of mere philosophical speculation. Similarly, the systems approach to a perception of the world which embraced its evident and hidden relationships was discovered and formulated. An outstanding place belongs to Pythagoras and his School, which constructed an interpretation of natural science, analysis, the wholeness of the natural world and its relationships in mathematical terms. A present-day formulation might be: Pythagoras' School tried to create a mathematical model of the world and numbers took the place between things and ideas (Volkov, 1975).[1])

The use of feedback control in irrigation systems (Institute of Management, 1974) has a 4000-year-old tradition. By irrigation of the land with water from the Euphrates and Tigris Rivers the old Babylonians achieved such a high degree of efficiency in agricultural production that it provided for the highest population density at that time. Using feedback, they controlled the moisture of the soil by opening and closing irrigation channels. One of the 282 laws in the Hamurabi's Code (approx. 2100 BC.) stated: If anybody opens the irrigation channel but is negligent, and water flows to the land of his neighbour, he will pay for the loss of corn. This can be qualified as the first known statement of a legal fine for negligence on the part of a human operator in a feedback control system.

The realization of artificial, relatively complex engineering systems is related to machine production. The systems machine technology was accompanied by the development of systems characteristics in scientific research. As soon as the engineer, biologist, economist or any other expert begins to realize that the mutual structural and functional relationships between the subjects of investigation form a whole, he has to think in systems terms, taking into account the relationships in his own field. If he does not do so, his reasoning is not comprehensive, the conclusions may not be complete and, consequently, not correct.

The nature of the issues in different scientific spheres influenced the earlier or later use of the systems approach. For example, ponds are generally isolated elements and therefore, in their design and operation, the relationship between them need not be considered. However, the designers of the old *systems* of ponds in the 16th century had to take into account the structural and functional relationships between individual ponds. For example, the levels of the ponds in South Bohemia (Czechoslovakia) which were fed by the Golden Canal were dependent on the water level

---

[1]) An interesting study of the interpretation of the systems approach in scientific knowledge was carried out by Ogurcov, 1974. He dealt with the ancient concepts of the characteristics of existence and recognition, using contemporary systems ideas in ontology, and with the interpretation of systems attributes of scientific knowledge in the philosophy of the new era, especially in classical German philosophy.

in the Canal and changing the levels during fish-harvesting operations had to be coordinated.

Much closer interrelations exist in the systems and cascades of modern reservoirs. The first dams were designed as isolated items (although at the beginning of the 20th century a flood control system of reservoirs was designed and built on the Nisa River). The modern reservoirs are the elements of a WRS with cascades from reservoirs on the same river or on different watercourses (e.g. WRS for the water supply in the Ostrava region see Chapter 12).

Such an attempt at the comprehensive treatment of problems by considering interrelationships was repeated in other branches of science, too. Habr and Vepřek, 1972, cited a number of economists (Quesnay, Marx, Keynes, etc.) who were conscious of the necessity to treat problems comprehensively. They state that there is a gap between those attempts to use a systems approach and the possibility of applying it in economics; these earlier economists did not have the appropriate methods and technical means for real systems work. However, a comprehensive systems treatment of economic problems is still more of a wish than a reality. In water management the situation is similar, and technical parameters are often used instead of economic ones.

So far, the reservoir has been considered as an element of the system. If a more detailed discriminating level is used, the dam (i.e. a part of the reservoir) can itself be treated as a system. Its elements are the embankment of the dam, spillway, outlets, diversion channels, etc. If an engineer designs the reservoir as an isolated structure, i.e. without the necessity of the systems approach in relation to other reservoirs, he has to deal with a component part – the dam – as with a system of elements with structurally and functionally dependent variables (conception of design and construction; relationships given by the release of water and withdrawals).

Another example of WRS is the River Váh cascade in Czechoslovakia. There are clear structural relationships there (e.g. between the water levels in the reservoirs – the elements of the system) but functional relationships are also evident (e.g. the time-related operation of the subsystems). The exploitation of the River Váh for hydroelectric power generation would not have been possible without a systems approach.

The oldest WRS structures are the systems of ponds in Bohemia and Moravia[1]) and the water-supply system for mines in Slovakia[1]) (14th – 17th centuries).

The interrelationship between different forms of surface water (i.e. watercourses, lakes, etc.) and ground water required the systems approach very early in history for a comprehensive use of water resources and their conservation. A well-known historical system existed in Solomon's ponds in Jerusalem (10th century B.C.) and there were some older irrigation systems in ancient cultures.

[1]) Czechoslovakia (ČSSR) consists of three geographical units: Bohemia, Moravia and Slovakia. Bohemia and Moravia form one part of the federation (ČSR) and Slovakia the second part (SSR).

Although the systems approach has been applied in scientific and engineering problems since ancient times and the necessity for it has become more and more evident, the origin of "systems sciences" dates from about 50 years ago. The growing complexity of systems and their problems required the formulation of general methodological principles for their definition and treatment.

The *general systems theory* is one of the modern gnostic approaches to the investigation of complex subjects. The Austrian biologist Ludwig von Bertalanffy was the founder of this new branch of science. In the thirties he developed the theory of open systems in biology and, around 1950, the principles of the general systems theory. In 1954 he was one of the founders of the Society for the Advancement of the General Systems Theory (renamed the Society for General Systems Research, an affiliate of the American Association for the Advancement of Science). Every year, this society publishes thick yearbooks (Yearbook of the Society of General Systems Research, University of Michigan, U.S.A.) with contributions not only on the general systems theory but also on its application in many spheres.

Lektorsky and Sadovsky, 1960, considered the von Bertalanffy theory of open systems to be an important philosophical and methodological achievement, synthetizing the tendencies in modern science. In this respect, it can be compared with the work of other outstanding progressive scientists, e.g. Wiener, Ashby, etc.

As far as WRS is concerned, it is open systems which are significant, i.e. systems with input and output, with stimuli and response, as opposed to closed systems, which have no input and output. The Belgian Prigogine, 1947, 1955, classified systems in three categories – open, closed and isolated. Isolated systems will not be discussed here as they do not exchange either matter or energy with their environment and, in fact, do not exist in nature.

Von Bertalanffy, 1962, considered the theory of open systems and states of dynamic equilibrium as a generalization of physical chemistry, kinetics and thermodynamics. The general systems theory was a further generalization. He mentioned it for the first time at the philosophical conference in Chicago, 1937, but he decided to publish it only after World War II when he found that other scientists were also using this special way of reasoning.

The investigation of living organisms as an open system with processes of exchange of matter with the environment has resulted in a new concept, which is also used in WRS, i.e. the notion of "homeostasis". It appeared in the publications of the American physiologist Cannon, 1929, who studied the control mechanisms which maintain the blood composition of mammals. However, homeostasis is a set of processes which maintain an inner equilibrium state of the system. For all the phenomena which protect the cell and the living organism from the effects of the environment and keep them stable, Heilbrunn, 1956, introduced the concept of "biostasis".

The beginnings of systems science can be found in the thirties. The rapid methodological development of this science occurred in the fifties and the main achievements of its applications at the beginning of the sixties when computers provided the necessary technology for the solution of complex system problems.

At the present time, there is continuous development of a number of relatively independent scientific fields relating to the treatment of systems problems, sometimes with varying definitions of the subjects[1]).

---

[1]) Let us try to sum up the development of the methods used in the treatment of systems up to the time of publication. A list of selected publications follows. Classification is according to the fields comprised in the systems science, its adjoining or related spheres, or its applications (in this part the name of the author and the year of publication do not necessarily mean a quotation or reference):

a) *In the field of general systems theory*:
- general systems theory (von Bertalanffy, 1950, 1956, 1958; Boulding, 1956; Ashby, 1958; Mesarović, 1963, 1964; Ackoff, 1963; Rapoport, 1963),
- general systems theory in scientific methodology (Boulding, 1956; Mesarović, 1964),
- mathematical systems theory (Wymore, 1967; Kalman–Falb–Arbib, 1968),
- theory of open systems (von Bertalanffy, 1950),
- theory of open systems in biology (von Bertalanffy, 1950),
- theory of control (Mesarović, 1963; Godunov, 1967),
- theory of optimal control (Kalman, 1958; Neustadt, 1960; Lee–Markus, 1961; Pontryagin *et al.*, 1961),

b) *In the field of cybernetics*:
- cybernetics (Wiener, 1948; Ashby, 1956; von Bertalanffy, 1962; Maruyama, 1963),
- cybernetics in air-defence artillery (Wiener, 1961),
- algebraic theory of machines (Riguet, 1951; Krohn–Rhodes, 1965, 1968; Hartmanis–Stearns, 1966),
- theory of automata (Rabin–Scott, 1959; Arbib, 1965, 1969; Dakajev *et al.*, 1970),
- theory of abstract automata (Arbib, 1969),
- theory of adaptation (Ashby, 1940; Mesarović, 1963),
- theory of regulation (automatic control) (Ashby, 1956; Kalman, 1958),
- theory of groups (Kurosh, 1953; Hall, 1959),
- topology (Bourbaki, 1951; Leeman, 1962; Langefors, 1966),
- information theory (Shannon, 1948; Goldman, 1953; Khinchin, 1954; Kolmogorov, 1956),
- information theory in psychology (Attneave, 1959),
- mathematical theory of communication (Shannon–Weaver, 1949),
- communication theory in secret systems (Shannon, 1948),

c) *In the field of systems engineering*:
- systems engineering (Goode–Machol, 1957; Hall, 1962; Chestnut, 1967),
- systems programming (Habr, 1965),
- decision theory (Chernoff–Moses, 1959; Raiffa–Schlaifer, 1961),
- engineering psychology (Lewin, 1936),

d) *In the field of operations research*:
- operations research (Morse–Kimball, 1951; Churchman–Ackoff–Arnoff, 1957),
- graph theory (Euler, 1736),
- network analysis of complex activities (Berge, 1958; Ford–Fulkerson, 1962),
- structural analysis (Vepřek, 1970),
- econometrics (Tinbergen, 1957),
- mathematical economy (Baumont, 1961; Nemchinoff, 1964; Mansfield, 1966),
- the theory of queues (Got–Smith, 1961; Saaty, 1961; Kaufman–Cruon, 1962),
- the theory of replacement (Cox, 1962),

Systems engineering, for example, was developed from industrial engineering with specific objectives and methods of organization of work. According to the definition of its subject by the committee for standardization of A.S.M.E. in 1943, it was close to the contemporary economics of industrial sectors. It also included optimization of benefits and minimization of risk in business.

With the growing complexity in business economics the systems approach to business firms and their environment was applied. As a result of this modernisation of views and methods, industrial engineering is approaching systems engineering which uses this new scientific method to assist in the management of complex industrial organizations. New systems are being modelled, old systems have been analysed and improved. In addition, the time intervals between a scientific discovery and its application and between the origin of the demand and its satisfaction have been reduced (Hall, 1962).

The concept "complexity", as it appears in different branches of human activity, has become a problem of integration:

– in transportation (the coordination of enormous quantities of goods by different means of transportation with large-scale capacity on an international scale),

– in communications (the automatic long-distance and international telephone systems),

– in water management (the long cascades of extensive multi-purpose WRS, with different water demands and different realiability requirements creating many interrelationships in the system),

– in industry (the big trusts, monopolies, etc. as complex and comprehensive economic production systems with combinations of automatic control and operation by human beings),

– in trade (large-scale information systems with modern computer hardware and software),

– in military contexts (the extensive military systems provided an impetus to the application of new scientific disciplines to complex problems),

---

- the theory of games and strategy (von Neumann–Morgenstern, 1944, 1947; McKinsey, 1952; Blackwell–Girshick, 1954),
- inventory theory (Moran, 1959; Heady–Within, 1963; Morse, 1958),
- factor analysis (Spearman, 1904).

e) *In the field of systems analysis*:

- systems analysis (Hitch, 1955; Hare, 1964, 1967; Vlček, 1968),
- systems analysis in military research (Hitch, 1955; Hoagh, 1956),
- systems analysis in the air force (Churchman *et al.*, 1957),
- systems analysis in the field of management and control (Johnson *et al.*, 1963; Modin, 1963; Hare, 1967),
- systems analysis in data processing (Langefors, 1966),
- systems analysis in computer science (McMillan–Gonzales, 1968),
- systems analysis in effective planning (Rudwick, 1969).

– in science (modern science is often associated with high costs but also with complexity in organization).

This integration of problems creates a demand for a new type of experts (systems engineers) who are able to synthetize the design and operation of large-scale and complex systems.

Science itself is, in fact, a complex and dynamic system and can therefore be the subject of systems analysis. Its elements may be chosen according to the standpoint of investigation (Kedrov, 1974):

– objects if gnostic objects matter,

– methodology if the way of cognitive analysis is most important,

– subjective goals if specific practical aims of scientific research are discussed.

The gnostic *object* is formed by a system of natural, and other sciences: the first group comprises the subsystems of mathematics, physics, mechanics, chemistry, biology, etc.; the second group consists of the social and economic sciences, philosophy, psychology, linguistics, etc. In addition, interdisciplinary sciences have been established including systems science (cybernetics, etc).

In the course of investigation not only objective but also subjective views occur. The abstract reasoning used in methods of investigation depends on the subject employing it.

The achievement of the *goal* of investigation and its *practical* utilization is even more closely related to the human being in question – the human subject.

Pure ("classical") sciences correspond more closely to the objective aspects of science, whereas applied sciences correspond to the subjective ones (e.g. purpose, goal).

The technical sciences straddle the line between the natural and social sciences as their goals are formed by social conditions but the achievement of these goals depends on natural laws. The new science of science (metascience) deals with science as an object of investigation. Of course, in this science the systems approach is utilized with the following concepts: the model of science, structure, mathematical models of scientific research, etc.[1])

The survey of scientific fields and references shows that extensive development of systems science began in the fifties. In Czechoslovakia systems references have been appearing since the sixties (Habr, Beneš, Kamarýt, Kubík, Kotek, Korda, Dráb, Walter, Vlček, etc.) and a reference to its application in water management appeared in 1967 (Kos, Partl). The development of systems science is characterized by special organizational and professional activities.

In 1954, as stated, the Society for the Advancement of General Systems Theory was founded in the U.S.A. The International Association for Hydraulic Research

---

[1]) Systems analysis issues featured prominently in the yearbook "Sistemnyye issledovaniya" (System research) published by Nauka, Moscow.

(IAHR-AIRH) was founded in 1935, and one of the eight committees under its direction is the Committee on Water Resources Systems, set up at the end of the sixties. WRS problems were also included among the main topics of the 15th Congress of IAHR in 1973.

The issues of water management are the concern of IWRA (International Water Resources Associations), which in 1973 organized its First Congress in Chicago under the slogan: "*Water for the Human Environment*" and a Second Congress in 1975 in New Delhi under the slogan "*Water for Human Needs*". The worldwide problems of water management were raised as early as in 1967 by the United Nations, which organised a conference "*Water for Peace*", which was attended by the U.N. member countries.

Since 1973, there has been working International Institute for Applied Systems Analysis (IIASA) in Laxenburg (near Vienna, Austria) (preparatory work started in 1967); the founders of the institute were institutions in the Soviet Union, Canada, Czechoslovakia, France, German Democratic Republic, Japan, Federal Republic of Germany, Bulgaria, the United States of America, Italy, Poland, the United Kingdom. By 1977, Austria, Hungary, Sweden, Finland and The Netherlands had all joined. In its Charter, IIASA laid down the following objectives:

"The Institute shall initiate and support collaborative and individual research in relation to problems of modern societies arising from scientific and technological development. To this end, the Institute shall undertake its own studies into both methodological and applied research in the related fields of systems analysis, cybernetics, operational research, and management techniques.

The Institute shall encourage and reinforce national and international efforts in corresponding fields of inquiry; devise means of enhancing appreciation of this type of research among scientists from all nations; and attempt to increase understanding through the development of uniform standards and terminology. Its work shall be open to all experts in conformity with the normal practice of international scientific co-operation.

The Institute's work shall be exclusively for peaceful purposes."

IIASA has directed its research program towards methodology, big organizations, integrated industrial systems, urban and regional systems, ecological systems, biological and health problems, computers, energy, water, etc. IIASA uses a matrix organization with four divisions: Resources and Environment; Human, Settlement and Services; Management and Technology; System and Decision Sciences; and two programs: Energy Systems and Food and Agriculture.

In Resources and Environment comprehensive research of WRS was carried out at three levels: *Level I* – the problems of local water reservoirs and lakes, water storage and distribution, operation of WRS, short-term operation; *Level II* – the problems of large regions, long-term operation, e.g. large-scale investment in water resources and long-term development plans; *Level III* – global problems, e.g. global air and water pollution, the global scarcity of fresh water resources.

In Czechoslovakia, since 1971, a WRS subcommittee of the Committee for Water Management of the Czechoslovak Academy of Sciences has systematically investigated the problems of WRS and has helped in the systems approach in all the main tasks.

In the Water Management Association of the Czechoslovak Association of Engineering and Science a new group has been established for WRS.

The concentration of Czechoslovak water resource engineers and scientists on the problems of WRS resulted in numerous studies, publications dealing with WRS, case studies, a special chapter in the General Water Plan, 1976, and the organization of five special symposia on WRS.

2 SYSTEMS IN WATER RESOURCE MANAGEMENT

In developed countries and countries with limited water resources, comprehensive and rational water management is a necessary condition of social and economic development. The bodies of water form extensive natural systems and these natural elements and hydrological relationships are supplemented by artificial relationships, so that the natural watershed ceases to be the boundary of the system.

Water management thus becomes a typical sector where the rational and purposeful utilization of water resources cannot be performed without multi-purpose water resource systems (WRS), including water resources, water users, and further factors related to or influenced by water management.

The optimal solution of problems involving competitive requirements for water needs the systems approach as a methodological procedure that takes all the internal and external relationships into account and utilizes the new theories of systems and modern computer hardware and software.

2.1 WATER RESOURCE SYSTEMS

The concept "water resource systems" can be treated as a real system since its elements are real objects. The system is defined according to the chosen objective and degree of abstraction.

WRS can be defined as a set of water resource elements linked by interrelationships into a purposeful whole.

The elements of the system can be either natural (precipitation, watercourses, lakes, ground water, etc.) or artificial (water management articles and facilities, reservoirs, channels, barrages, weirs, hydroelectric power plants, pumping stations, etc.).

The relationships between the elements are either real (e.g. water diversion) or conceptual (e.g. organization, information). WRS are "open systems" i.e. their elements bear some relation to the environment of the system.

If the link between some elements within the system is relatively closer than that between other elements, inside the system a relatively independent whole exists, which is called the *subsystem*. The selection of the discriminating level for the concept "water resource system (WRS)" is a matter of convention. An example of the discriminating level chosen in Czechoslovak conditions is given in Table 2.1; WRS comprises the total extent of the basins of the main rivers and their tributaries, the supersystem (supreme system) is water management of the whole country (and its supersystem is the whole economy of the country).

These systems are largely artificial and therefore purposeful. Each element, subsystem or system that consists of only one artificial component designed or used for some goal, is a purposeful object.

However, there exist natural WRS, formed by the hydrological networks, without any objectives, and these are not the objects of purposeful human activities (e.g. the system of karst ground water). Only exceptional examples of these purposeless WRS are investigated in this book.

Table 2.1 Hierarchical pattern of WRS (example)

Name	Object
Supersystem	Water management of the country
System	Basin of the main river
Subsystem of the 1st order	Municipal water supply
Subsystem of the 2nd order	Water supply of a town
Subsystem of the 3rd order	Water supply of a big factory
Subsystem of the 4th order	Water supply of a workshop

The interaction between the elements and subsystems of WRS can be local (topographical), hydrological, technological, engineering, etc.

A purposeful WRS is formed by a set of physical (material) items (elements), goals (which the WRS is to achieve) and some operational rules.

On the basis of the goals, WRS can be classified as:

– irrigation and drainage systems, hydroelectric power plant systems, water supply systems, fish-breeding systems, navigational systems, etc.,

– single-purpose or multipurpose systems.

A *single-purpose system* has various physical (material) items (elements) but only one purpose, e.g. a hydroelectric power plant system, or a flood control system, etc. Its goal is defined at the beginning of the investigation, mainly in technical units, e.g. to supply the discharge Q ($m^3 . s^{-1}$) with reliability p_o.

A *multi-purpose WRS* has various elements and a number of goals. The main aim in identifying it is to determine what combination of goals is optimal and what criteria are needed for the evaluation of this optimum.

As the goals are often competitive, optimization is difficult. The optimum-seeking task (determination of the objective function) is simplified if one particular goal (e.g. a municipal water supply) has higher priority than the others. Then this multi-purpose WRS can be treated as a single-purpose system and the "additional" goals are used as constraints.

Together with the basic attributes of a system, i.e. wholeness, unity, etc. the multipurpose WRS also possesses all the signs of complexity so that a very complicated system can be defined in the following manner:

– it is large-scale in its dimensions, number of components (subsystems and elements), the number of input and output parameters and the number of relationships (e.g. WRS in Chapter 12),

– change in one quantity causes a change in a number of other variables (e.g. a change of hydrological input data),

– it is dynamic (both the water resources and the water demands change with time),

– the nature of hydrological input data and some water demands is random (the values of flows, withdrawals, the indices of water quality, etc. are not known in advance; it is a stochastic process and only its statistical characteristics may be known),

– the demands on the multi-purpose WRS are often competitive (e.g. requirements for higher flood control storage and active storage of reservoirs; the increase in one withdrawal limits the achievement of other goals, etc),

– operation and control of WRS requires many devices for monitoring and information processing, together with computer hardware and software and automation of some activities,

– the multi-purpose WRS requires personnel who measure, monitor, evaluate and control the operating process to ensure the faultless functioning of the WRS,

– the nature of the multi-purpose WRS requires not only technical and economic parameters, but also environmental and intangible aspects for the evaluation of their objectives as they have a profound impact on the environment.

First of all, the WRS has to be analysed in all its complexity. We should not underestimate the danger of oversimplification and try to create a basis for the evaluation of the validity of results of simplified systems (models). A comprehensive analysis of the structure and function of the WRS facilitates the scientifically approved abstraction and simplification.

2.2 WATER RESOURCE SYSTEMS IN THE GENERAL WATER PLAN OF CZECHOSLOVAKIA

"Water Resource Systems" is the heading of an important chapter of the General Water Plan of ČSR and SSR[1])

The General Water Plan (GWP, 1976) states that the creation of WRS in the basins

[1]) Czechoslovakia – the Czechoslovak Socialist Republic (ČSSR) is a federation of two countries, the Czech Socialist Republic (ČSR) and the Slovak Socialist Republic (SSR).

of the main rivers is an objective necessity for the development of water management in the natural, social, and economic conditions of the country. The utilisation of water resources will double by the year 2000, the annual withdrawals will increase to 33% and consumption to 12% of the mean annual flow; in drought years these figures will approximately double. The quantity of waste flow will be approximately proportional to this increase, and requirements concerning conservation and the protection of surface and ground water from pollution will also increase.

Table 2.2 Survey of WRS in basins of General Water Plan

No.	WRS (basin)	Area [10^3 km^2]	Number of inhabitants (1970) [Thous.]	Mean annual flow $\left[\frac{10^9\ m^3}{year}\right]$
1	Upper and middle reaches of the Labe River	14.37	1738	3.34
2	The Vltava River (without the Berounka River)	18.30	2330	3.35
3	The Berounka River	9.28	725	1.26
4	Lower reaches of the Labe River	9.54	1183	2.11
5	The Odra River	6.25	1212	1.95
6	The Morava River	21.11	2630	3.14
7	The Danube River and the lower reaches of the Morava River	5.39	739	0.36
8	The Váh River and the Nitra River	16.77	1761	5.31
9	The Hron, the Ipel and the Slaná Rivers	12.30	814	2.92
10	The Bodrog, the Hornád, the Bodva and the Poprad Rivers	14.62	1174	3.84

The traditional classification of the water management sector into divisions (viz. water supply, irrigation and drainage, hydroelectric power generation, etc.) is not comprehensive as the different requirements concerning water in a geographical and administrative area (country, region, basin or parts of them) cannot be coordinated. The methodological prerequisite of the comprehensive and optimal utilisation of water resources and release operation is given by the multi-purpose WRS.

The treatment of the problems of the main WRS consists of the processing of demands and prediction of their growth, and of the estimation of the capacity of water resources and their operation, and it is supplemented by the design of water resource conservation and protection measures. A schematic representation of the

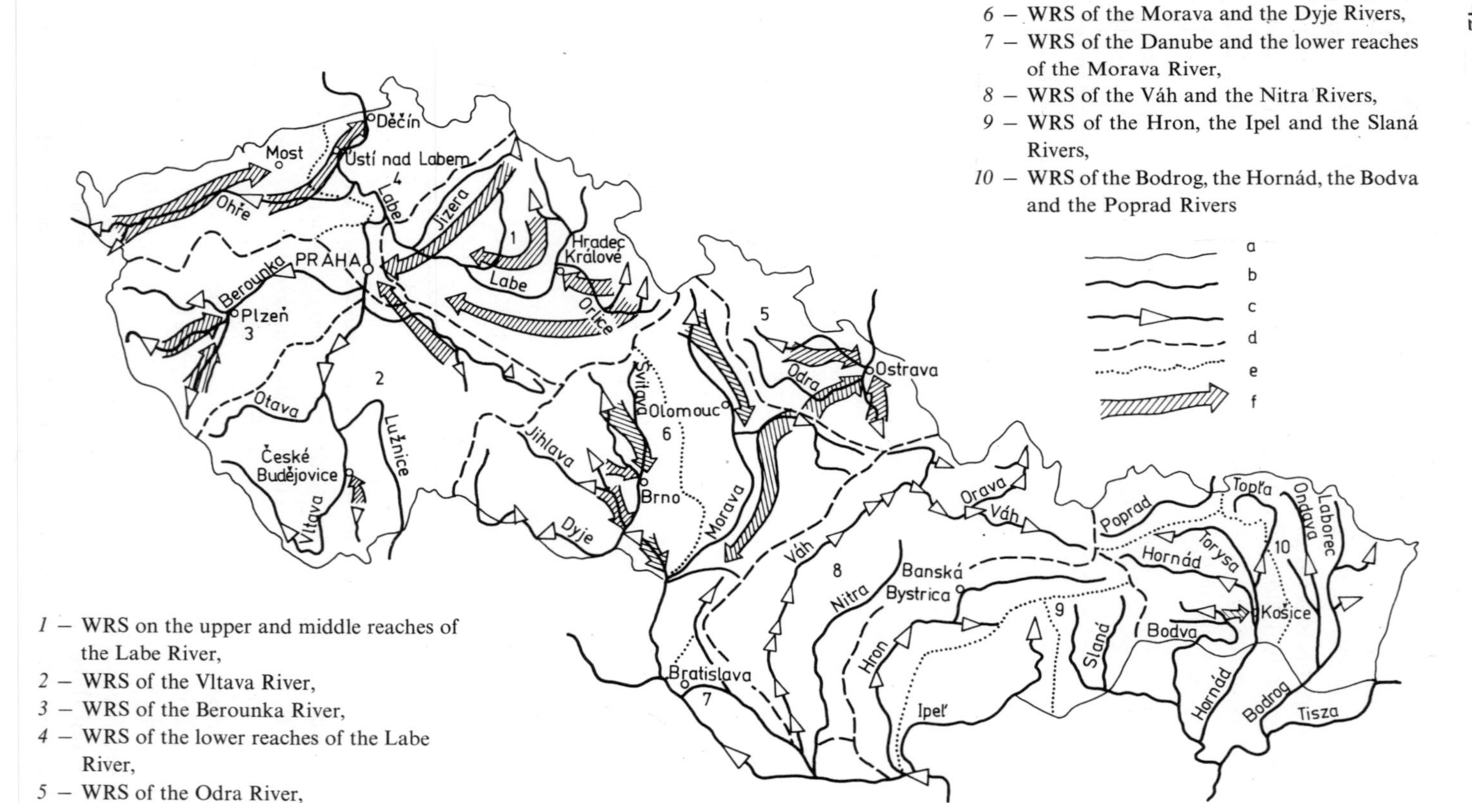

Fig. 2.1 Schematic representation of WRS in the General Water Plan

a – state frontier, *b* – water courses, *c* – key reservoirs of the WRS, *d* – boundaries of WRS, *e* – boundaries of the subsystems, *f* – the main directions of the influence of the key reservoirs.

extent of WRS in the GWP that cover approximately the area of the main basins, is given in Table 2.2 and Fig. 2.1.

In the first part of the chapter entitled "Water Resource Systems", water management in the main basins is described in the following sections:

– the main tasks of water management in the basin,
– procedures for meeting the main objectives,
– relationships with other basins of the GWP.

In the second part the individual main WRS are described. The comprehensive utilisation of water resources is treated in systems with two discriminating levels – the overall (rough) and the detailed level. The following goals were included in the systems with the rough discriminating level:

– municipal water supply,
– industrial and irrigation water supply,
– flood control and its relationship to water resources,
– water resource conservation, water pollution prevention and environmental improvement,
– hydroelectric power generation,
– inland waterways and navigation,
– water-related recreation and fishing,
– river treatment,
– low-flow augmentation, minimum-flow maintenance.

The following main goals were included in multi-purpose WRS with a detailed discriminating level:

– municipal water supply,
– industrial and irrigation water supply,
– low-flow augmentation, minimum-flow maintenance,
– flood control,
– water quality improvement,
– water-related recreation.

A simulation model was the main tool for analysis of WRS with the detailed discriminating level. The operational rules of WRS are coded as logical and arithmetical statements that determine the conditions for the storage and release of water from the reservoirs, the allocation of water between reservoirs and for different goals and time periods. This model was used for the investigation of the behaviour of WRS under different conditions. Each change in the operational rules requires a corresponding change in the program.

The input data of simulation models include the parameters of WRS, hydrological time series (i.e. uncontrolled monthly time series in the periods 1931 – 1960 and 1931 – 1970 respectively) and the demands on WRS at control points. These demands are expressed as withdrawal (without consumption), consumption and minimum acceptable flow. The capacity of WRS was tested by data projected for the year 2000 and preliminary data for the year 2015.

In the simulation model the operation of WRS was frequently required without deficits for the period of the observed time series. In the stochastic simulation model, this assumption corresponded to a reliability of 95–97%.

The basic civil engineering activity in WRS in the construction of reservoirs related to water supply and pollution prevention, river training, transbasin diversion and navigation. Water-related recreation was taken into account in the operational rules of reservoirs.

The WRS were subdivided into several subsystems for modelling and evaluating the results. As all the WRS are closely related to the system's environment a water resource supersystem (supreme system) in ČSR has been defined with the following goals:

– the determination of the transbasin diversion to meet water deficits in some WRS,

– the coordination of the operation of the main reservoirs in the supersystem for hydroelectric power generation and flood control on the lower reaches of rivers,

– acceptable water quality at points on the country's borders.

The water resource supersystem was subdivided into three subsystems:

– the basin of the River Labe,

– the basins of the Rivers Morava and Odra linked via the Teplice reservoir by a transbasin diversion channel,

– a water resource system connected to the Danube–Odra–Labe canal.

In the third part of the "Water Resource Systems" chapter in the GWP there is a description of the areas of important water resource conservation defined under the Water Act (No. 138/1973 of the Collection of Acts). This act defined new concepts such as: the conservation of areas with natural water storage (areas of water resource conservation), watercourses and basins for municipal water supply. Conservation also concerns areas with planned utilisation of water resources. Similarly, areas for the conservation of natural ground water resources have been designated.

In a special section of this chapter there is a description of the conservation of watercourses and basins for municipal water supplies and the flooded area of the planned reservoirs.

Preventive water resource conservation is very important for the present utilisation of water and for future trends in water resource development. The resulting constraints are also included in the design of WRS.

The GWP was used as an illustration of the problems of multi-purpose WRS and their importance in dealing with the growing difficulties of water management, and as an illustration of the extent of multi-purpose WRS with a discriminating level in the GWP. The incorporation of systems science in WRS is indispensable, if their design is to be optimized.

2.3 WATER RESOURCE SYSTEM DEFINITION

Concerning water management objects, an infinite number of systems can be defined, which differ in objective, discriminating level, degree of simplification, etc. They are defined as real objects, i.e. the systems are real and not conceptual.

In natural and social sciences, a simplified abstraction (system, model), reflecting only the principal characteristics, is frequently investigated instead of the very complex reality.

The development of scientific knowledge and scientific methods has facilitated the setting-up of systems of ever-increasing complexity in a way that comes closer to reality. However, the simplification of nature, i.e. its adaptation to the possibilities of computer software and hardware capacity (however limited) can exceed the limits of acceptability with results remote from reality. The interpretation of the results of models and systems must be based on the standpoint that not "nature", but its simplified reflection was dealt with directly.

Therefore, in the operation of WRS human control cannot be omitted, since:

– their operation is even more complex and modelling cannot replace reality,

– the behaviour of WRS is not deterministic; it can be predicted with only some degree of reliability.

The human operator remains the principal element even in automated control systems and his profession can be called the "profession of the century" in view of its expected development.

WRS with artificial elements are created with the objective of meeting certain fundamental requirements in an optimal way (if possible), i.e. satisfying some economic or other criteria of optimality.

The task of WRS optimization may be placed in three contexts:

– the WRS exists and its *function* is to be optimized by changes in its *operation*,

– the WRS exists and it is to be *enlarged* (some elements added) in order to meet its present functional requirements better or to fulfil some new functions,

– the WRS does not exist; it is to be *designed* and *constructed*; existing elements can be used along with new ones.

A simplified system is defined for the required objective. This process has three main stages:

– the definition of the WRS units and its schematic representation,

– analysis of the structure and behaviour of the system,

– the design of changes in the system (if any) and their implementation.

The first two stages are the subject of systems analysis, the third requires a synthesis of the system.

A proper definition of the system is the basic prerequisite of the successful solutions of problems in WRS. The correct choice of the main characteristics is indispensable for the achievement of the required objective (or several objectives), and the use of incorrect ones in the system (model) obstructs all efforts, even if the best methods are used. The system in that case does not represent reality, and the results cannot be used in practice since they might lead to incorrect decisions. If the incorrectness of the results is obvious or can be revealed by common sense, no wrong decision is taken and no losses occur. However, the most dangerous case is when an incorrect result seems plausible; then the error is not corrected in time and the wrong decision may be taken.

In defining a WRS it is necessary:

– to verify the correct and accurate formulation of the objectives of the system; this is particularly important if the objectives of the system and the goals of its elements are competitive,

– to choose the discriminating level of the system and by that means to determine its elements, subsystems and environment and their functions in the system,

– to determine the relationships between the parts of the system and between the system and its environment,

– to decide upon a schematic representation of the system (e.g. a graph, matrix, flow-chart, etc.).

The definition of the system is the top creative design activity where human reasoning and intuition are indispensable. The choice of important elements and significant relationships and the adequate reduction of the number of constraints and variety of feasible alternatives determines the validity of system definition but also the adequacy of the simplification and the mathematical tractability of the model.

Although systems analysis includes many methods of system simplification, it cannot guarantee the correct choice of elements and their relationships.

The definition of the system consists of two principal procedures:

– subdivision into parts with defined interrelationships,

– simplification.

For easier definition of a large-scale system a working procedure was worked out which mechanized some of the steps (Habr and Vepřek, 1972). Hare, 1967, recorded the elements on cards on which the main data were inscribed, e.g. the name and code of the element, the description of the transformations in the element, the relationship of the element to the system and to the system environment, etc. These cards are the basis for the mechanization of the matrix and graphical representation of the system and for the analysis of its structure.

An example of a graphical and matrix representation of a general system is given in Fig. 2.2. The nodes of the graph denote the elements of the system and the directed edges (arcs) represent relationships that are generally material, energy, and information flows. In the matrix representation the starting node is in the row and the final node in the column.

In a simple system with few elements graphical representation and matrix representation are equally suitable; in large-scale systems the information can be more easily extracted from a matrix representation.

In some graphical representations (with the exception of the logical analysis – see 9.4.2), the critical path method (or its modification PERT) can be used for analysis of the time dependence and determination of critical points in the system.

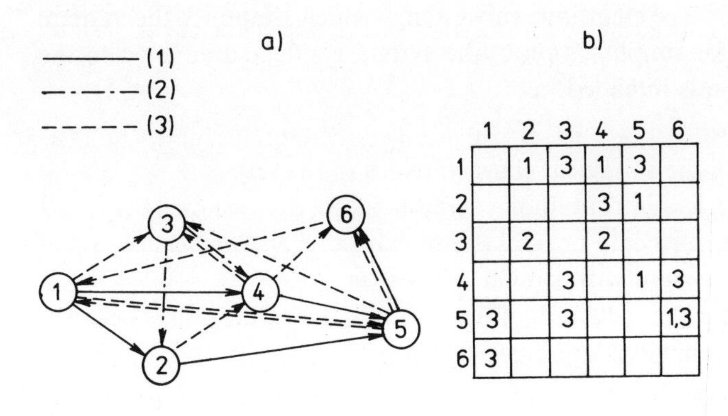

	1	2	3	4	5	6
1		1	3	1	3	
2				3	1	
3		2		2		
4			3		1	3
5	3		3			1,3
6	3					

Fig. 2.2 Schematic representation of the system
a – graphical (flowchart),
b – matrix
Notation of relations:
(*1*) material, (*2*) energy,
(*3*) information

A *simplification* (reduction) of complex and large-scale systems is frequently a compromise between two extreme situations:

– the system is not simplified (accurate results, no costs for simplification, but a very expensive treatment),

– the system is oversimplified (great reduction in costs for treatment but part of the costs is used for simplification; the results, however, may be unreliable).

The increase in the speed and capacity of facilities for information processing makes the choice of less reduced systems possible.

A simpler system can be reduced directly. For simplification of the WRS the following direct methods can be used:

– elimination or aggregation of elements and relationships in the system,
– substitution and transformation of variables and relationships,
– subdivision into subsystems.

Elimination can be performed by "filtering" the components of the system through the given constraints. The following procedures can be employed:

– a reduction of the acceptable range of variables (from among the whole range of the set, values with probability $0.1 \leqq p \leqq 0.9$ are considered),

– reduction of the area of interest; the boundary of the system with the environment is reduced;

– reduction of the number of possible combinations by logical reasoning, or by statistical sampling,

– selection of the inter-element relationships (e.g. by quantity or by frequency of occurence) can eliminate relationships that are above or below a certain value, i.e. the threshold (e.g. withdrawals under a certain limit are omitted); the decrease or increase of the frequency or the number of relationships among the elements can help in the subdivision of the system into subsystems, which simplifies the system.

The use of elimination for simplification of the system is often suitable when the optimization of its function is intended.

Aggregation can be done:

– statistically (i.e. using the statistical characteristics of the set),

– logically (e.g. by change of a continuous variable into a discrete one),

– using prototypes, i.e. elements or subsystems whose properties enable them to represent a group in a system with similar properties.

Aggregation is suitable for the reduction of the system if the relationships between the subsystems and their behaviour are being tested.

Substitution and transformation of variables and relationships are to simplify the description of the system or transform it into a known form. The transformation has to be unique (i.e. possible inversion transformation of the reflection to the original system); the properties of the system that are invariable with respect to the applied transformation should be known.

Subdivisions into subsystems, which are relatively closed, is suitable for large-scale systems; the subsystems are then treated as separate systems and the treatment of the original system is simplified.

All the direct methods of system reduction are based on the assumption that the system is stationary (i.e. with time independent statistical properties). In grouping and selection, the consistency of the system should be maintained (i.e. the same approach by different researchers, the same methods of data collection and processing, etc.).

In the *analysis of structure and behaviour of existing systems* the following items are investigated:

– the principles and objectives of the system,

– the structure of the system and its adequacy for the given objectives,

– the behaviour of the system (whether it corresponds to the assumptions or why not),

– the means of system improvement (if any) with the expected impact on the behaviour of the system.

As the main purpose of this analysis is the diagnosis of weak points in the structure and behaviour of the system, this analysis is called *diagnostics*. It can be classified as (Habr and Vepřek, 1972):

– *qualitative* – the weak points of the present organization are detected, the goals for its improvement are determined, the principles of perfect system are stated, the introduction of new methods and facilities is planned and approved,

– *testing* – the basic information about the analysed system is obtained from by internal and external tests: in *external* tests the system is taken as a black box, the input parameters are varied and the response is investigated (the response test is used mainly in real systems with a flow of material); in *internal* tests discontinuities are detected, and the interface relationships between the elements, the impact of feedback and the alternative function of some parts of the system are investigated.

Simulation of the system is often a successful way of system testing. In principle, it provides the possibility to evaluate a real system before putting it into practice, often evaluation of real systems is not, in fact, possible in complex WRS.

The multi-purpose WRS are often simulated mathematically by the Monte Carlo technique, i.e. the technique of random processes. Computers are necessary and modelling is expensive. Therefore it is necessary to find out if:

– the conditions are suitable for simulation,

– there is a cheaper and easier method for solving the problem than simulation,

– the human role in the control process does not eliminate the possibility of using the simulation technique,

– the low frequency of task computation does not prevent the setting-up of a special simulation program,

– the conditions in the simulation model and in the real system are the same.

2.4 METHODOLOGY OF WATER RESOURCE SYSTEM DEFINITION

From among the many possibilities of defining a WRS, an example of a multi-purpose WRS where one goal has priority is presented.

2.4.1 Task Formulation by Water Resource System Analysis

A system of reservoirs in a basin is to be completed and its operation optimized. The system is multi-purpose with the following objectives:

– municipal and industrial water supply,

– flood control,

– low-flow augmentation,

– hydroelectric power generation,

– water-related recreation.

In Fig. 2.3 there is a schematic representation of
- water resources and existing and designed reservoirs,
- larger, suitably aggregated withdrawals of water,
- watercourses and canals.

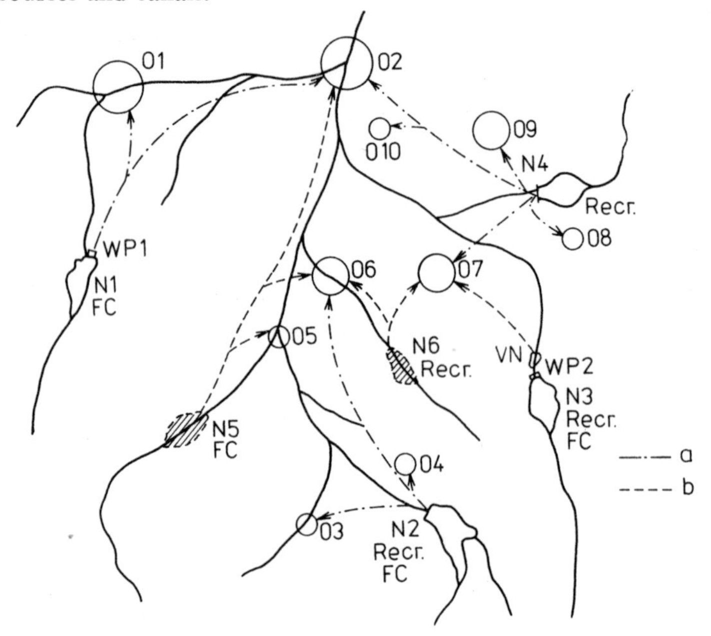

Fig. 2.3 Schematic representation of a multipurpose WRS
N1 to *N4* – existing reservoirs, *N5*, *N6* – designed reservoirs, *VN* – buffer storage reservoir, *01* to *010* – withdrawals of water, *a* – existing water mains, *b* – designed water mains, *Recr.* – water related recreation, *FC* – flood control, *WP* – water power plant

At present, there are four reservoirs in operation. Their goals, together with the water supply, are given in Table 2.3, the designed reservoirs are denoted *N5* and *N6*.

Since the aim was to obtain only a reflection of the relationships among the elements of the WRS, the simplest representation by a zero-one matrix was used.

A system with six water reservoirs and ten aggregated water withdrawals is relatively simple; the matrix representation is more suitable than the graphical one (Fig. 2.3) even if it is simplified.

One can read directly from the matrix representation that:

– the existing reservoirs are mainly used as isolated reservoirs (i.e. independent, isolated subsystems) with the exception of reservoirs *N1* and *N4* whose purpose is to supply water for withdrawal *O2*,

– water for withdrawal *O5* is not supplied from the reservoirs,

– reservoir *N3* is not utilized for the municipal water supply yet; its utilization

Table 2.3. The purpose of reservoirs in the system

Reservoir	Purpose of the reservoir				Withdrawals of water									
	Water supply	Flood control	Power generation	Recreation	*O1*	*O2*	*O3*	*O4*	*O5*	*O6*	*O7*	*O8*	*O9*	*O10*
N1	1	1	1		1	1								
N2	1	1		1			1	1		1				
N3	(1)	1	1	1								(1)		
N4	1			1			1				1	1	1	1
N5	1	1				1			1	1				
N6	1			1						1	1			

for this purpose (see (1) in Table 2.3) can relieve water resource *N4* which is the only resource for withdrawals *O8, O9* and *O10*,

– the recreational function is combined with the municipal water supply and hydroelectric power generation,

– reservoir *N5* is an important complementary resource for WRS for the following reasons:

a) it supplies water for withdrawal *O5*,

b) it increases the quantity and reliability of the water supply for *O2* and *O6*,

Table 2.4. Matrix of the competitive purposes of the system

Purpose	*Z*	*O*	*H*	*R*
Z – water supply	–	1	1,3	2,3
O – flood control	1	–	1	–
H – hydropower	1,3	1	–	3
R – recreation	2,3	–	3	–

1 – Competition in reservoir storage requirements

2 – Competition in water quality

3 – Competition in operation requirements

c) via *O2* and *O6* the system is linked with *N1*, *N2* and *N4*, which forms a compact water supply system,

d) it increases the degree of flood control (flood alleviation).

– reservoir *N6* is a planned versatile element of the water supply system. If it delivers water for large withdrawals *O6* and *O7*, water resources *N2*, *N4* and *N5* can be relieved; its location makes it suitable for recreation.

To simplify the defined WRS and the treatment of the problem, a logical analysis is carried out:

The system has four objectives, i.e. it is a multi-purpose system. Firstly, a *reduction in the number of aims* should be achieved, since the multi-purpose nature of the system greatly complicates its optimization, especially if the aims are competitive; a matrix of competitions has been set up (Table 2.4):

This matrix is, of course, symmetrical around the main diagonal. It is clear that all the four purposes are to some degree competitive, with the exception of recreation and flood control (in this example). Competitions (1) and (3) are approximately equally frequent, competition (2) occurs only once.

The respective priorities of the purposes can be determined in this relatively simple case by a comparison of their social and economic impacts in the context of the plan for economic and water management development of the area, e.g. in this way:

The most important demand on water resources in this region, as compared with other requirements (represented by three purposes), is the *municipal and industrial water supply*; the question is, if the other three purposes can be considered as the constraints on the main demand.

Flood control: if the necessary flood control volumes A_r in the reservoirs are determined, the single existing conflict (1) is thus resolved; if the reservoir has a carry-over volume (i.e. large active storage A_z) and storage A_z is also used for a reduction in flood discharges, conflict (3) can be resolved by unconstrained requirements for the water supply. If there is no provision for the construction of the storage volumes of reservoirs for both purposes *Z* and *O*, some new alternative serving both these purposes will have to be found.

Hydroelectric power generation: the conflict between purposes *Z* and *H* concerns requirements on storage of the reservoirs (1) and their operation (3). As the reservoirs are on small watercourses equipped with hydroelectric power plants of limited capacity and of minor significance in the power system, the hydroelectric power requirements can be subordinated to purpose *Z*; in this way the conflict has been resolved. The economic efficiency of purpose *H* should be determined under these constraints; e.g. the adverse impact of peaking of hydroelectric power plants on the water quality would restrict their peaking operation.

Recreation: the conflict between purposes *Z* and *R* concerns the water quality requirements (2) and operation (3). In the present situation, in principle, the combination of recreation and municipal water supply is possible. A detailed analysis is to be used to decide if this combination of purposes Z and R in a reservoir is effective or if it would be more advantageous to separate these two important purposes (both have a high environmental impact).

Further relationships between the *various purposes* are treated, e.g. between purposes O and H: the conflict between purposes O and H is of type (1) and it has been dealt with by resolving the conflict between purposes Z and O, since purpose H is subordinated to Z. If there is a conflict between O and H, the social and economic analyses will, under the given conditions, give higher priority to flood control O.

The conflict between purposes H and R concerns operation (3); for R a constant water level is advantageous, H requires a variable water level due to peaking operations which are, however, constrained by Z.

The result of this short analysis is the transformation of a system with four purposes into a system with a single main purpose, with the remaining three purposes being

Table 2.5 The relation among the resources and withdrawals of water in WRS

Element of WRS	*N1*	*N2*	*N3*	*N4*	*N5*	*N6*	*O1*	*O2*	*O3*	*O4*	*O5*	*O6*	*O7*	*O8*	*O9*	*O10*
N1	–	0		*A*	*A*	0	1	1			0	0	0	0	0	0
N2	0	–	0	0	*A*	*A*		0	1	1	0	1	0			
N3		0	–	*A*	0	*A*		0				0	1	0	0	0
N4	*A*	0	*A*	–	*A*	*A*	0	1			0	0	1	1	1	1
N5	*A*	*A*	0	*A*	–	*A*	0	1	0	0	1	1	0	0	0	0
N6	0	*A*	*A*	*A*	*A*	–	0	0	0	0	0	1	1	0	0	0
O1	1			0	0	0	–	*a*								
O2	1	0	0	1	1	0	*a*	–			*a*	*a*	*a*	*a*	*a*	*a*
O3		1			0	0			–	*a*		*a*				
O4		1			0	0			*a*	–		*a*				
O5	0	0		0	1	0		*a*				*a*				
O6	0	1	0	0	1	1		*a*	*a*	*a*	*a*	–	*a*			
O7	0	0	1	1	0	1		*a*				*a*	–	*a*	*a*	*a*
O8	0		0	1	0	0		*a*					*a*	–	*a*	*a*
O9	0		0	1	0	0		*a*					*a*	*a*	–	*a*
O10	0		0	1	0	0		*a*					*a*	*a*	*a*	–

Notation of relations:

1 – direct relation between the water resource and withdrawal

A – direct relation among the water resources supplying the same withdrawal

a – direct relation among the withdrawals from the same water resource

0 – indirect relation among the water resources and withdrawals via one intermediate withdrawal

used as constraints on this main one. The conflicts among these three subordinate purposes have been analysed.

The *discriminating level* in the definition of the WRS was expressed in Fig. 2.3 where the elements and relationships (directed edges giving the flow of water) were drawn.

The graphical representation in Fig. 2.3 together with Table 2.5 can be used for a possible subdivision of the WRS into *subsystems*. The direct relationships, denoted in the matrix as 1, *A*, *a*, will play the decisive role. The indirect relationships, denoted as 0, will be taken into account in the second order as a characteristic of the interrelationships between the elements of the system.

A look at Table 2.5 shows:

– a high degree of interrelatedness among the elements of the system; the definition of WRS was therefore necessary and the systems approach is therefore methodologically correct,

– the relationships involving the functions of water resources are mainly direct or via one intermediate withdrawal (with the exception of *N1* and *N3*),

– water resources *N5* and *N6* were well chosen; they are directly or indirectly (0) related to all the other elements of the system,

– the following subsystems can be defined: *N1* and *O1* (the environment of this subsystem is the constraint on *O2*), *N2*, *O3*, *O4* (the environment of this subsystem is the constraint on *O6*).

The *relationship between the system and its environment* is reduced to O_N, i.e. the reliability of low-flow augmentation in the river that leaves the system under withdrawal *O2*.

This analysis having been carried out, as was briefly discussed, the *formulation of the task* is possible: Design an optimal completion of the multi-purpose WRS by additional water resources, considering primarily sites *N5* and *N6*. The system is to serve four main purposes (*Z*, *O*, *H*, and *R*), purpose *Z* being the main one and the other three constraints on it. The relationship to the system environment is given by reliability requirements for discharge O_N under withdrawal *O2*. In the treatment of the system it is necessary to delimitate two subsystems of the WRS, (*N1*, *O1*) and (*N2*, *O3*, *O4*), which are relatively loosely related to the other parts of the system.

2.4.2 Water Resource System Definition

In a multi-purpose WRS different subsystems can be defined:

– a subsystem of watercourses and reservoirs,
– a subsystem of drinking-water consumers,
– a subsystem of gauging stations,
– a subsystem of power transmission,
– a subsystem of communication, etc.

On the basis of the preliminary analysis the multi-purpose WRS was transformed into a single-purpose system with some constraints. The main purpose was water supply. Therefore this WRS was defined mainly as a water supply system.

The problem was precisely formulated to allow definition of the system, the system was quantified, i.e. subdivided into parts (viz. elements, subsystems), the relationships were determined, as given by the flow of water (Fig. 2.3). In the graphical representation a desirable and reasonable simplification was made and a system abstraction prepared.

The *elements and subsystems* are (*N1, O1*), (*N2, O3, O4*), *N3, N4, N5, N6, O2, O5, O6, O7, O8, O9, O10*; they include water resources and water withdrawals (canals and pipelines that facilitate linkage are considered as relationships).

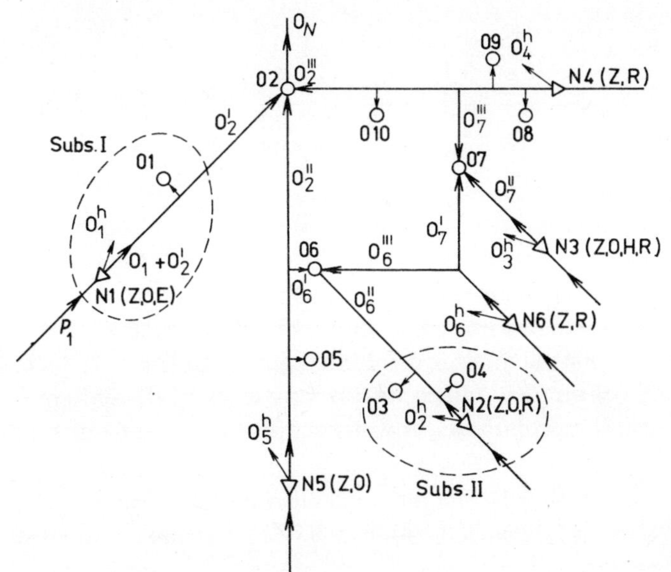

Fig. 2.4 Graphical representation of the water supply system

The river downstream of site *O2*, where O_N is to be maintained, is considered as an *element of the environment*.

The relationships between the elements of the system and with the environment are graphically represented in Fig. 2.4.

Further treatment is only outlined roughly; a detailed explanation is more suitable for a concrete case (see Chapter 12). The schematic representation in Fig. 2.4 or in Table 2.3 is not sufficient for an unambiguous (unique) procedure as it does not include e.g. the capacity of water resources, the dimensions of storage volumes of reservoirs, the quantity and importance of the withdrawals, water quality, etc.

The water supply system has dynamic inputs and outputs. It is an extensive system with six reservoirs so that further simplification would be necessary for optimization by dynamic programming:

– alternatively, the system is treated with different time horizons, (e.g. for the time level 1990, and 2000 with consideration of further development),

– for each time level a number of alternatives are dealt with and from the chosen criteria an optimum alternative is selected.

Firstly, both subsystems are considered (Fig. 2.4):

a) Subsystem (*N1, O1*): The relationships between the stochastic inflow $P_1 = P_1(t)$ and its statistical characteristics are known, assuming a similar knowledge of the municipal water supply $O_1 = O_1(t)$ and a given minimum flow O_1^m. Using these data and assumptions the storage volume A_{z1} of the reservoir *N1* can be calculated for a given reliability p_1:

$$A_{z1} = f(P_1, O_1, O_1^m, p_1) \tag{2.1}$$

It was assumed $O_2' = 0$. In addition, the flood control storage A_{r1} and dead storage A_{s1} are determined and then (omitting the hydropower generation) the total storage volume of the reservoir *N1* is calculated (A_{p1} is the volume corresponding to spillage):

$$A_{1\,\mathrm{total}} = A_{s1} + A_{z1} + A_{r1} + A_{p1} \tag{2.2}$$

The possibility of allocating this volume in the given geomorphological conditions is determined and the technological and economic indices etc. are calculated.

If the hydrological and geomorphological conditions are suitable for $O_2' > 0$, then the corresponding indices are determined for several values of O_2' and compared with indices for O_2'' and O_2''' to optimize the water supply for O_2 from different water resources.

b) Subsystem (*N2, O3, O4*) is dealt with in a similar manner to subsystem (*N1, O1*) as the two cases are equivalent in principle, if withdrawals (*O3* + *O4*) are combined and constraints *H* and *R* are omitted.

In this subsystem a relationship between O_6'' and the water reservoir *N2* is obtained and expressed in appropriate indices.

We could further simplify the system by forming a subsystem (*N4, O8, O9, O10*), which has two relationships with the environment: *O2* and *O7*.

Having sufficiently and acceptably simplified the system, we can start to deal with the system in technological and economic units.

The optimum alternative is determined, using economic indices relating to the main goal and taking into account the secondary goals and intangibles as constraints.

In this way the problems of the system are solved for one time level. The system is then treated similarly for the second time level, using the same intermediate results from the first time level (e.g. the technical and economic parameters for O_2', etc.).

The results of both time levels are used in the choice of a final alternative.

2.5 WATER RESOURCE SYSTEM EVALUATION

The evaluation of a WRS and the decision-making concerning the optimum alternative for the system's development are difficult and possess some special characteristics, so that the usual ("classical") methods may fail. The decision-making process involves various institutions which prefer different criteria for evaluation. Even one decision-maker has a difficult task in a multi-purpose WRS with incompatible goals or in evaluating a single-purpose WRS using various criteria. In addition, such decisions are often irreversible, cause considerable impact, and involve great uncertainty in input data and prediction of development. A special, and in WRS a typical, problem is the combination of economic and intangible criteria.

In spite of many references to the subject (see e.g. Haimes, Hall, and Freedman, 1975), and attempts to formalize the decision-making and optimization process, subjective heuristic reasoning is still often an indispensable part of it.

The "Recommendation" of the Ministry of Forestry and Water Management of the ČSR on the evaluation of the effectiveness of structures in water management, 1976, includes methods for and examples of the evaluation of alternatives for water engineering structures by a procedure which includes relative and absolute efficiency, an evaluation of construction by stages, and the allocation of costs in a multi-purpose WRS. For the evaluation of WRS the method of decision analysis can be used as it is useful for the evaluation of alternative structures when a significant proportion of the requirements and consequences cannot be expressed in terms of money. This method reduces the subjective part of the evaluation (Novotný and Polenka, 1973).

2.5.1 Water Resource System Evaluation by Decision Analysis

The decision-making process consists of four steps:

1) definition of the criteria for the evaluation of the extent to which the alternative fulfils the purpose,

2) a simple evaluation of alternatives on a point scale,

3) mutual comparison of criteria in pairs using Fuller triangle,

4) a weighted evaluation of the alternatives by combining steps (2) and (3).

The *criteria* should express the principal purposes of the system and should not overlap. They are chosen according to the case being analysed:

a) *Cost* criteria include the initial costs, or a comparison of initial costs with a given limit. An alternative that requires lower initial costs may be preferred, using the objective function

$$Y = \sum_{i=1}^{H} I_i \overset{!}{=} \min \qquad (2.3)$$

where H is the time horizon under consideration (e.g. Five-year plan), I_i – initial costs in time interval i.

b) *Production* criteria are concerned with the quantity and quality of water management production and services. The quantity is stated in technical units or in dollars, if adequate prices or unit costs are available. The quality is evaluated separately if price cannot express it (e.g. reliability of water supply, increase in water temperature in summer, etc.).

c) *Environmental* criteria are used for

– conservation and enhancement of the environment (the positive and negative effects of the alternatives on the present state of the environment and possibilities for reducing negative effects are evaluated),

– town planning and architectural evaluation of the water engineering structures,

– impact on the working and living environments,

– the use of water engineering structures for water-related recreation and sports,

– changes in land use (it is assumed that the economic value of the ground is expressed in costs for land, which are included in the initial costs),

– the extent of displacement of the inhabitants,

– preservation of buildings of cultural or historical value.

d) Criteria of *economic effectiveness* can include:

– the present value of initial costs and OMR costs (operation, maintenance and repair), if the outputs of the alternatives do not differ too much,

– the average discounted net benefits (if the outputs of the alternatives differ significantly and if it is possible to express the output in monetary units).

e) *Construction* criteria apply to the evaluation of positive and negative factors relating to construction that are not included in the initial costs such as the following items:

– duration of construction work,

– uncertain foundation conditions which have not yet been described by a geological survey,

– the dependence of the construction work on parts imported from abroad,

– the precision requirements for products which the producer can guarantee only with great difficulty,

– local climatic, transportation, manpower, etc. conditions,

– the use of standardization and prefabrication,

– the possible use of free construction capacity of the contractor.

f) Criteria of *optimization* include:

– the reliability of operation,

– the ability to rectify operational failures,

– the frequency of maintenance and repairs,

– the quality requirements for the supply of fuel, energy, and material, which are not included in their costs,

– the climatic conditions of operation,

– the operational constraints imposed by repair and replacement of construction parts.

g) *Development* criteria include:

– coordination with water resource development of the region, e.g. with respect to the development of superior WRS, General Water Plan, etc.

– development adaptability to changed conditions,

– conceptual reliability with reference to possible failures,

– the degree of continuous utilisation of the designed capacities with possibilities for construction in stages.

h) *Other* criteria, e.g.

– conflicts with activities of other sectors, which are mutually exclusive or constitute certain constraints,

– preparation for construction (design, land use, projects, operation, etc.),

– international conditions of construction in cases involving extention across borders into neighbouring countries, national defence, etc.

– coordination with the political interests in the region.

Table 2.6 Fuller triangle

Evaluated criteria	Criteria	Number of preferences
1 1 1 1 / 2 3 4 5	1	3
2 2 2 / 3 4 5	2	4
3 3 / 4 5	3	0
4 / 5	4	2
	5	1

a – with couples of rows

2 3 4 5	Criteria	Number of preferences
2 1 1 1	1	3
2 2 2	2	4
4 5	3	0
4	4	2
	5	1

b – with a single row

From the above-mentioned criteria those items are chosen that best characterize the effectiveness of the designed alternatives (or some new criteria are added).

A *simple evaluation of alternatives* is carried out by awarding each alternative and each chosen criterion a number of points, e.g. 0 (does not satisfy) to 5 (satisfies best). The total number of points gives the simple score of the individual alternatives.

The *mutual evaluation of the criteria* is carried out by comparison in pairs of all the criteria used and it is represented in a tabular form called Fuller triangle.

The Fuller triangle is composed of pairs of rows; the first row of each pair contains the serial numbers of the criteria being processed and the second row contains the numbers of other criteria that are compared with the criteria in the first row. The pairs in the column are compared and the preferred criteria are underlined. The number of preferences (underlined numbers) of the criteria score in the whole triangle determine their respective weights. To eliminate the case of a zero weight, all the weights are increased by one.

The Fuller triangle is substantially simplified (to half the rows without underlining) if instead of double rows only a single row with preferred criterion is used (see Table 2.6 for five criteria).

The *weighted evaluation of alternatives* is obtained by multiplying the figures resulting from the simple evaluation of alternatives in the second step by the weights of the corresponding criteria in the third step. The sum of all the weighted values of the criteria for each alternative yields the total weighted score, the highest of which determines the most advantageous alternative from the standpoint of the criteria used.

Evaluation of the method of decision analysis

The whole procedure is formalized and therefore objective. The choice of criteria and their evaluation is, however, heuristic. Both these properties tend to be advantageous but are sometimes disadvantageous.

The advantage stems from the fact that the procedure is unambiguous, and that the calculation is preceded by reasoning and deliberation that leads to a deeper understanding of the individual characteristics of the alternatives, which can be numerous and would otherwise be hardly commensurable.

The disadvantages pertain to the fact that the formalized procedure need not express in the way described the proper proportions of the relative advantages of the alternatives, and that in a choice of alternatives and their evaluation, a subjective standpoint might prevail.

Therefore the chosen criteria should exhaust all the important properties of the alternatives investigated. In particular, properties that differentiate the alternatives must not be omitted. On the other hand, criteria which are not important for the investigated structure can be omitted.

The subjective element can be diminished by first having the evaluation performed by several researchers individually. Their different views are then discussed, the effect of different conclusions on the results is determined and the evaluation is repeated.

The result of this analysis is not identical with the final decision of the decision-maker, but such an analysis is a very important decision-aiding tool.

An example of decision analysis

We are to decide between two alternatives using the following steps:

1st step – definition of criteria: they are listed in Table 2.7. (in this case, eight criteria).

2nd step – determination of the number of points (scale 0 to 5 points), the simple evaluation of both alternatives being given by the total number of points (Table 2.7).

Table 2.7 The list of criteria and the simple evaluation of alternatives

Criterion	Alternative	
	I	II
1 Economic effectiveness	5	4
2 Quantity and reliability of water supply	4	3
3 Flood control	2	4
4 Environmental impact	3	1
5 Land use and its change	1	2
6 Requirements of construction	3	1
7 Operational reliability	4	2
8 Adaptability to changed conditions	2	3
The number of points by simple evaluation	24	20

It is clear that the evaluation of some criteria by points and their interrelationships depend on the expert's judgement. In a real case, a greater number of criteria would often be used. Adding more alternatives causes no trouble. In our case, the simple evaluation seems to indicate that the first alternative (I) is the more advantageous, even if it is less advantageous from the standpoint of flood control, land use and change and adaptability to changed conditions.

3rd step – the comparison of pairs in the Fuller triangle (Table 2.8) is used in a simplified form (see Table 2.6.b). A subjective element also appears in this comparison in pairs.

4th step – the weighted evaluation of alternatives (Table 2.9) is obtained by multiplication of the values from the second and third steps (see Tables 2.7 and 2.8), the weight being the number of preferences increased by one.

Table 2.8 Evaluation in pairs in the Fuller triangle

2	3	4	5	6	7	8	Criterion	Number of preferences
2	1	1	1	1	1	1	1	6
	2	2	2	2	2	2	2	7
		3	3	3	7	3	3	4
			4	4	7	4	4	3
				6	7	8	5	0
					7	8	6	1
						7	7	5
							8	2

Table 2.9 The weighted evaluation of alternatives

Criterion	Weight	I		II	
		simple	weighted	simple	weighted
1	7	5	35	4	28
2	8	4	32	3	24
3	5	2	10	4	20
4	4	3	12	1	4
5	1	1	1	2	2
6	2	3	6	1	2
7	6	4	24	2	12
8	3	2	6	3	9
Total			126		101

It is clear that in a weighted evaluation a difference in the effect of the individual criteria is more pronounced than in a simple evaluation.

In this case also, alternative (I) was more advantageous.

In view of the nature of the evaluation the result must be considered as a preliminary one. It is necessary to return to the decision criteria, to the uncertainty in the preferences, and to determine the effect of possible changes on the resulting final evaluation. (Sýkora, 1980). However, only one part of the decision-aiding information will be obtained in this way[1]).

The weights of criteria are very important in the methodology applied. The deterministic approach using the Fuller triangle in fact does not determine the "weights" according to the meaning of experts but their significant order based on mutual comparison. Bečvář, 1986, suggested to use instead of the point evaluation a linguistic one based on the evaluation of criteria and their values for the individual alternatives analysed.

The experts are given a sufficient number of statements describing the quality (e.g. 25 statements for evaluation from "excellent" or "extraordinary" to "catastrophic") and approximately the same number of statements for expression of the quality (from "giant" to "zero") and about nine sentences for different combinations of both individual evaluations. The variability of the possible evaluations is sufficiently rich.

For an objectively justified transition of the verbal statements to the numerical evaluation about two hundred experts were interviewed by Bečvář in order to transfer the imagination of the numerical evaluation of the statement into a 50-point scale. The graphical evaluation by a subjective function of the relationships of the point scale elements to the statements and the sentences was added. The central "gravity" points and the "width" of each statement were the results of these data processing.

---

[1]) The method of decision analysis described above was used by a team of eleven members of the research institute, the university of technology and the water board responsible for WRS operation, for the determination of the optimal alternative for the capacity of the Slezská Harta reservoir (see Chapter 12). The capacity designed by various projects varied from 150 to 330 mil. m^3. The table shows the results of six typical respondents, including the whole capacity range. About half of the thirteen criteria show profound differences. Nevertheless, the weighted evaluation tends to favour the lowest capacity A or the middle capacity B. The results of respondents 4, 5, 6 reveal little difference between alternatives A, B and C:

Respondent		1	2	3	4	5	6
Weighted	A (150 mil. m^3)	1.00	1.00	1.00	0.99	1.00	0.99
score of	B (210 mil. m^3)	0.74	0.86	0.91	1.00	0.94	0.99
alternatives	C (250 mil. m^3)	0.64	0.80	0.88	0.99	0.93	1.00
(final values):	D (290 mil. m^3)	0.35	0.69	0.71	0.77	0.71	0.88

This method was used with a good result in comparing four alternatives of a big pumping water plant in Czechoslovakia where the environmental ciriterion of an ecological area protection was applied. Using the Fuller triangle the ratio of the lowest criterion to the highest would be 1 : 11, according to the verbal evaluation of the experts this ratio was approximately 1 : 2 only.

2.5.2. Economic Evaluation of Water Resource Systems

The economic evaluation of WRS is always a fundamental one and no decision-making process can do without it. Economic and other evaluations are used in the rational design of WRS not only in the final stage of the project but during all its stages. Only in this way can a project that is feasible and sufficiently close to the optimal alternative be obtained.

The evaluation of effectiveness from the standpoint of society as a whole is more difficult than a mere cost-benefit analysis. A cost-benefit analysis using only market prices for monetary evaluation of effects is not acceptable as it does not include all those social consequences which cannot be expressed as relationships between supplier and consumer.

Therefore, not only market prices but also "shadow" prices (Krutilla, 1962; Maass *et al.*, 1962; Vitha and Doležal, 1975; Recommendations, 1976) are recommended for the evaluation of products and services, which are not used in relationships between supplier and consumer.

Vitha and Doležal, 1975, describe the method of calculation of these shadow prices (price equivalents). For products where the amount of human labour involved can be calculated, the two prices must be the same. For multi-purpose production an iterative procedure is recommended.

In the first step of the calculation, it is assumed that the total value of the benefits, CV, of Water Boards and engineering organisations all over the country has to correspond to the total costs, SNN, that must be paid by society over the given period (e.g. five years plan or an annual average of this five year period), then

$$CV = SNN \tag{2.4}$$

and

$$CV = V_1 + V_2 + \dots V_n = \sum_{i=1}^{n} V_i \tag{2.5}$$

where V_i are benefits from the individual water-related products and services, for which the following relationships is valid

$$V_i = SNN_i \tag{2.6}$$

Since, in this step, the prices for the determination of the benefits are not known, they are determined on the basis of replaceable products (e.g. electric power from thermal power plants, transportation by railways, etc.) or by prices that the consumer is willing to pay (e.g. the shadow price of water-related recreation). These benefits are denoted by $\bar{V}_i$ and $\overline{CV}$; it is possible that

$$\overline{CV} = \sum \bar{V}_i \gtreqless CV \tag{2.7}$$

If $\sum \bar{V}_i$ is greater than CV, the multi-purpose treatment is effective; if $\sum \bar{V}_i$ is less than CV, this treatment is not advantageous.

Since equations (2.4) and (2.5) must be satisfied, the benefits $\bar{V}_i$ may be reduced to $\bar{V}_{ir}$, as the total cost of replaceable products and services should be greater than their socially necessary costs:

$$\sum \bar{V}_{ir} = CV = SNN \tag{2.8}$$

Since up to this point an equal effectiveness for all the products was assumed,

$$\bar{V}_{ir} = k\bar{V}_i \tag{2.9}$$

where

$$k = \frac{CV}{\sum \bar{V}_i} < 1.0 \tag{2.10}$$

The calculated price of the product or service in the first step will be

$$C_i = \frac{\bar{V}_{ir}}{P_i} \tag{2.11}$$

where P_i is the quantity of the product or service for one year (in technical units).

The prices obtained in this way in the first step are corrected by common and individual costs of the corresponding common and individual parts of the multi-purpose water engineering construction.

There are many methods for the allocation of costs of a multi-purpose project among the individual benefits (Votruba and Broža, 1966). "Recommendations", 1976, allocates the costs in proportion to the benefits and side effects expressed in monetary units. If there are significant side effects among the inputs and outputs of the multi-purpose water engineering project, which cannot be evaluated by shadow prices, these effects may be evaluated by the initial and OMR costs of the substitute or they may be determined by agreement among the users. The result depends on the precision of the econometric measurement of products and services.

2.5.3 Multi-Objective Optimization

In the optimization of multi-purpose WRS several criteria are applied and the problem of multi-purpose optimization arises. Studies on such subjects have been published since the early seventies[1].

Treatment of multi-purpose optimization by means of a mathematical model requires a definition of the individual objectives, the relationships between them and the goals of the system and, based on them, a global objective of the system leading to a unique algorithm for the selection of the optimum alternative. Often, ideas of desirable and feasible values of the criteria are modified during this treatment; this leads to a compromise solution, which takes into account the interrelationships of criteria in feasible solutions and the requirements of the decision-makers.

The partial goals require their maximum or minimum values (extreme goal) or they are expressed in the form of the constraint (constrained goal). The analysis of their relationships to the objectives of the system should be followed by a synthesis to form a selection method.

The multi-objective task can be treated in such a way that the most advantageous is selected from among several solutions, or that this task is transformed into a single-purpose problem. Molnár, 1976, considers as promising the development of heuristic convergent models which combine both these approaches. Their principle is iterative and the decision-maker reduces, in a series of steps, the set of feasible solutions on the basis of information about the criteria gained in the previous steps. Each objective in this procedure is interpreted as a constraint (principle of satisfaction), and step by step the individual constraints are loosened until the resulting value of the objective is unaffected by the loosening in the previous step. The values of the objectives obtained in each step express the "degree" of satisfaction of the decision-maker. The iterative treatment gives a sequence consisting of the non-decreasing "values" of satisfaction so that, at the end of this sequence, a solution with a maximum "value" of satisfaction (utility) is obtained. The treatment can be applied to the model of the theory of games (Belenson and Kapur, 1973), the parametric model (Fandel, 1972), the eliminating model (Banayoun and Tergny, 1970) and others.

If in WRS one goal (e.g. economic effectiveness) is ranked above the other goals, a single objective can be used (e.g. the standpoint of the national economy), and the optimization can be performed using mathematical decision-optimization models; the secondary goals can be used as constraints with estimated values. Lately, however, there has been a tendency to give intangible goals the same priority and the same weight as economic efficiency.

---

[1]) They include the following: Banayoun and Tergny, 1970; Roy, 1970; Lebedyev *et al.*, 1971; Fandel, 1972; Belenson and Kapur, 1973, etc. The multi-purpose optimization of WRS: e.g. O'Riordan, 1973; Monarchi *et al.*, 1973; Cohon and Marks, 1973; Haimes and Hall, 1974; Haimes *et al.*, 1975.

The application of a vector of objective functions creates a new quality in modelling, mathematical programming and optimal control, especially if there is no numerical estimation of optimal solution. In water resource analysis, a number of studies of systems with several goals have been carried out. O'Riordan, 1973, when planning basins in Canada, introduced the objective of economic growth and an environmental objective. Monarchi *et al.*, 1973, developed an iterative method that leads to a solution on the principle "not less than". Cohon and Marks, 1973, tried to find the best solution from the standpoint of the gross national product and regional reallocation of the product in a developing country. Haimes *et al.*, 1974 and 1975, developed a new approach, viz. the surrogate worth trade-off method. This method is based on the assumption that in the optimization theory the relative values of increments of different intangible goals, given the value of each objective function, are more important than their absolute values. The decision-maker can more easily estimate the relative values of the increments and decrements between any two goals than their absolute values.

A great advantage of the surrogate worth trade-off method is that intangible goals can be evaluated quantitatively. Therefore, they can be used for the multi-purpose analysis of water resource utilisation when questions of the design, construction and operation of reservoirs or water quality are involved.

Multi-purpose optimization, although not yet completely worked out, is one of the most important issues of systems analysis, including the analysis of WRS. In this book some literature references are given and some approaches are discussed which are promising and which should be developed in general and in application to WRS.

2.6 APPLICATION OF HEURISTIC METHODS IN WATER RESOURCE SYSTEM ANALYSIS

Heuristics[1]) is an interdisciplinary science that deals with the investigation of methods of treatment and reasoning that optimize (e.g. by reduction) the solution of problems as compared with less effective random procedures or highly specialized methods (Linhart, 1976).

Heuristics (i.e. the theory of heuristic methods) takes into account logical, mathematical, cybernetic, psychological and philosophical aspects. It investigates the psychological and logical conditions necessary to make the treatment of a problem

---

[1]) Definition of heuristics in Webster's Third New International Dictionary (G & C Merriam Co., Springfield, Mass. U.S.A., 1963): heuristic is the science or art of heuristic procedure valuable for stimulating or conducting empirical research but unproved or incapable of proof – often used for arguments, methods and constructs that assume or postulate what remains to be proven or that lead a person to find out for himself.

by an individual or by a team more effective, it rationalizes mental activity and helps in its planning by heuristic programs.

The *heuristic process* treats a problem in terms of a certain probability and the researchers use analogy, intuition, additional questions, etc.

The new method that can help to reduce the number of alternative approaches to the problem under consideration, with unknown procedures leading to the solution, is often called heuristic. On the other hand, algorithmic methods can be used if the method leading to the solution of the problem is known (an algorithm is a unique, precisely determined procedure of problem solving).

The selection of the method depends on the type of problem: If the problem is well defined and thoroughly logically described, it can be solved by some algorithm on a computer; the vaguely defined problem, however, can be solved only by a human being, since the treatment includes some gnostic activity that is, by definition, a non-algorithmic process. The heuristic treatment of a problem or control of activities starts at the moment the goals and methods used are inadequate for the requirements and a new plan of activities needs be set up.

The heuristic method consists of three main stages (Linhart, 1973):

- formulation of the problem and analysis of the conditions and constraints,
- treatment, i.e. generation of hypotheses, selection of strategies and operations,
- verification of the correctness of hypotheses and adequacy of the method, implementation of the results in practice.

Six main heuristic methods can be distinguished:

- "generate and test" (GT),
- method of adaptation (MA),
- "hill-climbing" (H),
- heuristic search (HS),
- inductive method (IM),
- hypothetical-deductive method (HD).

These methods can be combined and new ones added (Newell *et al.* 1958, 1959, 1969; Pushkin *et al.* 1969, 1971, etc.). All these methods involve the creation of an internal model of the problem by a hypothetical-deductive process (Rubinstein, 1960), when the hypothetical causes of the phenomena appear in the mind of the researcher and he makes deductions from them. In this creative thinking, which is one of the basic heuristic approaches, there is a dialectical unity of analysis and synthesis, generalisation and abstraction. Depending on the state of the treatment, synthesis can procede analysis or vice versa, so that synthesis is an indispensable tool for analysis and analysis for synthesis.

The six above-mentioned heuristic methods were arranged according to the degree of activity of the hypothetical-deductive model, complexity, strength, and generic properties of the methods. GT is the simplest method, as appears from its flow-chart (Fig. 2.5, Linhart, 1976). The researchers generate at random an element x

from the set X; it becomes an input of a decision test which shows whether x has the property P; if it has, the loop ends and a further element can be generated; if it does not, the element is omitted. The GT method is used for the determination if the property P is satisfied in the set X. It is clear that the heuristic strength of this trial and error search of the GT method is low.

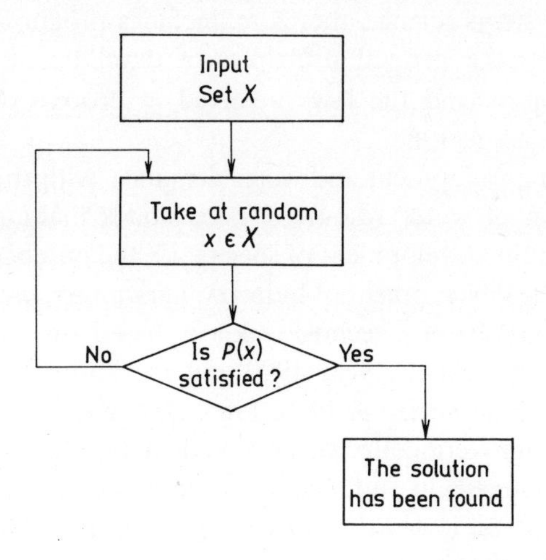

Fig. 2.5 Flowchart of the method "Generate and test" (GT)

The strength of the method is evaluated from three standpoints:

– what is the probability that it will yield a solution?

– what is its quality? how does it optimize the solution?

– what variability of procedures does it have (e.g. different methods of calculation, etc.)?

The universality of the method is given by the variety of problems that can be solved in this manner. Frequently new information that increases the strength of the method decreases its universality.

Experiments to integrate the logical – formal procedures of mathematical models and human activities on the subject of decision-making in heuristic models produced the heuristic convergent models used for multi-objective optimization (see section 2.5.3). The main working tool is the mathematical model. Comparison of alternatives is possible if there are measurable consequences of the decisions taken in accordance with the given goals. In the final estimate of the consequences in complex technical and economic systems, e.g. WRS, a decisive role is played by the "individuality" of the decision-maker with his knowledge and his priorities. The grading of goals is significantly influenced by the subjective attitude of the decision-maker.

2.7 PROGNOSTICS IN THE DESIGN AND OPERATION OF WRS

The nature of water management (i.e. limitation of water resources, the long-term consequences of water engineering structures for man and the environment, their irreversibility and long-term economic life, etc.) forces the water resource planners to deal with the distant future, which is not possible without long-term, medium-term and also short-term predictions. Systems science, therefore, includes prognostics as a scientific field that investigates the general principles of development, methods of prediction for arbitrary objects and the laws involved in process of making predictions (Gvishiani and Lishichkin, 1968).

The causal dependence between social development and water demands, with the necessary development of the utilisation of water resources, necessitates making water management predictions related to the development of society, i.e. the number of inhabitants, their standard of living, the development of industry, agriculture, etc.

Therefore, the General Water Plan (GWP) of Czechoslovakia is based on the development plan of the national economy till the year 1990 and the known or predicted number of inhabitants for the time horizons 1970, 1985, 2000 and 2015. For these time horizons the following data were collected and predicted:

– structure of the settlement (i.e. the movement and concentration of the inhabitants, the proportion of the population living in towns and villages of up to 2000, 5000, 10 000, 100 000, and above 100 000 inhabitants),

– the standard of living and the facilities available (e.g. category of flats, municipal water supply, sewage, etc.).

Trends in industrial production, allocated to the individual sectors, were given till 1985, the long-term trends in agriculture and environment till the year 2000. The general evaluation includes the prediction that the greatest industrial and urban development will be concentrated in the present industrial and population centres. In other regions the development of recreational utilisation can be envisaged.

Water resource predictions are related to social and economic predictions that are typical for centrally planned economies. Specific predictions precede the planning stage, but they need not be incorporated in the plan[1]).

Predictions are integral components of every theory, particularly of systems theory. Management and control (the subject of cybernetics) include two integral parts: the forming of predictions and plans. Prognostics as a scientific field is in a state of rapid development. The stage of creating working methods has not yet been completed; Pearce, 1973, enumerates eighteen possibilities.

Prediction stems from past development; therefore a thorough knowledge of that is a necessary but not the only condition of reliable prediction. In social, econ-

---

[1]) In this sense the GWP is more of a prediction than a plan, as it does not determine the tasks that have to be performed but rather provides an outline ("scenario") of the further development of water resources in the whole country and in its individual regions.

omic and natural phenomena the deterministic components are accompanied by many random factors; even the best prediction, therefore, has a limited degree of accuracy.

Even if a diagnostic analysis of the past and present states has been performed from a prognostic standpoint, the bounds of variability of the future development may be found to increase with the distance of the time horizon of the prediction.

In economics, predictions are often classified according to the distance of the time horizon into short-term (1 – 3 years), medium-term (3 – 7 years), long-term (10 – 20 years), and perspective (30 years and more). The definitions of these terms are not clearcut (Kozák and Seger, 1975) and in water management and in GWP mainly long-term and perspective predictions are used. A great problem in water management, therefore, is the danger of extrapolation of trend curves into the distant future.

Prediction, including its exactness, depends on three sources:

– the results of observation and monitoring of the past period and their correct processing (*input data*),

– *prognostic theory* (model), describing the effect of individual factors,

– *predictor*, i.e. the method of prediction determination used for calculation of the prediction from the input data.

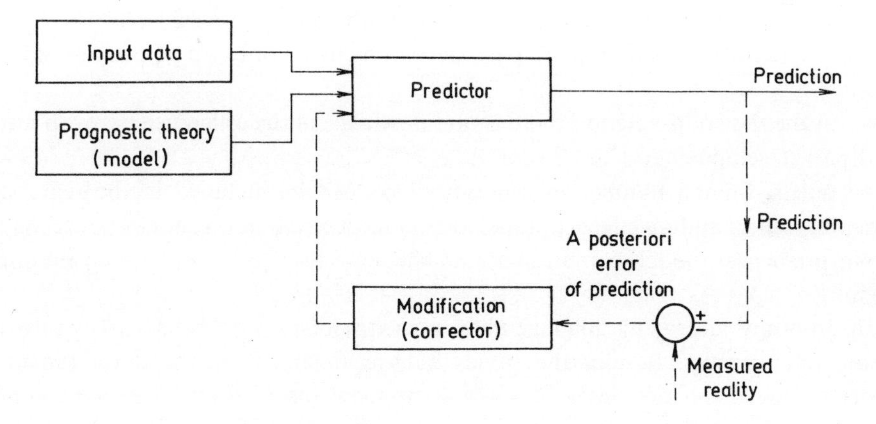

Fig. 2.6 Schematic representation of the prediction process

The flowchart in Fig. 2.6 shows the basic scheme of the predictive process marked by a solid line (Kozák and Seger, 1975). This flowchart indicates that prediction is the estimation of the future level of an unknown quantity according to a model which uses a certain predictor and observed input data.

The prediction may be *quantitative* (numerical) or *qualitative*. In technical and economic problems, quantitative predictions are fundamental; they are expressed by single values (*point* prediction) or by ranges (*interval* prediction). The difference between the predicted value and the real value is the *error of prediction.*

Together with general scientific methods and methods from other fields, interdisciplinary methods are used in prognosis.

The predictions in water management employ mainly the following methods:

- interpolation and extrapolation methods,
- methods based on the probability theory and mathematical statistics,
- methods of analogy,
- modelling methods,
- methods based on the estimation of experts.

Interpolation and extrapolation methods

The elements of a dynamic sequence are determined by *interpolation* using the analysis of relationships between the existing elements. The problem of interpolation is often solved by the determination of a regression function which includes a degree of deviation of empirical values from theoretical ones.

Extrapolation serves for the estimation of the values of the selected function outside the known interval based on the known values inside this interval. In technical and economic predictions, errors of prediction always occur because:

– the regression function of past development is only an approximation to reality,

– at the time of prediction there is no knowledge of the difference between future and past development,

– only a limited number of random effects can be included in the regression function; other random effects not included in its determination cannot be estimated.

No prediction model can eliminate all the sources of error but it can minimize them.

In economy and water management, an *extrapolation of trend curves* into the future often occurs. Besides the trend (long-term tendency), the development in water management includes a *seasonal* component (oscillation), often with annual periods, and a *random* component that is not a function of time. The choice of the trend has the greatest impact on the predicted values, and this influence grows with the distance of the time horizon.

The applied trend curves have three forms:

– monotonically increasing, unbounded for $t \to \infty$ (e.g. linear trend $y = a + bt$, quadratic trend $y = a + bt + ct^2$, exponential trend $y = a + b\,e^{ct}$, logarithmic trend $y = a + b \log t$);

– monotonically increasing or decreasing (without inflection point) to a non-negative asymptote for $t \to \infty$ (e.g. a simple exponential trend: $y = b\,e^{-ct}$, hyperbolic trend $y = a - b/t$; Törnquist' curve $y = at/(b + t)$ etc.);

– having one inflection point and a non-negative asymptote (S – shaped curve, e.g. Gauss–Laplace integral curve $y = \int_{-\infty}^{t} F(x)\,dx$, logistic curve $y = a/(1 + b\,e^{-ct})$ etc.).

The shapes of the trend curves are given in Fig. 2.7a. The trend functions are compared by means of the functions of growth rate that are given by the first derivative of the trend function $y' = dy/dt$ (Fig. 2.7b).

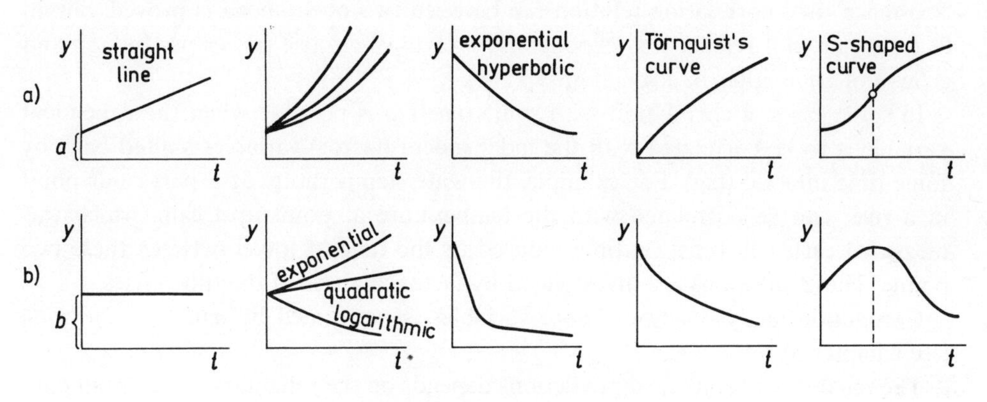

Fig. 2.7 The forms of functions
a – trend function, b – growth rate

The choice of the trend and the growth rate function need to be verified by analysis of the nature of the extrapolated phenomenon, and empirically (i.e. statistically); the former is the basic verification, the latter is an auxiliary one.

The final verification of the exactness of extrapolation (and other) predictions is a comparison of reality with prediction. The errors of prediction can be clearly seen in the Theil P-S diagram (prediction – reality) as deviations from the line of "errorless predictions" that crosses the origin of axes and has a 45° slope – on the abscissa the predictions and on the ordinate the observations are drawn to the same scale. Using errors thus revealed, the prediction model can be modified (corrected) step by step and the adapted prediction can be attained. The basic flowchart of prediction can therefore be supplemented by modification or by feedback (in Fig. 2.6 drawn by a dotted line).

Methods based on probability theory and mathematical statistics

Although probability calculus is used in other methods, too, statistical estimation methods are classified as an individual group since they have a common basis in a correlation relationship between real phenomena (that need not necessarily be causal). The correlation between two phenomena may be caused by the common influence of a third factor; e.g. the correlation between the time series of flows in two gauging stations is caused by similar meteorological and morphological properties of both basins.

The correlation relationship can be described by some statistical characteristics, or by regression equations that can be used for the determination of data concerning one phenomenon on the basis of data for a second one.

Formal mathematical correlation analysis must always be combined with physical analysis to avoid the danger of *spurious correlation*, i.e. when, numerically, the "existence" of a correlation relationship between two phenomena is proved, but, in fact, they are not related (e.g. exponential growth of electric power production and growth of criminality in a social group etc.).

In some cases, a correlation with some *time lag* is possible when the dependent variable is to be correlated with the independent (factor) variables shifted back by some time interval (lag). For example, the water temperature at a particular point in a river can be correlated with the temperature at point upstream, which was measured earlier in time, i.e. time reduced by the time of travel between these two points. These questions are investigated by factor analysis of the time series.

Correlation analysis among more variables is performed by *multiple regression* (see Chapter 3).

The reliability of statistical predictions depends on the reliability of the input data and the logically approved validity of the probability method (model).

Method of analogy

Methods using analogy have often been used in water management problems (e.g. hydrological analogy, electronic and hydrodynamic analogy, etc.). Analogous phenomena are governed by the same laws. Physically they need not be related, e.g. electrical current and water infiltration are both described by the relationship $z = k(\partial y/\partial x)$, the increase of some biological parameters and the growth of scientific information both by an S-shaped curve, etc. In view of the physical independence of analogous phenomena, the reliability of prediction by analogy has to be proved by an alternative method.

In management, sometimes, the *method of historical analogy* is used when the prediction of future development is based on the development of an analogous past phenomenon. For example, the development of water demand in an area with a designed municipal water supply is derived from the growth of demand that occured in another area where the municipal water supply was installed earlier.

The basis of the success of this method is the proof of the likelihood and validity of the analogy under consideration (e.g. analogy of social, local, historical or other conditions).

Modelling method

With the increasing complexity of the problems and the phenomena being predicted, the importance of modelling is increasing. Instead of real objects, models of them are investigated and the results are then transferred from the model to reality. Modelling (i.e. mathematical modelling) of WRS is an integral part of the methods with which they are treated (see Chapters 5, 6, 7, 8).

Technical, scientific and economic predictions often require computer-based models. Four steps have been worked out for such models (Gvishiani and Lishichkin, 1968):

a) Representation of a future reality in technical, economic, social or other areas with regard to national requirements; the trends are extrapolated at the upper and lower levels,

b) probability calculation of relative frequencies,

c) systems correction of errors,

d) integration of the previous stages by computer modelling (sometimes without computer); the result is the prediction of development trends and alternatives for planning strategies.

Methods based on the estimation of experts

Predictions of phenomena that have no analogy in the present or in the past and that cannot be formalized are often based on the predictions of experts, who are assumed to be able to estimate the trends of future development in their own fields. These estimates can be obtained in different ways:

a) *Brainstorming* aims to reveal new ideas through the intuition of experts "in the process of brain tension". It is assumed that among the many ideas of the individuals and groups of experts some important ones may be discovered. For the success of the method the following points are recommended:

– the basic features of the problem should be delineated from one preferred standpoint;

– no idea should be omitted as being incorrect; it should be developed, even if it does not seem realistic;

– open discussion should be stimulated and the intellectual tension should be maintained.

b) The method of *group tension* is based on the consensus of views on some question in a group of experts,

c) The method of *operation creative activity* aims at the most probable solution in the situation where only the chief officer of the project is thoroughly informed about its nature and possible methods of treatment.

d) The *Delphi* method tries to predict future development by a statistical evaluation of the views of the experts on expected development in their own and marginal fields. The experts receive a questionnaire on these issues which is (a) formulated in such a way that the answer may be quantitative, and (b) used in iterative cycles with increasing precision of questions and answers.

Every expert with a specific point of view has to explain and justify it. The method is carried out in writing which excludes direct discussion but facilitates a continual exchange of views. The result combines the prevailing views on the given question expressed quantitatively in terms of means, quantiles and correlation between the time horizons of the events (e.g. discovery).

Computers are used for the prediction of long-term development trends. The computer prediction programme of the British chemical concern ICI (Pearce, 1973) includes six basic trend curves: straight line, parabola, simple exponential curve, logarithmic parabola, modified exponential curve, and the Gompertz curve, where the growth rate decreases to the final value given by the user.

In the light of ICI's experience, these trend curves are more suitable for the prediction of total consumption in a country than for the sales of any particular firm. For a good prediction the past series has to be longer than 7 years; better 12 to 15 years, and, for very variable data, 25 years give reliable results.

Non-linear extrapolation is sensitive to the most recent values. In short-term predictions over one or two years, deviations from the trend have a decisive influence. The selection of input data and the interpretation of the results require a good insight into the problem.

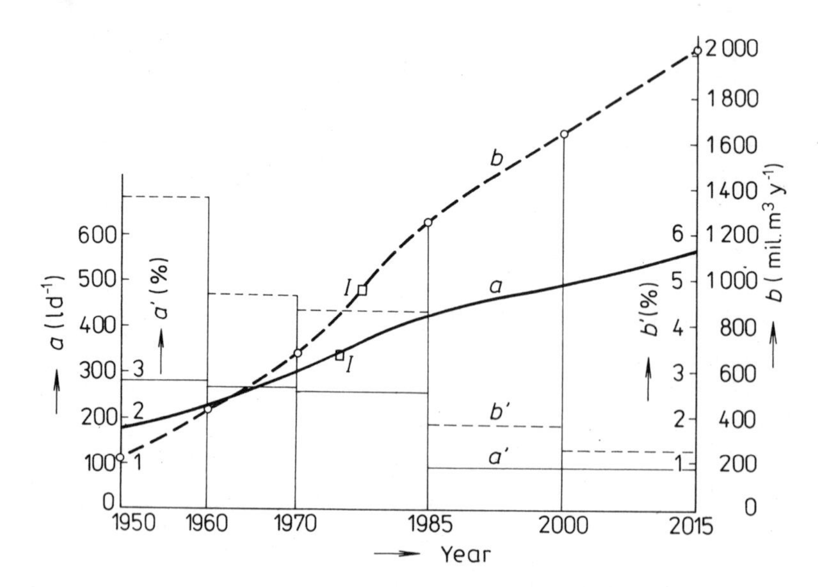

Fig. 2.8 The development and prediction of municipal water supply in ČSR
a – specific demand (litres per day per inhabitant), b – annual prediction of drinking water ($10^6 . m^3$ per year), a' – average annual increment of specific demand (%), b' – average annual increment (%)

The prediction of water resource demands in the GWP was derived from the prediction of social development and from other research papers. In alternatives obtained from processing the data from all over the country, the differences between the extreme values were so small that, with the exception of water pollution, the trend of the development of all the basic requirements could be characterized in each time horizon by a single value. Regional correction of the requirements is part of the continuous prediction-conception activity that follows the GWP.

The prediction of the development of the municipal water supply in ČSR in Fig. 2.8, is expressed in specific demand, its average annual increment, total production of drinking-water, and its average annual increment. The trend curves are flat S-shaped curves with the inflection point expected in the years 1976 to 1980: the average annual increments (growth-rate function) have an irregular but steadily decreasing tendency.

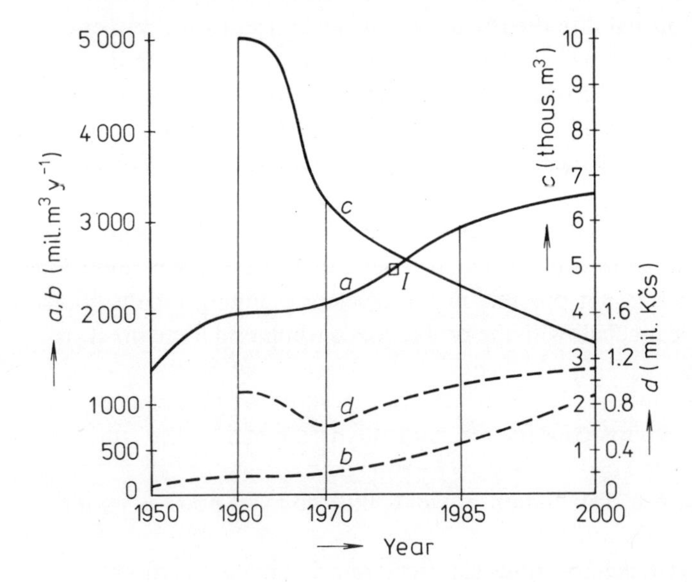

Fig. 2.9 The development and prediction of trends of withdrawals and consumption of water in industry of ČSR
a – withdrawals of water resources, *b* – consumption of water, *c* – withdrawal of water per 1 mil. Kčs of production, *d* – consumption of water per 1 mil. Kčs of production

Similar trend curves of withdrawals and consumption of water in industry are given in Fig. 2.9 (curves *a*, *b*). The further development curve (*a*) has an expected inflection point in the years 1975 to 1980; curve (*b*) monotonically increases. The same figure (curves *c*, *d*) shows the predictions of the development of withdrawals and consumption of water per 1 mil. Kčs (monetary unit in Czechoslovakia) of industrial production in ČSR.

The specific withdrawals decrease monotonically while specific consumption, after decreasing until the year 1970, increases again, and it will double in 30 years.

Similarly, trend curves can be analysed for water supply for agriculture, development of water quality, flood control, etc. Plans for the development of water management activities and the requirements of other sectors that form the necessary conditions were based on these predictions.

Similar analyses were carried out for the main basins which approximately coincide with the areas where WRS were defined on the discriminating level of GWP.

2.8 FUNCTION OF CREATIVE TEAMS AND WORK GROUPS IN THE DESIGN AND OPERATION OF WRS

The origin and development of WRS and the necessity of their rational design and operation require a qualitatively new activity that goes beyond the boundaries of any one discipline. In the treatment of interdisciplinary problems, the coordination of experts on different subjects is necessary; e.g. the problems of WRS require experts in the following fields: hydrology, hydrogeology, water quality and water pollution, environment and wildlife, probability theory and mathematical statistics, numerical analysis and computers, systems analysis, economics, operation of water engineering structures, town planning, etc. According to the nature of the problem and the relative importance of different fields, the set of experts may form either a creative team (also called a consulting panel, task force group, advisory committee) or a work group.

A *creative team* is a group of highly qualified experts from selected backgrounds whose aim is to tackle and solve a particular problem, or to design a method for a given task, a set-up which is not possible in a project or planning organization. Unlike a work group, this team deals with the problem as a whole and therefore its responsibility is collective.

In a *work group* (e.g. of authors, department) work is allocated to members, the results are put together, perhaps coordinated, and this is the end of the work on a given task.

In a creative team there is a distribution of work, but the main point is its integration.

In homogenous as well as in heterogenous teams, no one discipline is predominant, all are equally important. The team is free to determine the content of the work and the direction the investigation of a roughly outlined problem should take. Its success depends on team interaction, the exchange of information inside the team and a climate of friendship and trust, where everybody can express his own views and influence or oppose the views of the others.

When there are different views, decision by voting is not acceptable; the problem has to be discussed and explained and the alternatives optimized, using the approved criteria, until agreement is reached. If it is not possible to decide which of two alternatives is the better, both must be followed up.

Sládeček, 1973, describes typical approaches to creative team activities:

– *brainstorming*, in order to generate, in a short time, as many alternatives for the treatment of a given problem as possible,

– a *morphological* approach when the problem is split into various elements, after which relationships and mutual combinations for the set-up of the new functions, relationships, etc. are derived,

– *synthesis* when a special method of connecting seemingly unrelated elements is used, together with analysis and analogy,

– the use of *control lists*, i.e. listing the interrelationships of particular elements and their transformation in accordance with certain rules,

– the *optimization* method, generating alternatives and selecting the best one,

– the use of *lists of properties*, suitable for tasks which aim to improve the functioning of the system,

– the use of *models*,

– an *operations research* approach,

– a *systems engineering* approach,

– *heuristic* methods.

The probability of finding a successful solution to the problem depends on the quality of the methods applied and on the number of original ideas which can be generated.

The team cannot do without a leader, whose authority has to be based on the trust of all the members of the team and not on his power or social position. A good leader must be outstanding in his own field and he has to have original ideas; he has to be able to differentiate between good and bad suggestions, he has to stimulate work and create a favourable atmosphere. The success of the work of the team will be endangered if its composition is unsuitable, if the task is beyond its professional or time capacity or if its style of work is not effective.

The work on WRS problems can be performed by a creative team and a work group: the former for scientific research and the latter for the design and operation of concrete WRS, using known methodology. The creative teams are often composed of employees of different organizations and institutes, whereas the members of the work groups come from the same organization. The head of the team is usually a well-known expert with informal authority; the head of the work group is the formal head of the organization or one of its departments. In the team there is no hierarchy, but in the work group this does exist. The team is led, the work group is managed or controlled. There are, therefore, better conditions for the development of creativity in the team than in the work group. However, neither type of internal organization is in itself a guarantee of a good working style and good results. There are work groups where the management, style of work, and results are on the level of a good creative team and, conversely, there are creative teams full of "comedies of errors" (viz. administrative, personal or actual errors) which hinder the success of the work.

In the selection of the team members, not only professional orientation should be taken into account but also personal character. Egocentric individuals (who are far from scarce in research or science) are not suitable people for team work. Team cooperation is not a temporary fashion but an integral part of the scientific and technological revolution, and as such it should be promoted and its foundations laid during school education.

An example of a creative team in WRS

The investigation is related to the nature of the task and the problem is outlined roughly so as not to restrict the creative freedom of the team – e.g., "Set up a methodology for the design and evaluation of a WRS with water supply its first priority".

Table 2.10 Formation of creative team for the design of WRS with water supply having priority

1 Conception of the structure of system and definition of WRS	Experienced water resource planner with education in systems sciences	Water resource research institute
2 Determination of municipal water supply subsystems	Experienced water resource engineer with education in water supply and systems analysis	Hydroproject
3 Water balance of the system	Hydrologist and water resource engineer or planner 1 or 2	Hydrometeorologic institute
4 Assembly of mathematical models of WRS	Mathematician with education in systems science or systems engineer with mathematical education	Research institute of computer science or water management
5 Optimization of alternatives	Economist with systems education	Water management or economy research institute
6 Consultative cooperation of experts	– for land use plan – for environment – for the objectives of WRS – for operation of WRS	Terplan (territorial planning) various organisations various organisations Water Board

A schematic representation of the development of the team is given in Table 2.10.

First, a list of activities necessary for the solution of the problem is compiled. Then, the types of experts are added to this list and possibly the organisations and institutes where these experts are to be found. To facilitate cooperation and communication inside the team, the number of members should be kept low.

For this reason, the types of activities and types of experts are divided into two groups:

– basic permanent activities which participate substantially in the problem solving during the whole period of the work,

– auxiliary activities necessary for dealing with the problem but not on a permanent basis.

The experts in the first group should be members of the team, the experts in the second group should be consultants. However, the formation of the team is ultimately adjusted according to the ability and character of the individuals involved.

In our case, a team with five members, supported by some consultants, will be formed. The natural head of the team should be an experienced water resource planner with a training in systems sciences. The table shows that every member of the team for WRS should have some knowledge of systems sciences (analysis); the applied systems sciences should be a normal part of the basic training in water resource engineering and planning.

2.9 AUTOMATIC CONTROL OF WATER RESOURCE SYSTEMS

2.9.1 Subjects of the Automatic Control Theory and Its Relationship to Cybernetics

Control as a specific human activity is understood as a purposeful impact on an object with the aim of attaining a given objective (Kubík *et al.*, 1972).

The principles of control, its forms and technical means of implementation are related to the nature of the controlled system.

Scientific research of the control system requires the analysis of the controlled system, an important and integral part of it being the control function.

The general principles of control systems and processes are investigated by cybernetics and originated on the borderlines between mathematics, logic, electronics, physiology and other fields. The theory of automatic control and computers contributed to the formation and development of cybernetics as a new field of science using the results of the theories of communication and information and their investigation by the theory of mathematical probability. These fields provided cybernetics and its components with its basic method and procedures.

The theory of control and decision-making is sometimes considered as an extension of operations research, which is, in this respect, interpreted as a set of specific ways of decision-making (Sládek, 1974).

New branches of technology came into being, together with the development of the theory of cybernetics, e.g. automatic control that investigates the use of technical facilities and devices to free man from the tedious mental and physical work involved in control.

Although automatic control has been developing since the end of the last century, the systems approach has been stressed only recently. Its basic feature is a comprehensive understanding of the system with all its internal and external relationships, with the objective of optimum effectiveness on the basis of a comparison of alternatives. The theory of automatic control was involved and has become an important and promising branch of technical cybernetics.

Initially, the theory of automatic control dealt with the problems of automatic protection of technical devices, automatic signalling and automatic regulation. The intensive development of the theory of automatic control in recent years, together with the intensive development of computer hardware and software and methods of mathematical programming, led to the formation of the theory of optimal control systems.

This theory investigates systems of automatic control with different properties and under different conditions of technical realisation. For example, the optimal control of systems with complete and incomplete initial information, deterministic and stochastic systems, adaptive, learning systems, etc.

The theory of optimal control systems has penetrated into various human activities. It can be predicted that automatic control systems will develop in all branches of the national economy as subsystems of the central economic control systems. In water management, too, suitable conditions for their implementation exist. The multi-purpose WRS with their characteristic properties (e.g. territorial extensiveness, dynamic and stochastic character of input data, their great number, their reliance on modern computer hardware and software for dispatching operations, etc.) represent a typical example of possible implementation of the optimal control theory, which can be considered as a significant improvement on the previous approaches to control.

Further basic information is available concerning the principles and methods of automatic control, the theory of optimal control of automatic systems and information about the possibilities of their implementation in WRS problems. Development is far from complete but there is plenty of literature on the subject (Švec and Kotek *et al.*, 1969; Kubík *et al.*, 1972) with many references.

2.9.2 Principles and Methods of Automatic Control Theory and Optimal Systems Theory

Automatic protection, signalling, control and regulation were mostly applied in the comprehensive automation of technological processes in machinery, energetics, steelworks and chemistry. Automation includes the issues of control in the given technological process; a prerequisite is a knowledge of the properties of the controlled technology and objects and the control functions, i.e. a knowledge of the effects of the control action on the process.

An especially important kind of control is regulation, which maintains some physical quantities at predetermined values. The actual values are monitored during the regulation process and compared with the required ones; the goal of regulation is the attainment of minimum deviations.

The automatic regulation theory has been extensively and broadly developed. The subject includes the design of elements and loops of automatic control, investigation of their static and dynamic properties in connection with the problem of stability, investigation of special regulation circuits (e.g. multi-branch, multivariable, non-linear control). Impulse and digital control is studied as the basis of the control of complex systems by computers, and the problems of identification of the regulated systems, which is the basis for the construction of mathematical models of them, are also important.

In the theory of automatic control an important place belongs to the methods of the theory of probability and the theory of random processes. They are used for the investigation of dynamic characteristics of control systems, which are influenced by random noise, and the theoretical relationships from which the effective design of these circuits are derived. Research in this area is not complete: important applications of the theory of automatic control can be dealt with by statistical methods, e.g. statistical prediction, filtering (of the information signal from noise), statistical linearisation of non-linear systems of the problems of adaptive control systems.

The application of the methods of the theory of probability and the theory of stochastic processes in tasks of automatic control has been developed in recent years and has generated a new extension of the theory of automatic control, which is now called statistical dynamics of regulation circuits (Beneš, 1961; Švec and Kotek *et al.*, 1969).

The requirement of optimization of automatic control systems is often combined with the problems of their control. It is concerned with that form of automatic control which has the best properties of control as evaluated by some optimization criterion. It can, for example, be the requirement of the optimal form of the transients of the controlled quantity or other quantities in the control loop caused by a change in the control or noisy variables (i.e. dynamic optimization) or the requirement to attain other optimum values of quantities in the steady state to optimize the technological and economic results of control, e.g. maximum energetic effectiveness, the best quality of production etc. (i.e. static optimization). In optimization control some extreme value of the optimization criterion is frequently required (maximum or minimum).

The theory of optimal control now covers a more extensive area than the treatment of optimal regulation loops, which had been the main subject of the traditional theory of automatic regulation. Nowadays, the general problems of complex systems are investigated, under different conditions, and control computers are used for automatic control. Control is carried out by a system of programs of varying importance. An adequate program is selected depending on the behaviour of the system and the momentary values of the parameters monitored. The control computer should include in its software several programs with different priorities. The requirements of real-time systems control implies requirements for the speed and reliability of the control computer.

The development of mathematical methods facilitated the treatment of deterministic and stochastic problems. Frequently, the basis is formed by the methods of mathematical programming, but other systems fields are promising, too, e.g. the theory of games, the decision theory, etc. For WRS contexts, the theory of statistically optimal systems and adaptive systems are important since the properties of the system can be adapted to changing conditions. These properties can be maintained or improved on the basis of experience gained in the search for optimum behaviour (e.g. adaptive learning systems).

Adaptive systems can be classified into several types according to the aim of the activity and the realization of change in it. A change in activity can be derived from change in the external signals influencing the system or from changes in the parameters of the system and it can be realized by a change in the structure of the control system. However, both the structure and the parameters of the controlled system may change.

A characteristic feature of algorithms of adaptivity and learning is the ability to find the optimum solution under conditions of uncertainty in initial input data. The control system must continually monitor the behaviour of the controlled system, influence it and decide on the further strategy of control.

Design of a control system which realizes optimum control involves the performance of a relatively extensive set of tasks which that can be subdivided as follows (Kubík *et al.*, 1972):

– the problem of optimal identification of the control goals, the solution of which provides an optimum formulation of the control objective,

– the problem of optimal identification of the controlled object, the solution of which provides the mathematical model,

– the problem of optimal identification of the constraints which are related to the conditions of the object,

– the problem of the determination of the optimum algorithm of control and its realization.

One of the most difficult tasks is the determination of the optimum control algorithm, which has to satisfy the criteria not only of optimum control but also of all the constraints.

2.9.3 Prospects of Application of Optimal Systems of Automatic Control in WRS

The scientific and technical problems of automatic control systems are complex and difficult to solve. In their implementation in the operation of WRS the following fundamental problems of a general and applied nature need to be taken into account: the design of the control centre, a system of computer software, a monitoring system, a system of data collection, processing and communication with the centre including remote transmission and control and the problems mentioned above concerning the design of the control centre. The experience of other technical sectors can be put to advantage. In water management the development of these systems has just started. It can therefore be assumed that their realization and full operational functioning is a long-term task for the institutions responsible for water research design and operation.

The first conditions for optimal control of WRS in the main basins have been satisfied. In accordance with the general trends, the Water Boards prepare the fundamental conditions for the comprehensive water resource dispatch centres, which are to be in charge of the scientifically based operation and control of water resources.

In WRS the main problems of the design of optimal control include the construction of a mathematical model (or a set of models), its verification under real operational conditions and implementation on a control computer. It is an extraordinarily complicated task whose solution has, as yet, no precedent. It is clear that the process of construction of systems software includes the problems of systems analysis and synthesis; the design of the system, the construction of models and control algorithms and their verification in operation all have to be developed.

The difficulties that arise in the development of mathematical control models are related to the specific conditions of implementation of systems and stochastic methods in WRS. Actual problems are to be treated, e.g. multi-objective optimization of the competing interests of water users, classification of decision (operation) situations in non-stationary conditions, integration of continual and discrete simulation, etc. For WRS control systems, particularly applicable are the automatic adaptive models with random input, which keep the important control parameters within certain limits, derived from optimization analysis of WRS objectives.

In the main basins, water resource control by control computers in real-time operation is an important task, which can be considered a final stage in the implementation of automatic operation of WRS. It is assumed that the control computer in a dispatch centre will produce the control instructions in a system of control algorithms, based on concrete, monitored data describing the situation in the basin (i.e. flows in watercourses, levels in reservoirs, water and air temperatures, precipitation, water quality at selected points, etc.).

It is a favourable circumstance that an investigation of all the main tasks and problems associated with the application of cybernetics in WRS was initiated in the Czechoslovak State Program of Task of Basic Research and Applied Technical Development.

The Water Board in the Ohře River basin is an example of a highly developed water resource dispatch centre (Chomutov, ČSSR). Here, the aspect of WRS operation control involves the following devices:

– automatic monitoring stations in the basin,

– transmission devices for the transmission of data from the monitoring stations to the regional dispatch points and to the dispatch centre,

– communication devices linking the monitoring stations with the regional dispatch points and dispatch centre,

– regional dispatch points in Cheb and Karlovy Vary (ČSSR),

– a central dispatch centre in Chomutov (ČSSR) with a control computer RPP-16-S.

3 MATHEMATICAL METHODS USED IN SYSTEMS THEORY

The mathematical methods used in the theory of systems are related to the nature and function of the systems analysed.

WRS have, in principle, a probability character. Therefore probability methods, and methods of mathematical statistics and the theory of stochastic processes are used in their investigation. The methods of Fourier, Laplace and z-transformation are used for the description of the dynamic complex behaviour of these systems, and the methods of the calculus of variations are applied for optimization in tasks where the structure and behaviour of systems are described and investigated.

A selection of the basic mathematical methods related to these issues is presented in this chapter; a more detailed treatment can be found in the specialist literature referred to in the text.

3.1 PROBABILITY THEORY

3.1.1 Basic Notions

Probability theory deals with investigation of *mass random events*; it is based on a large number of observations or measurements of phenomena occurring under a given set of conditions. For the determination of the statistical properties of events, many elementary events (viz. sample points) have to be determined (e.g. by measurement, observation etc.), if that is possible in the given case. In general, a large sample of these elements can be realized taken for random events from a sample space (e.g. possible outcomes of an experiment; all possible outcomes form the sample space). The occurrence of a random event can be predicted with some probability, but not exactly, even if the set of conditions governing its occurrence is maintained. If this probability is equal to one, the event is certain (sure), if it is equal to zero, the event is impossible.

The term probability can be more easily explained than defined. Often, the classical definition of probability is used. It says: If event A can be split into elementary events X_i ($i = 1, 2, \ldots, m$) (i.e. events that cannot be further split into particular events), the probability of its occurrence in the given complete system of n mutually exclusive and equally probable events (where the sum of all the elementary events is a certain event and each pair of elementary events has a zero product) is

$$P(A) = \frac{m}{n} \qquad (m \leqq n) \tag{3.1}$$

In the case of a complete system of n equally probable events $A_1, A_2, \ldots, A_n$ which the event A is composed of and if

$$P(A) = \sum_{i=1}^{n} P(A_i) \tag{3.2}$$

then

$$P(A) = \sum_{i=1}^{n} p = np = 1 \tag{3.3}$$

so that

$$p = \frac{1}{n} \tag{3.4}$$

The *unconditional probability* of occurrence of event A was related only to a system of conditions. If event A also depends on the occurrence of an other event, B (given that the event B has occurred), then

$$P_B(A) = \frac{P(AB)}{P(B)} = \frac{P(A)\,P_A(B)}{P(B)} \tag{3.5}$$

where $P(AB)$ is the probability of the product of two events (occurrence of both events A and B) and $P_B(A)$ is called the *conditional probability* of event A, provided that event B has taken place.

For independent events A and B,

$$P_B(A) = P(A) \quad \text{and} \quad P_A(B) = P(B) \tag{3.6}$$

so that

$$P(AB) = P(A)\,P(B) \tag{3.7}$$

Assume that event A can occur together with one and only one of the n mutually exclusive events B_i ($i = 1, 2, \ldots, n$) with the known probability $P(B_i)$. The formula for total probability is

$$P(A) = \sum_{i=1}^{n} P(B_i)\, P_{B_i}(A) \tag{3.8}$$

The events B_i are often called hypotheses of event A and at least one of them must occur for event A to take place. The probability of occurrence of individual hypotheses if event A has occurred, is given by Bayes' formula:

$$P_A(B_i) = \frac{P(B_i A)}{P(A)} = \frac{P(B_i)\, P_{B_i}(A)}{P(A)} = \frac{P(B_i)\, P_{B_i}(A)}{\sum_{j=1}^{n} P(B_j)\, P_{B_j}(A)} \tag{3.9}$$

where $P_A(B_i)$ is the probability of occurrence of the i-th hypothesis together with the occurrence of event A.

3.1.2 Theoretical Probability Distributions of Random Variables

3.1.2.1 Random Variables

The result of a random event is quantitatively described by the value of a *random variable*. The exact value of the random variable cannot be ascertained; only the probability of occurrence of its values can be determined. Random variables are classified as follows:

– *discrete* variables that have, in the given interval, a finite, discrete number of different values,

– *continuous* variables – the set of their values is infinite and continually fills up part of the axis of real numbers.

The *probability distribution* of a random variable describes the allocation of some probability of occurrence to each value of a discrete random variable, or some probability of occurrence of a continuous random variable in a certain interval of possible values.

The probability distribution of a random variable can be described by the *probability density* function (p.d.f.) or by the cumulative distribution function (c.d.f.).

The p.d.f. (Fig. 3.1.a) of a continuous random variable is expressed by function $f(x)$, which is the first derivative of the c.d.f. of the random variable $F(x)$ with respect to x.

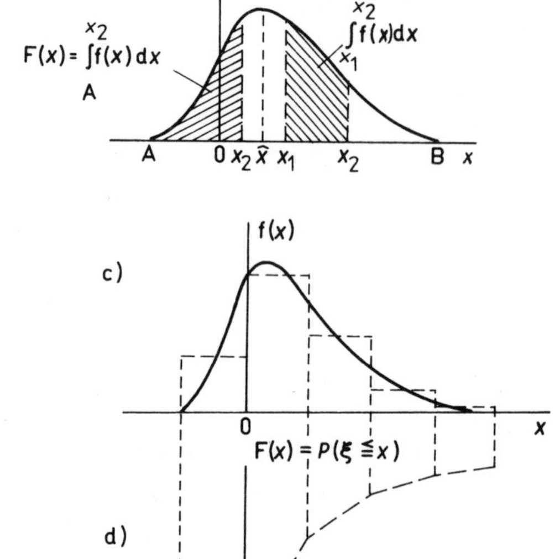

Fig. 3.1 Graphs of the p.d.f. and the c.d.f. of a continuous random variable
a – p.d.f. of a continuous random variable,
b – c.d.f. of a continuous random variable,
c – relative class frequencies of a random variable, *d* – its distribution function

The value of the integral

$$\int_{x_1}^{x_2} f(x)\,\mathrm{d}x = P(x_1 < \xi \leqq x_2) \tag{3.10}$$

gives the probability of occurrence of the values of the random variable ξ in the interval $(x_1; x_2\rangle$.

The c.d.f., $F(x)$, of a random variable ξ gives the probability that the value of the random variable ξ will not exceed the selected value x, i.e.

$$F(x) = P(\xi \leqq x) \tag{3.11}$$

The c.d.f. (Fig. 3.1.b) of a continuous random variable is continuous, that of a discrete random variable is continuous with the exception of a finite number of points. The c.d.f. of a continuous random variable is a continuous, monotonically increasing function (represented graphically by a curve)

$$F(x) = \int_{-\infty}^{x} f(x)\,\mathrm{d}x \tag{3.12}$$

The c.d.f. of a discrete variable is a step function that changes its values only at a finite (or countably infinite) number of points that are the possible values of the random variable ξ and where the jumps occur.

If the sample of discrete random variable values is very large, it is helpful to group the values into classes of equal size. The centre of the class (viz. class mark) represents all the values of the class. It is associated with the class frequency, i.e. the frequency of occurrence of the discrete random variable in the given interval. For each class frequency, n_i, the relative class frequency n_i/n is calculated. The p.d.f. is a step function and the corresponding c.d.f., which is the cumulative function of the relative class frequencies, has the form of a broken line.

The general property of the p.d.f. is

$$\int_{-\infty}^{\infty} f(x)\,\mathrm{d}x = 1 \tag{3.13}$$

3.1.2.2 Numerical Characteristics of Random Variables

Assume a random sample of a certain population of a discrete random variable and a continuous one. There are many problems in mathematical statistics in which it is difficult, or at least not feasible, to determine completely the c.d.f. of a random variable. In such cases it is often possible to describe the distribution of a random variable partially by its moments (i.e. moments of probability distribution). They are classified as a) general (or merely moments) and b) central moments.

a) Moments (*general moments*) are defined by

$$m_k(\xi) = \int_{-\infty}^{\infty} x^k\, f(x)\,\mathrm{d}x \tag{3.14}$$

where x is the value of the random variable, $f(x)$ is its p.d.f., and k is the order of the moment (k is an integer; this moment is often called the k-th moment of a continuous random variable); or by

$$m_k(\xi) = \sum_{i=1}^{n} x_i^k p_i \tag{3.15}$$

which is the k-th moment of a discrete random variable, where P_i are the probabilities allocated to the values x_i;

b) *Central moments* about the mean μ are defined as the mean (expected) value of $(x - \mu)^k$ (in a mechanical interpretation, the moment is related to the axis which includes the centre of gravity of the graph of the p.d.f.) of a continuous random variable

$$M_k(\xi) = m_k(\xi - \mu) = \int_{-\infty}^{\infty} (x - \mu)^k\, f(x)\,\mathrm{d}x \tag{3.16}$$

Table 3.1 The main moments of the random variables

Type of moment	Order of moment	Expression for the continuous random variable	Expression for the discrete random variable		Meaning of the moment
			for $p_1 \neq p_2 \neq \ldots \neq p_n$	for $p_1 = p_2 = \ldots = \text{const}$	
General	I.	$m_1(\xi) = \int_{-\infty}^{\infty} x\, f(x)\, dx$	$m_1(\xi) = \sum_{i=1}^{n} x_i p_i$	$m_1 = \frac{1}{n} \sum_{i=1}^{n} x_i$	mean $\mu(x)$ [$\bar{x}$]
Central	I.	$M_1(\xi) = \int_{-\infty}^{\infty} (x - \mu)\, f(x)\, dx$	$M_1(\xi) = \sum_{i=1}^{n} (x_i - \mu)\, p_i$	$M_1(\xi) = \frac{1}{n} \sum_{i=1}^{n} (x_i - \bar{x})$	–
Central	II.	$M_2(\xi) = \int_{-\infty}^{\infty} (x - \mu)^2\, f(x)\, dx$	$M_2(\xi) = \sum_{i=1}^{n} (x_i - \mu)^2\, p_i$	$M_2(\xi) = \frac{1}{n} \sum_{i=1}^{n} (x_i - \bar{x})^2$	variance $\sigma^2(x)$ [s_x^2]
			$s = \sqrt{M_2(\xi)}$	$s = \sqrt{M_2(\xi)}$	standard deviation $\sigma(x)$ [s_x]
				$C_v = \frac{s}{\bar{x}}$	coefficient of variation C_v [C_{vx}]
Central	III.	$M_3(\xi) = \int_{-\infty}^{\infty} (x - \mu)^3\, f(x)\, dx$	$M_3(\xi) = \sum_{i=1}^{n} (x_i - \mu)^3\, p_i$	$M_3(\xi) = \frac{1}{n} \sum_{i=1}^{n} (x_i - \bar{x})^3$	–
				$C_s = \frac{M_3(\xi)}{\sqrt{M_2^3(\xi)}}$	coefficient of skewness $C_s(x)$ [C_{sx}]
Central	IV.	$M_4(\xi) = \int_{-\infty}^{\infty} (x - \mu)^4\, f(x)\, dx$	$M_4(\xi) = \sum_{i=1}^{n} (x_i - \mu)^4\, p_i$	$M_4(\xi) = \frac{1}{n} \sum_{i=1}^{n} (x_i - \bar{x})^4$	–
				$\gamma = \frac{M_4(\xi)}{M_2^2(\xi)} - 3$	coefficient of kurtosis $E(x)$ [γ]

or of a discrete random variable

$$M_k(\xi) = \sum_{i=1}^{n} (x_i - \mu)^k p_i \tag{3.17}$$

The meaning of the first moment and the first to the fourth central moments with the expressions for continuous and discrete random variables and for p = const are given in Table 3.1.

3.1.2.3 Application of Theoretical Probability Distributions

Theoretical probability distributions are used for the evaluation of the type of probability distributions of random events and the values that describe them, and for fitting the empirical probability distributions corresponding to random samples

Table 3.2 Main theoretical probability distributions

Group	Type	Modification (alternative)
Normal		
Normalizing	log-normal	Gauss-Gibrat Galton Ven-Te-Chow
Gamma-distribution		
Beta-distribution		
Pearson	Type I Type III	
Extremal	exponential Weibull Gumbel	
Discrete	geometrical hypergeometrical binomial Poisson	
Involved in analysis of variance	Student – t distribution – χ^2 Snedecor – F	 Fisher

of the population (see 3.1.4). In WRS, *exceedance probability functions* (e.g. flow duration curves) are often used and they are related to the c.d.f. by the formula

$$P(x) = 1 - F(x) = 1 - \int_{-\infty}^{x} f(x)\,\mathrm{d}x \tag{3.18}$$

The empirical frequency function cannot reliably determine the probability of extreme values, even if relatively long time series of observed data are available. Fitting them to the theoretical c.d.f. facilitates extrapolation to the extreme values. The quality and reliability of the results depend on the goodness-of-fit of the selected theoretical c.d.f. with the unknown actual distribution.

Table 3.2. contains the main theoretical c.d.f. used in water management and related tasks.

In fitting the data to the theoretical c.d.f. the following methods are used:

1) By analysis of the characteristics and properties of the random variable and its origin (analysis of the boundary conditions, skewness of the empirical c.d.f., etc.) a feasible type, or several types of theoretical c.d.f. are determined.

2) By means of the parameters of the empirical c.d.f. (sample values) the parameters of the theoretical c.d.f. are estimated.

3) The goodness-of-fit between the empirical and theoretical c.d.f.s is evaluated.

3.1.2.4 The Normal Probability Distribution

The normal (Gaussian) probability distribution is the most important one in probability theory and in mathematical statistics, as it is suitable for the approximation of empirical p.d.f. of quantities observed in different phenomena, both continuous and discrete, and also for the approximation of other, more complicated, theoretical probability distributions. A random variable, the values of which are the sum of a large number of mutually independent effects, each having a small impact, has the normal probability distribution.

A further condition for normal distribution is a large number of elements in a sample. It is necessary to differentiate between a large number of elements in a sample and a large number of independent effects. In general, it can be said that a large number of elements in the sample contributes to the goodness-of-fit between the empirical and theoretical distributions, whereas the number of independent effects determines the type of theoretical distribution adequate for the given random variable.

The normal distribution of a continuous random variable has two parameters: the mean μ and the standard deviation σ. It can also be used for a discrete random variable.

The p.d.f. and c.d.f. of the normal distribution are presented in general and standardized forms in Table 3.3.

Table 3.3 Normal probability distribution

random variable	x
mean	$\mu(x)$
standard deviation	$\sigma(x)$
p.d.f.	$f(x) = c\,\mathrm{e}^{-[x-\mu(x)]^2/2\sigma^2(x)}$; $c = \dfrac{1}{\sigma(x)\sqrt{2\pi}}$
c.d.f.	$F(x) = \dfrac{1}{\sigma(x)\sqrt{2\pi}} \displaystyle\int_{-\infty}^{x} \mathrm{e}^{-[x-\mu(x)]^2/2\sigma^2(x)}\,\mathrm{d}x$
standardized variable	$t = \dfrac{x-\mu(x)}{\sigma(x)}$
p.d.f. of the standardized variable	$f(t) = \dfrac{1}{\sqrt{2\pi}}\,\mathrm{e}^{-t^2/2}$
c.d.f. of the standardized variable	$F(t) = \dfrac{1}{\sqrt{2\pi}} \displaystyle\int_{-\infty}^{t} \mathrm{e}^{-t^2/2}\,\mathrm{d}t$

It is convenient to refer to the normal distribution with the general form of p.d.f. (Table 3.3) as the distribution $N(\mu, \sigma^2)$, and the standardized form of p.d.f. as $N(0, 1)$.

The graph of the p.d.f. and c.d.f. for some selected values of σ and $\mu = 0$ is given in Fig. 3.2.

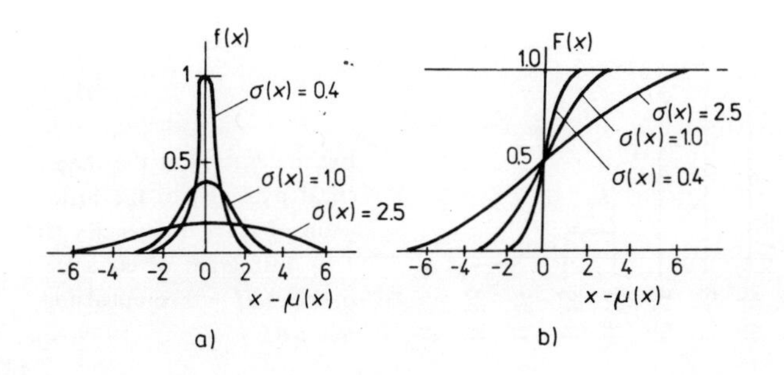

Fig. 3.2 Normal probability distribution of the random variable for different values of $\sigma(x)$
a – p.d.f., *b* – c.d.f.

In WRS contexts the probability distributions are frequently not symmetrical and they are bounded on one or both sides. The normal distribution, however, is an approximative basis used for the derivation of other convenient distributions and for the development of methods of probability and the statistical treatment of random samples.

Table I.1 Class frequencies

i	Class interval	x_i'	n_i
1	8.88 – 9.12	9.00	1
2	9.13 – 9.37	9.25	3
3	9.38 – 9.62	9.50	11
4	9.63 – 9.87	9.75	24
5	9.88 – 10.12	10.00	42
6	10.13 – 10.37	10.25	62
7	10.38 – 10.62	10.50	58
8	10.63 – 10.87	10.75	44
9	10.88 – 11.12	11.00	19
10	11.13 – 11.37	11.25	9
11	11.38 – 11.62	11.50	4
12	11.63 – 11.87	11.75	2
13	11.88 – 12.12	12.00	1
			280

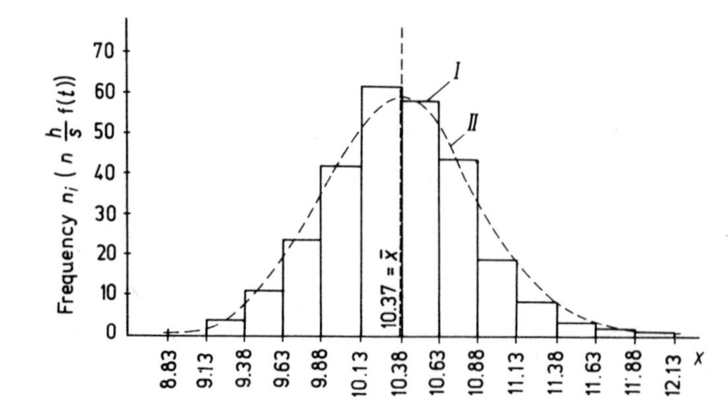

Fig. I.1 Fitting of the theoretical p.d.f. (II) to the histogram of class frequencies (I) *I* – histogram of class frequencies, *II* – theoretical normal p.d.f.

Example I

A sample of 280 elements is given by its class frequencies (Table I.1) with the sample mean $\bar{x} = 10.37$ and the sample standard deviation $s = 0.475$. A theoretical normal p.d.f. is to be fitted to it.

Solution

The empirical p.d.f. is clearly close to the normal p.d.f. The values of the theoretical normal p.d.f. are calculated in Table I.2. The values $t = (x_i - \bar{x})/s$ are the values of the standardized variable where x_i are the lower class limits for which the theoretical values of the p.d.f. are calculated.

Table I.2 Computation of the theoretical normal p.d.f.

i	x_i	$x_i - \bar{x}$	$t = \frac{x_i - \bar{x}}{s}$	$f(t)$	$\frac{nh}{s} f(t)$
1	8.88	−1.493 2	−3.14	0.002 9	0.426
2	9.13	−1.243 2	−2.61	0.013 3	1.95
3	9.38	−0.993 2	−2.08	0.046 0	6.77
4	9.63	−0.743 2	−1.55	0.120 2	17.5
5	9.88	−0.493 2	−1.03	0.234 8	34.5
6	10.13	−0.243 2	−0.51	0.350 2	51.5
7	10.38	+0.006 8	+0.014	0.394 0	58.0
8	10.63	+0.256 8	+0.54	0.344 5	50.6
9	10.88	+0.506 8	+1.06	0.227 5	33.5
10	11.13	+0.756 8	+1.59	0.112 8	18.2
11	11.38	+1.006 8	+2.12	0.041 5	6.10
12	11.63	+1.256 8	+2.64	0.012 2	1.79
13	11.88	+1.506 8	+3.17	0.002 7	0.397

The values of $f(t)$ corresponding to values t are obtained from tables (e.g. Reisenauer, 1970). These values are multiplied by the value of the following expression

$$\frac{nh}{s} = \frac{280 \cdot 0.25}{0.475}$$

where h is the class length.

The values of the class frequencies and theoretical p.d.f. are given in Table I.1.

3.1.2.5 Normalizing Probability Distributions

The normal probability distribution has a simple mathematical representation and tables on the p.d.f. and c.d.f. of standardized variables are available (the computer programs for their calculation are included in the software of each modern computer). Therefore, there have been many attempts to transform skew probability distributions of a random variable x to a variable y that is normally distributed.

The principle of this transformation is the determination of a function φ for *normalizing the transformation* $y = \varphi(x)$.

The simplest transformation function is $y = x^a$. In hydrology, C. K. Stidd used the exponent $a = \frac{1}{3}$, which proved to be helpful for various types of time series of precipitation (e.g. annual, monthly and daily precipitation) (Kendall, 1970).

The *logarithmic transformation* expressed by Johnson in a general form is often used:

$$y = a + b\left(\log \frac{x - m}{l}\right) \tag{3.19}$$

This formula is the most general and includes other formulae (e.g. Galton's) as a special case. An apparent disadvantage is its large number of parameters (a, b, l, m), but on the other hand it has the advantage of good adaptability to a wide range of skew distributions.

Modifications of the log-normal distribution applied in meteorology, hydrology and water management have been developed by F. Galton, R. Gibrat and Ven Te Chow.

The log-normal probability distribution of a random variable x is such that the variable, after the logarithmic transformation (e.g. (3.19)), has the normal probability distribution.

Table 3.4 Log-normal probability distribution

p.d.f.	$f(x) = \frac{1}{(x - x_0)\,\sigma(y)\sqrt{2\pi}} \exp\{-[\lg(x - x_0) - \mu(y)]^2/2\sigma^2(y)\}$
form of logarithmic transformation (Galton)	$y = \lg\lvert x - x_0\rvert$
mean	$\mu(y) = \lg\sigma(x) - \lg\lvert C_\gamma\rvert - \frac{1}{2}\lg(1 + C_\gamma^2)$
variance	$\sigma^2(y) = \lg(1 + C_\gamma^2)$
coefficient of skewness	$C_s(x) = C_\gamma^3 + 3C_\gamma$

The log-normal distribution proved to be adequate for many skew distributions that correspond, for example, to maximum flows (frequently $C_s > 3C_v$, i.e. the coefficient of skewness is greater than three coefficients of variation) and to monthly flows.

The analytical expression of the p.d.f. and relationships describing the basic statistical characteristics are given in Table 3.4 for Galton's transformation of the random variable

$$y = \lg |x - x_0| \tag{3.20}$$

Further transformations were used by R. Gibrat (Gauss–Gibrat distribution)

$$y = a \log (x - x_0) + b \tag{3.21}$$

Ven Te Chow

$$y = \ln x \tag{3.22}$$

and F. Galton (a new formula)

$$y = \alpha(\lg x - \beta) \tag{3.23}$$

The Gauss–Gibrat distribution fitted the probability distributions of daily flows and annual maximum flows well (Votruba and Broža, 1966). Gumbel's extreme probability distribution is a special case of the log-normal probability distribution (Ven Te Chow, 1964).

3.1.2.6 Gamma Probability Distribution

The mathematical expression of the p.d.f. of the gamma distribution uses the gamma function, i.e. Euler's function of second order $\Gamma(x)$. It is defined by the integral

$$\Gamma(x) = \int_0^\infty e^{-t}\, t^{x-1}\, dt \tag{3.24}$$

that is convergent for $x > 0$. Its form can be described by the expressions

$$\lim_{x \to \infty} \Gamma(x) = +\infty \quad \text{and} \quad \lim_{x \to 0+} \Gamma(x) = +\infty$$

and it has a local minimum for $x \simeq 1.46$.

Further properties of the gamma function are

$$\begin{aligned} &\text{a) } \Gamma(x + 1) = x\,\Gamma(x) \\ &\text{b) } \Gamma(n) = (n - 1)! \text{ for positive integer } n \end{aligned} \tag{3.25}$$

The gamma function has been tabulated (e.g. Abramowitz *et al.*, 1964; Rektorys *et al.*, 1963), and it is used for the calculation of many difficult integrals that can be transformed to it.

Pearson and Weibull used the gamma function for the expression of their distributions. The gamma distribution (or more precisely the incomplete gamma distribution) was used by G. L. Barger and H. C. S. Thom in processing precipitation time series (Kendal, 1970). Due to the possibility of selections many combinations of its two parameters with a pronounced effect on the shape of the p.d.f. (Fig. 3.3), further applications in water management are envisaged.

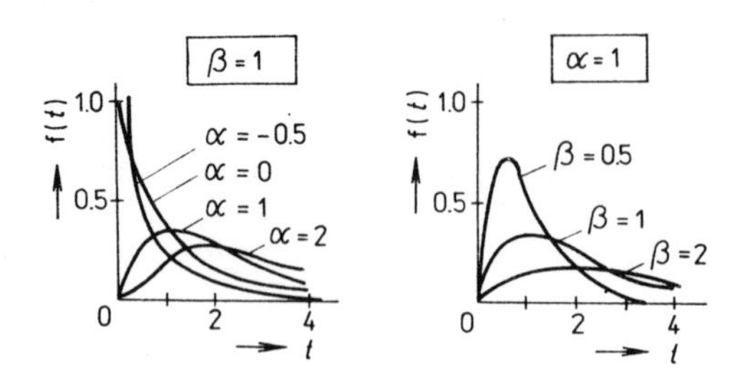

Fig. 3.3 P.d.f. of the gamma-distribution for different combinations of α and β

The general mathematical form of the p.d.f. of the random variable x, i.e. mean, standard deviation and mode, is given in Table 3.5. This distribution has been tabulated by K. Pearson, and it is published not only in specialist mathematical literature but also in water management references. For the estimation of parameters the method of maximum likelihood is recommended and preferred to the method of moments.

Table 3.5 Gamma probability distribution

p.d.f.	$f(x) = \begin{cases} \dfrac{1}{\alpha!\,\beta^{\alpha+1}}\, x^{\alpha}\, e^{-x/\beta} & \text{for } x > 0 \\ 0 & \text{for } x \leq 0 \end{cases}$
c.d.f.	$F(x) = \dfrac{1}{\alpha!}\, \Gamma_{x/\beta}(\alpha + 1)$
mean	$\mu(x) = \beta(\alpha + 1)$
standard deviation	$\sigma(x) = \beta\sqrt{\alpha + 1}$
coefficient of skewness	$C_s(x) = \dfrac{2}{\sqrt{\alpha + 1}}$

α, β – parameters of the gamma probability distribution in a general form

$\Gamma_{x/\beta}(\alpha + 1)$ – gamma-function

Table 3.6 The Pearson probability distribution – type III

Quantity		General form	Standardized form	Geometrical interpretation
random variable		x	$t = \dfrac{x - \mu(x)}{\sigma(x)}$	ξ
mean		$\mu(x)$	0	$\mu(\xi) = \xi_0 + a + d$
standard deviation		$\sigma(x)$	1	$\sigma(\xi) = \sqrt{d(a + d)}$
coefficient of skewness		$C_s = \dfrac{2}{\alpha}$	$C_s = \dfrac{2}{\alpha}$	$C_s = 2\sqrt{\dfrac{d}{a + d}}$
p.d.f.		$f(x) = \dfrac{\alpha^{\alpha^2} \exp\left(\dfrac{\alpha\mu(x)}{\sigma(x)} - \alpha^2\right)}{\Gamma(\alpha^2)\,\sigma^{\alpha^2}(x)} \cdot$ $\cdot \left[\alpha\sigma(x) - \mu(x) + x\right]^{\alpha^2 - 1} e^{-(\alpha/\sigma(x))x}$	$f(t) = \dfrac{\alpha^{\alpha^2} e^{-\alpha^2}}{\Gamma(\alpha^2)} \cdot$ $\cdot (\alpha + t)^{\alpha^2 - 1} e^{-\alpha t}$	$f(\xi) = \dfrac{1}{d(a/d + 1)\,\Gamma(a/d + 1)} \cdot$ $\cdot (\xi - \xi_0)^{a/d} e^{(-\xi - \xi_0)d}$
the range of the random variable	$\alpha > 0$	$[\mu(x) - \alpha\sigma(x); +\infty]$	$(-\alpha; \infty)$	
	$\alpha < 0$	$[-\infty; \mu(x) - \alpha\sigma(x)]$	$(-\infty; -\alpha)$	
schematic geometrical representation of the p.d.f.		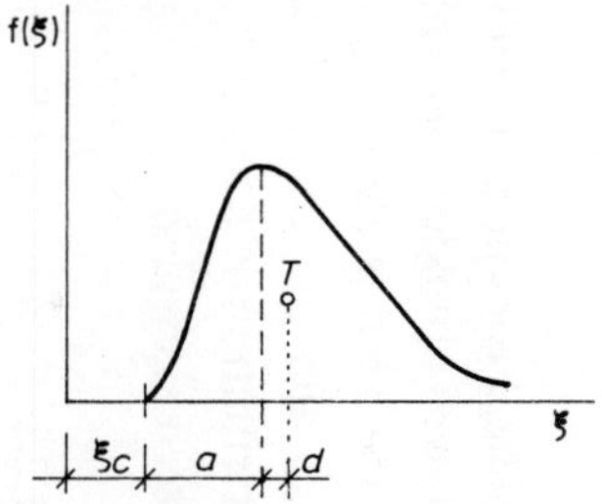		

3.1.2.7 The Pearson Probability Distribution

K. Pearson derived the p.d.f. on the basis of a general differential equation

$$\frac{\mathrm{d}y}{\mathrm{d}x} = \frac{x + a}{b_0 + b_1 x + b_2 x^2}\, y \tag{3.26}$$

with coefficients a, b_0, b_1, b_2, which are real numbers. According to the values of these coefficients and their combinations, he described 12 types of theoretical distributions, derived, in principle, from the general beta distribution. These functions have a purely theoretical basis; however, in many cases they can be successfully used to fit empirical distributions. Sometimes they are considered a good universal type of skew distributions, adequate for discrete random variables (e.g. approximation of the binomial distribution).

The general form of the p.d.f. of the Pearson distribution is given by the expression

$$f(x) = \exp\left(\int_{-\infty}^{x} \frac{x + a}{b_0 + b_1 + b_2 x^2}\, \mathrm{d}x\right) \tag{3.27}$$

The Pearson probability distribution type III is a type of skew distribution with a lower bounded, bell-shaped form. From the analytical description of the p.d.f. in Table 3.6 it is clear that it has three parameters: the mean $\mu(x)$, standard deviation $\sigma(x)$, and the constant α, which defines the coefficient of skewness. The normal distribution is a special case of the Pearson probability distribution type III with a zero skewness.

3.1.2.8 Exponential Probability Distribution

Exponential probability distribution is a single-parameter probability distribution, and it is given in Table 3.7 by its p.d.f. and c.d.f. It can be considered a special case of the gamma distribution, possibly also of the Weibull distribution.

This function is widely used in reliability theory. However, in hydrology and in water management it has been successfuly applied in the statistical treatment of the maximum flows in a given period (Votruba and Broža, 1966).

For application to water management the c.d.f. can be written in a general form

$$F(z) = 1 - \mathrm{e}^{-z} \tag{3.28}$$

where

$$z = \varphi(x) \tag{3.29}$$

Goodrich (Votruba and Nacházel, 1969) used the transformation of the type

$$z = k(x - x')^{1/\alpha} \tag{3.30}$$

with three parameters (constants) k, x', α.

Table 3.7 Exponential probability distribution

p.d.f.	$f(x) = \begin{cases} \Theta^{-1}\, e^{-x/\Theta} & \text{for } x \geqq 0 \\ 0 & \text{for } x < 0 \end{cases}$
c.d.f.	$F(x) = \begin{cases} 1 - e^{-x/\Theta} & \text{for } x \geqq 0 \\ 0 & \text{for } x < 0 \end{cases}$
mean	$\mu(x) = \Theta$; Θ – parameter of distribution
standard deviation	$\sigma(x) = \Theta$
coefficient of skewness	$C_s(x) = 2$

3.1.2.9 The Gumbel Probability Distribution

The Gumbel probability distribution is a type of extreme probability distribution, which is suitable for the maximum values of samples. In hydrology and in WRS problems it has been used for processing flood flows (Coutagne, 1951).

Table 3.8 The Gumbel probability distribution

p.d.f.	$f(x) = \exp(-z) \exp[-\exp(-z)]$
c.d.f.	$F(x) = \exp[-\exp(-z)]$
equation of the exceedance probability curve	$P(x) = 1 - \exp[-\exp(-z)]$
transformation of random variable	$z = \alpha(x - \beta)$
constants in transformation	$\alpha = \dfrac{\pi}{\dot{\sigma}(x)\sqrt{6}}$ $\beta = \dot{\mu}(x) - \dfrac{\gamma}{\alpha}$

α, β – constants in transformation
$\pi = 3.14 \ldots$ – Ludolf's number
$\gamma = 0.577 \ldots$ – Euler's number
$\dot{\mu}(x), \dot{\sigma}(x)$ – point estimates of the mean and standard deviation of the population

Table 3.9 The method of fitting theoretical exceedance probability functions by quantiles method

Probability distribution	Quantiles estimated from empirical probability curve	Auxiliary parameters	Estimation of parameters of the exceedance probability curve	Equation of the theoretical exceedance probability function
log-normal	x_5, x_{50}, x_{95}	$x_0 = \frac{x_5 x_{95} - x_{50}^2}{x_5 + x_{95} - 2x_{50}}$	$s_y = 0.304 \log \frac{x_5 - x_0}{x_{95} - x_0}$	$\log(x_p - x_0) = \log(x_{50} - x_0) + s_y\, \Phi_B(P, C_s)$ for $C_s = 0$
binomial	x_5, x_{50}, x_{95}	$S = \frac{x_5 + x_{95} - 2x_{50}}{x_5 - x_{95}} = \frac{\Phi_5 + \Phi_{95} - 2\Phi_{50}}{\Phi_5 - \Phi_{95}}$	$\bar{x} = x_{50} - s_x \cdot \Phi_{50}$ $s_x = \frac{x_5 - x_{95}}{\Phi_5 - \Phi_{95}}$; $C_v = \frac{s_x}{\bar{x}}$	$x_p = \bar{x}[1 + C_s\, \Phi_B(P, C_s)]$
Gumbel	x_5, x_{95}	$\alpha = \frac{4.067}{x_5 + x_{95}}$ $\beta = 0.27x_s + 0.73x_{95}$	$\bar{x} = 0.412x_5 + 0.5x_{95}$ $s_x = 0.315(x_5 - x_{95})$	$x_p = \beta + \frac{1}{\alpha} Z_p = \bar{x} + s_x\, \Phi_G(P, C_s)$
exponential	x_5, x_{50}, x_{95}	$S_z = \frac{2 \log x_{50} - \log x_5 - \log x_{95}}{\log x_5 - \log x_{95}}$ a) for $C_{sz} > 0$: $\alpha = \frac{\log k_5 - \log k_{95}}{\log x_5 - \log x_{95}}$ b) for $C_{sz} < 0$: $\alpha' = \frac{\log k_{95} - \log k_5}{\log x_5 - \log x_{95}}$ k_5, k_{95} tabulated for C_{sz}	$\log \bar{z} = \alpha \log x_{50} - \log k_{50}$ $\log \bar{z} = \alpha' \log x_{50} - \log k_{50}$	$\log z_p = \log \bar{z} + \log[1 + 0.5C_{sz}\, \Phi_B(P, C_{sz})]$ $\log x_p = \frac{1}{\alpha} \log z_p$

Under Czechoslovak conditions this distribution was applied in processing maximum precipitation (for *n*-day, *n*-hour values). In view of the analytical form of the p.d.f. (Table 3.8) it is also called double-exponential probability distribution, with the standardized deviation about the mode as the argument. E. J. Gumbel called it the law of maximum values as it was derived from an investigation of the highest age. It is skew (with one constant value of the skewness coefficient $C_s = 1.139$) and unbounded.

3.1.2.10 Estimation of Parameters of Cumulative Distribution Function

Analyses performed in WRS studies have shown that for the determination of the parameters of the c.d.f. (or exceedance probability curves) the method of moments need not necessarily be the best one. Therefore some other methods have been developed for a simpler, less cumbersome and often more reliable estimation of the parameters of the c.d.f. (Alexeyev, 1960). These methods use *quantiles*, i.e. values

Table 3.10 The used probability papers

Type of network	Scale of the abscissa (probability)	Scale of the ordinate
normal distribution	derived from the normal c.d.f. (symmetrically increasing on both sides)	linear
log-normal	derived from the c.d.f.	logarithmic
Gumbel	derived from the straight form of the relation between the standardized variable and the c.d.f. (skew, it increases to zero more than to one)	linear
Fréchet	-”-	logarithmic
Goodrich (exponential distribution)	log log $(1/P)$ (P ... derived from the c.d.f. of the normal distribution)	logarithmic
Brovkovich	derived from the c.d.f. of normal distribution (symmetrically increasing on both sides)	derived from the straight form of the standardized variable and the c.d.f. of binomial distribution with $C_s = 2C_v$

chosen in such a way that the observation less than this value forms a given (required or chosen) part of the sample (more correctly: for the number p the p-th quantile x_p of the continuous random variable x having c.d.f. $F(x)$ is defined by the smallest number for which $F(x_p) = p$). The method of quantiles is used for log-normal, binomial, Gumbel and exponential distributions. It is often based on values of two or three quantiles estimated from the smoothed empirical c.d.f. (exceedance probability function) and standardized values of the c.d.f. $\Phi\ (P, C_s)$ which have been tabulated for different types of probability distributions.

The equation for determining the basic statistical characteristics and the value of the theoretical exceedance probability function determined by the quantiles method and the corresponding standardized variables are given in Table 3.9.

The fitting of the theoretical c.d.f. (exceedance probability function) to the empirical one in hydrology is often performed in the following way: the values of the random variable are ranked in the ascending (descending) order, the empirical probabilities (probabilities of exceedance) are calculated by the plotting – position – probability formula $p = (m + a)/(n + b)$ where m is the rank, n is the number of ranked events, a and b are constants (e.g. Chegodayev has $a = -0.3$, $b = 0.4$, Hazen, $a = -0.5$, $b = 0$ or Weibull, $a = 0$, $b = 1$).

These values p and the corresponding values of stochastic variables are plotted on probability paper, where the scale of probability typically increases towards both sides. The exceedance probability function (or c.d.f.) is represented on this paper by a curve approaching a straight line, or by a straight line if the sample investigated has the theoretical c.d.f. used in the construction of the probability network. Therefore, the probability paper that seems to correspond best to the empirical sample is selected. Table 3.10 shows the main probability papers used in WRS with a description of the scales on the axes.

Example II

Earth samples were taken during the construction of an earthen dam to provide a sample of values for the density of wet earth. The frequencies (histogram) in Fig. II.1 show a negative skewness that suggests the use of binomial distribution. The parameters of the fitted theoretical c.d.f. will be determined using the method of quantiles.

Solution

The values of the quantiles are estimated from the empirical exceedance probability curve (Fig. II.2)

$$x_5 = 2.190 \quad x_{50} = 2.080 \quad x_{95} = 1.937$$

An auxiliary value S is calculated

$$S = \frac{x_5 + x_{95} - 2x_{50}}{x_5 - x_{95}} = \frac{2.190 + 1.937 - 2 \cdot 2.080}{2.190 - 1.937} = -0.130 \tag{II.1}$$

From the tables (e.g. Votruba *et al.*, 1973, appendix II) we get $S = -0.13$

$$C_s = -0.470; \quad \Phi_5 - \Phi_{95} = 3.260; \quad \Phi_{50} = -0.080$$

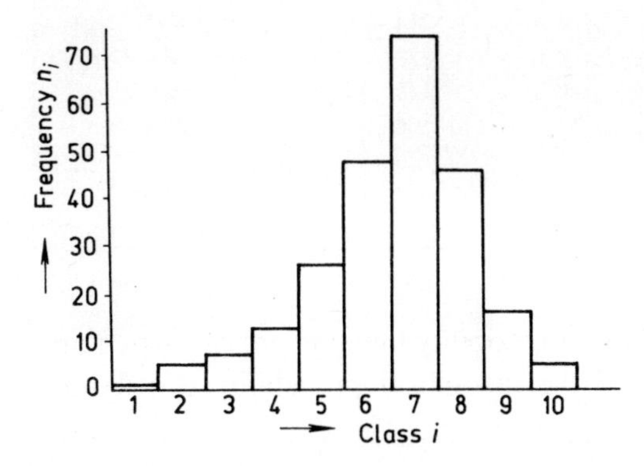

Fig. II.1 Histogram of class frequencies of earth density in a wet stage

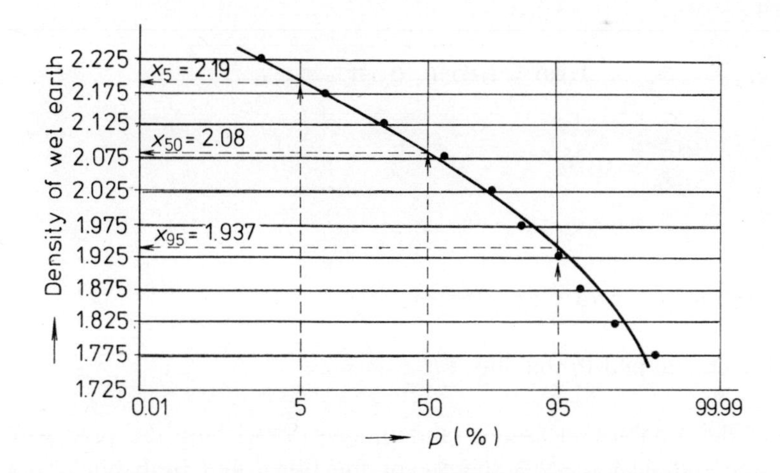

Fig. II.2 Smoothed empirical exceedance probability curve

Table II.1 The class frequencies of the density of wet earth

i	Value of the class mark $[x_i]$	Frequency $[n_i]$	Rank of the last element in each class $[m]$	Exceedance probability $p = \frac{m - 0.3}{n + 0.4} 100\%$
1	1.775	1	239	99.8
2	1.825	4	238	99.2
3	1.875	7	234	97.3
4	1.925	13	227	94.6
5	1.975	26	214	89.1
6	2.025	48	188	78.4
7	2.075	74	140	58.3
8	2.125	46	66	27.4
9	2.175	16	20	8.22
10	2.225	4	4	1.54

The sample parameters of the exceedance probability function are calculated and used as the estimators of parameters of the theoretical probability distribution (Table II.1)

$$s_x = \frac{x_5 - x_{95}}{\Phi_5 - \Phi_{95}} = \frac{2.190 - 1.937}{3.260} = 0.078 \tag{II.2}$$

$$\bar{x} = x_{50} - s_x \Phi_{50} = 2.080 - 0.078(-0.08) = 2.086 \tag{II.3}$$

$$C_v = \frac{s_x}{\bar{x}} = \frac{0.078}{2.086} = 0.038 \tag{II.4}$$

3.1.3 Dependence and Correlation

3.1.3.1 Functional and Probability Relationships

Among the variables (including random ones) describing the processes in the natural and technical sciences there exist functional and probability (stochastic) relationships; in water management a linear correlation is the most frequent one.

The investigation of the probability relationships of the random variables was facilitated by the following factors: the accumulation of sufficient information concerning the phenomena and describing random variables; a deeper understanding of the genesis of the processes investigated, which helped to reduce possible errors in the conclusions concerning the existence and nature of the probability relationships, improvement of methods of analysis; and, last but not least, the use of computers, for their evaluation.

Assume that the phenomena investigated have, in the simplest case, a single property that can be measured and expressed quantitatively. The quantity that describes this property is, in general, variable in time, t, and in space, s, and includes some random component ε, so that

$$x = f(t) + t_0 + \varepsilon_1 \tag{3.31}$$

and, at the same time,

$$x = g(s) + s_0 + \varepsilon_2 \tag{3.32}$$

or

$$x = h(s, t) + s_0 + t_0 + \varepsilon_3 \tag{3.33}$$

where s_0 and t_0 are constants

Similarly, a variable Y can be expressed. Assume that the variables X and Y are function of more random variables $Z_1, Z_2, \ldots, Z_n$, which are common to both variables, and that Y is also a function of $U_1, U_2, \ldots, U_r$ and Y a function of $V_1, V_2, \ldots, V_s$; then there is a probability relationship between the variables X and Y.

In WRS contexts, probability relationships between the variables investigated are quite common. If overlooked, this could lead to incorrect results with serious technical and economic consequences.

3.1.3.2 Multivariate Probability Distribution

In section 3.1.2, one-dimensional probability distributions were investigated where the probability of occurrence of the random variable can be described by a single value. In the investigation of probability relationships among random variables it is useful to start with the idea of multivariate probability distributions which take into account more factors of each element of the statistical sample. The probability of their occurrence is not given by a single value but by several values depending on the occurrence of a certain factor; in fact, this probability is a conditional probability in the sense of section 3.1.1.

Let us start with the two-variate probability distribution of the random variable (X, Y), where $X = x_i\,(i = 1, 2, \ldots, n)$ and $Y = y_j\,(j = 1, 2, \ldots, m)$, the matrix on which is given in Table 3.11. Each row of the matrix gives the conditional probability distribution of the random variable X, given the value Y, and each column the conditional probability of the random variable Y, given X. The last row and last column contain the values of the unconditional probability distributions of the values X and Y respectively.

Table 3.11 Schematic representation of the matrix of the probability distribution of a two-variate discrete random variable

$Y \backslash X$	x_1	x_2	...	x_i	...	x_n	
y_1	$p(x_1, y_1)$	$p(x_2, y_1)$	...	$p(x_i, y_1)$	...	$p(x_n, y_1)$	$p(y_1)$
y_2	$p(x_1, y_2)$	$p(x_2, y_2)$	...	$p(x_i, y_2)$	...	$p(x_n, y_2)$	$p(y_2)$
$\vdots$							
y_j	$p(x_1, y_j)$	$p(x_2, y_j)$	...	$p(x_i, y_j)$	...	$p(x_n, y_j)$	$p(y_j)$
$\vdots$							
y_m	$p(x_1, y_m)$	$p(x_2, y_m)$	...	$p(x_i, y_m)$	...	$p(x_n, y_m)$	$p(y_m)$
	$p(x_1)$	$p(x_2)$		$p(x_i)$		$p(x_n)$	

The k-th and c-th moments and central moments of the two-variate random variables (X, Y) are determined by the expressions

$$_{x,y}m_{k,c} = \frac{1}{n} \sum_{i=1}^{n} x_i^k y_j^c \tag{3.34}$$

$$_{x,y}M_{k,c} = \frac{1}{n} \sum_{i=1}^{n} (x_i - \bar{x})^k (y_i - \bar{y})^c \tag{3.35}$$

If $k \neq 0$ and at the same time $c \neq 0$, then the moments are called *mixed product moments*; the most important of these is the first central moment $_{x,y}M_{1,1}$ called the *covariance*

$$_{x,y}M_{1,1} = \frac{1}{n} \sum_{i=1}^{n} (x_i - \bar{x})(y_i - \bar{y}) \tag{3.36}$$

All the covariances of a n-dimensional sample of the random variables $X_1, X_2, \ldots, X_n$ can be arranged in a square matrix of the following form:

$$ {}_{i,j}\mathbf{M} = \left\| \begin{matrix} {}_{1,1}M & {}_{1,2}M & \ldots & {}_{1,n}M \\ {}_{2,1}M & {}_{2,2}M & \ldots & {}_{2,n}M \\ \ldots & & & \\ {}_{n,1}M & {}_{n,2}M & \ldots & {}_{n,n}M \end{matrix} \right\| \tag{3.37} $$

which is called a *covariance matrix*. It is symmetrical with respect to the main diagonal so that it can be written as the upper triangle matrix

$$ {}_{i,j}\mathbf{M} = \left\| \begin{matrix} {}_{1,1}M & {}_{1,2}M & \ldots & {}_{1,n}M \\ & {}_{2,2}M & \ldots & {}_{2,n}M \\ & & \ddots & \vdots \\ & & & {}_{n,n}M \end{matrix} \right\| \tag{3.38} $$

For the independent (uncorrelated) random variables X_1, X_2, ..., X_n this matrix is, in fact, a diagonal matrix with non-zero elements on the main diagonal only:

$$ {}_{i,j}\mathbf{M} = \left\| \begin{matrix} {}_{1,1}M & 0 & \ldots & 0 \\ & {}_{2,2}M & \ldots & 0 \\ & & \ddots & \vdots \\ & & & {}_{n,n}M \end{matrix} \right\| \tag{3.39} $$

Instead of the covariance matrix, a *correlation matrix* is often used for the standardized variables. Its elements are not the covariances but the *correlation coefficients* defined by

$$ r_{ij} = \frac{{}_{i,j}M_{1,1}}{S_i S_j} \tag{3.40} $$

so that (3.38) becomes (3.41)

$$ \mathbf{r}_{i,j} = \left\| \begin{matrix} 1 & r_{1,2} & r_{1,3} & \ldots & r_{1,n} \\ & 1 & r_{2,3} & \ldots & r_{2,n} \\ & & 1 & \ldots & r_{3,n} \\ & & & & \ldots \\ & & & & 1 \end{matrix} \right\| \tag{3.41} $$

3.1.3.3 Methods of Determination of the Linear Statistical Dependence Between Random Variables

These methods encompass the calculation of the coefficients of linear regression and correlation from the covariances, the coefficient of rank correlation, and also the estimation of the statistical dependence by graphical representation of the regression relationship by plotted points. The *coefficient of linear correlation, r*, is a measure of the statistical dependence between the random variables X and Y (their values x_i and y_i) and it is calculated from the covariances by the following equations

$$r = \frac{{}_{x,y}M_{1,1}}{s_x s_y} = \frac{\sum_{i=1}^{n}(x_i - \bar{x})(y_i - \bar{y})}{\sqrt{\sum_{i=1}^{n}(x_i - \bar{x})^2 \sum_{i=1}^{n}(y_i - \bar{y})^2}} \tag{3.42}$$

$-1 \leqq r \leqq 1$, and for uncorrelated (i.e. statistically independent) random variables X and Y, $r = 0$.

The statistical dependence can be either *direct* (i.e. positive for $0 < r \leqq 1$) or *indirect* (i.e. negative for $-1 \leqq r < 0$). The value of the correlation coefficient is invariant to linear transformation of random variables.

The coefficient of linear correlation is a good measure of the statistical dependence of random variables X and Y, if the relationship between them approaches the linear one. Therefore it is useful to evaluate the degree of linearity of the relationship between two random variables (e.g. by plotting of points).

The following formulation is recommended for the calculation of the correlation coefficient:

$$r = \frac{n \sum_{i=1}^{n} x_i y_i - \sum_{i=1}^{n} x_i \sum_{i=1}^{n} y_i}{\sqrt{\left[n \sum_{i=1}^{n} x_i^2 - \left(\sum_{i=1}^{n} x_i\right)^2\right]\left[n \sum_{i=1}^{n} y_i^2 - \left(\sum_{i=1}^{n} y_i\right)^2\right]}} \tag{3.43}$$

The *rank correlation coefficient* (Spearman's correlation coefficient) used for testing the linear dependence of two random variables, is given by the formula

$$r_p = 1 - \frac{6 \sum_{i=1}^{n} d_i^2}{n(n^2 - 1)} \tag{3.44}$$

For its calculation, the values x_i and y_i are replaced by their ranks (in descending order). The difference of ranks of the corresponding values v_{x_i} and v_{y_i} in both samples of random variables X and Y are denoted by d_i, n being the number of correlated pairs. To check the correct determination of the values d_i, the condition

$$\sum_{i=1}^{n} d_i = 0$$

may be used. The values of the rank coefficient have the same range and can be compared with the ordinary correlation coefficient determined by covariances.

Example III

Calculate the value of rank correlation coefficient of the mean annual flows Q_a in the period 1931 – 1960 in the town of Písek on the River Otava and compare it with the ordinary correlation coefficient.

Table III.1 Calculation of the rank correlation coefficient

i	Q_r (m^3 s^{-1}) a)	v_{x_i}	v_{y_i}	$d_i = v_{x_i} - v_{y_i}$	d_i^2
1	40.2	6	8	− 2	4
2	37.6	8	26	−18	324
3	35.2	26	28	− 2	4
4	31,0	28	22	+ 6	36
5	28.2	22	19	+ 3	9
6	27.5	19	15	+ 4	16
7	27.1	15	10	+ 5	25
8	26.4	10	3	+ 7	49
9	25.6	3	1	+ 2	4
10	24.9	1	2	− 1	1
11	23.8	2	14	−12	144
12	23.4	14	27	−13	169
13	22.7	27	5	+22	484
14	21.8 (2×)	5	11	− 6	36
15	21.5	11	12	− 1	1
16	21.1	12	23	−11	121
17	20.5	23	9	+14	196
18	20.3	9	16	− 7	49
19	19.7	16	27	−11	121
20	19.4	27	24	+ 3	9
21	17.8	24	25	− 1	1
22	17.2	25	21	+ 4	16
23	17.1	21	14	+ 7	49
24	16.3	14	4	+10	100
25	16.2	4	17	−13	169
26	14.1	17	13	+ 4	16
27	12.4 (2×)	13	7	+ 6	36
28	10.4	7	18	−11	121
29		18	20	− 2	4
$\sum$				×)	2314

a) ranked in descending order

×) the sum $\sum d_i$ is not equal to zero as in the column v_{x_i} in the computation of the autocorrelation coefficient the 30th year element is not included, which has $v_{x30} = 20$; similarly in the column v_{y_i} the 1st element is missing, which has $v_{y_1} = 6$. Then $(-\sum d_i = -14) + 20 - 6 = 0$

Solution

The calculation was arranged in a tabular form (Table III.I.). The coefficient of the rank correlation was calculated by formula (3.44)

$$r_p = 1 - \frac{6 \sum_{i=1}^{n} d_i^2}{n(n^2 - 1)} = 1 - \frac{6 \, . \, 2314}{29(29^2 - 1)} = 0.428 \qquad \text{(III.1.)}$$

The value of the ordinary correlation coefficient using formula (3.42) or (3.43) is a little higher; $r = 0.497$.

The correlation technique is often accompanied by regression analysis. Linear regression functions (geometrically represented by straight lines) relate a dependent variable X to an independent variable Y or vice versa. The plotted points of the relationship are concentrated along a straight line if the absolute value of the correlation coefficient converges to one (Fig. 3.4).

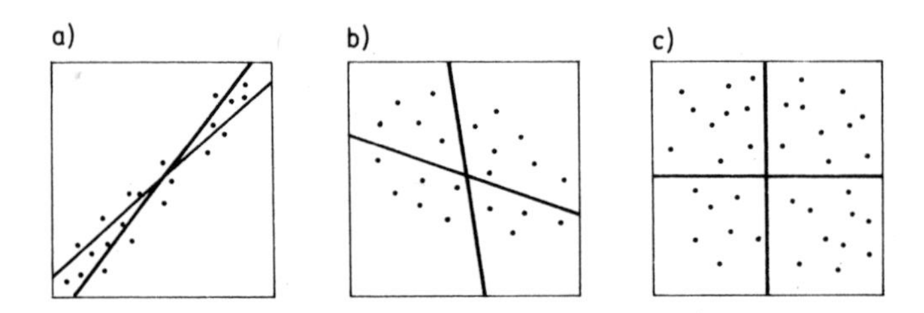

Fig. 3.4 Relative position of the regression straight lines
a – close direct linear regression, b – loose indirect linear regression, c – independence

The general equation for the linear regression (equation of the regression straight line) of the dependent random variable Y and independent random variable X is

$$Y = a + bX \qquad (3.45)$$

where a (intercept) and b (regression coefficient) are constants that are determined by the least squares method from the sum

$$\sum_{i=1}^{n} y_i = na + b \sum_{i=1}^{n} x_i \qquad (3.46)$$

The following expressions for these constants have been derived

$$a = \frac{\sum_{i=1}^{n} x_i^2 \sum_{i=1}^{n} y_i - \sum_{i=1}^{n} x_i \sum_{i=1}^{n} x_i y_i}{n \sum_{i=1}^{n} x_i^2 - \left(\sum_{i=1}^{n} x_i\right)^2} \qquad (3.47)$$

$$b = \frac{n \sum_{i=1}^{n} x_i y_i - \sum_{i=1}^{n} x_i \sum_{i=1}^{n} y_i}{n \sum_{i=1}^{n} x_i^2 - \left(\sum_{i=1}^{n} x_i \right)^2} \tag{3.48}$$

By substituting these expressions into eq. (3.45), the equation of the linear regression, the expanded function $Y = f(x)$ is obtained. Similarly, the contents a' and b' can be calculated for the function $x = g(Y)$ in the form

$$X = a' + b' Y \tag{3.49}$$

The constant b in equation (3.45) is the *regression coefficient* of the linear function $Y = f(X)$, and b' in equation (3.49) the regression coefficient of the linear function $X = g(Y)$. They can be related to the standard deviations s_x and s_y of the random variables X and Y in the equations

$$b = r \frac{s_y}{s_x} \tag{3.50}$$

$$b' = r \frac{s_x}{s_y} \tag{3.51}$$

and the regression function becomes

$$y - \bar{y} = r \frac{s_y}{s_x} (x - \bar{x}) \tag{3.52}$$

and

$$x - \bar{x} = r \frac{s_x}{s_y} (y - \bar{y}) \tag{3.53}$$

The correlation coefficient can be written in the form

$$r = \sqrt{bb'} \tag{3.54}$$

The graphic representation of pairs of regression straight lines showing high and slight degrees of dependence and independence, respectively, is shown in Fig. 3.4.

The value of the correlation coefficient between two variables is influenced by, among other things, autocorrelation in the samples of the random variables, a frequent occurrence in WRS contexts. Therefore, for individual cases under investigation, it is advisable to check the significance of autocorrelation in series of both random variables X and Y in relation to cross-correlation by means of the criterion

$$\alpha = \frac{1}{1 + 2r_1 r'_1 + 2r_2 r'_2 + \ldots} \tag{3.55}$$

where r_1 is the correlation coefficient between $x(t)$ and $x(t + 1)$,
r_2 – the correlation coefficient between $x(t)$ and $x(t + 2)$,
r'_1 – the correlation coefficient between $y(t)$ and $y(t + 1)$,
r'_2 – the correlation coefficient between $y(t)$ and $y(t + 2)$, etc.

If α converges to one the cross-correlation coefficient can be calculated as if X and Y were two samples without any autocorrelation (Solomon, 1970). Conversely, the autocorrelation may be analysed in relation to cross-correlation (Matalas, 1970; Cavadias, 1970).

3.1.3.4 Methods of Determination of Complex Statistical Dependence Between Random Variables

Non-linear and multivariate correlation and regression are the main examples of complex statistical dependence, which is important for WRS problems.

Non-linear regression is applied in cases where no other transformation to linear regression is possible. The plotted points are concentrated more or less closely along some curve. There are no simple and exact methods for the determination of this curve, and therefore the form of the regression curve must be estimated. An analogy of the coefficient of linear regression is the *correlation ratio* η and an analogy of the regression straight lines are the regression curves with the *correlation indices* $\boldsymbol{I}$ used as their parameters. They are determined by the equations

$$\boldsymbol{I}_x = \sqrt{1 - \frac{s_{yx}^2}{s_x^2}} \tag{3.56}$$

$$\boldsymbol{I}_y = \sqrt{1 - \frac{s_{xy}^2}{s_y^2}} \tag{3.57}$$

$$\eta = \sqrt{1 - \frac{\bar{s}_y^2}{s_y^2}} \tag{3.58}$$

where s_{xy}^2 and s_{yx}^2 are the residual variances of empirical points with respect to the estimated regression curve and

$$\bar{s}_y^2 = \frac{1}{n} \sum_{i=1}^{n} \left[\sum_{j=1}^{k} (y_i - \bar{y}_j)^2 \right] \tag{3.59}$$

where k is the number of classes in which the points have been grouped. If $\eta^2 - r^2 \to 0$ the correlation converges on the linear, if $\boldsymbol{I}_x^2 - \eta^2 \to 0$ and also $\boldsymbol{I}_y^2 - \eta^2 \to 0$ the regression function was well estimated. The range of the correlation ratio is the same as the range of the coefficient of linear correlation, i.e. $|\eta| \leqq 1$. In general $|\eta| > |r|$. The difference $\varkappa = \eta^2 - r^2$ is called the *criterion of linearity*.

The transformation of random variables from a non-linear to a linear relationship can have a logarithmic form. This *logarithmic anamorphosis* is very useful in many cases.

In hydrological and other WRS issues multivariate random variables are often investigated (e.g. multivariate regression models). The relationship of the random variables is described by equations of multivariate (in the following example by threevariate) correlation and regression. The regression function in the case of two dependent variables x_1 and x_2 (regression plane) has the following form

$$y = a + b_1x_1 + b_2x_2 \tag{3.60}$$

and the expanded form

$$y - \bar{y} = \frac{r_{x_1y} - r_{x_1x_2}r_{x_2y}}{1 - r^2_{x_1x_2}} \frac{\sigma_y}{\sigma_{x_1}} (x_1 - \bar{x}_1) + + \frac{r_{x_2y} - r_{x_1x_2}r_{x_1y}}{1 - r^2_{x_1x_2}} \frac{\sigma_y}{\sigma_x} (x_2 - \bar{x}_2) \tag{3.61}$$

with the usual meaning of the symbols used, and the multiple *correlation coefficients* is

$$r_{y.x_1x_2} = \sqrt{\frac{r^2_{x_1y} + r^2_{x_2y} - 2r_{x_1x_2}r_{x_1y}r_{x_2y}}{1 - r^2_{x_1x_2}}} \tag{3.62}$$

In this form the relationship between the independent variable Y and two dependent random variables X_1 and X_2 is tested. In hydrological applications the graphic coaxial correlation method is used for the multivariate correlation and regression, e.g. for the rainfall – runoff relationship (Dub and Němec, 1969).

3.1.4 Statistical Estimation of Probability Distribution Parameters

The purpose of the statistical estimation of probability distribution parameters is to provide information for scientific conclusions concerning mass reproducible phenomena and processes, based on samples of observed and measured data. The relationship between the population parameters and the sample estimator is investigated.

3.1.4.1 Estimators of Samples from a Population with Normal Distribution

The population is assumed to be infinite. Random or controlled samples from this population have a finite number of elements (size). The parameters of the population are related to the estimators derived from the sample. These estimators are known functions of the sample elements, which are used in place of the unknown true values of the estimated parameters. Therefore they vary to some extent from sample to sample. The main parameters and estimators are provided in Table 3.12.

Table 3.12 Main population parameters and sample estimators and their notation

Quantity	Population parameters	Sample estimators
size number of elements	N ($N \to \infty$)	n
mean	$\mu(x)$	$\bar{x}$
variance	$\sigma^2(x)$	s_x^2
standard deviation	$\sigma(x)$	s_x
coefficient of variation	$C_v(x)$	C_{vx}
coefficient of skewness	$C_s(x)$	C_{sx}
coefficient of kurtosis	$E(x)$	E_x
k-th moment	$m_k(x)$	$m_{x,k}$
k-th central moment	$M_k(x)$	$M_{x,k}$
mean of sample means	$\mu(\bar{x})$	
variance of sample means	$\sigma^2(\bar{x})$	
mean of sample variances	$\mu(s_x^2)$	
variance of sample variances	$\sigma^2(s_x^2)$	

Assume a population with normal probability distribution. The point and interval estimates of the mean, variance and standard deviation of sample mean and variances are given in Table 3.13.

A limited number of correlation and regression relationships for the other probability distributions of the population are cited in references (Šor, 1965). Their analytical expression is often complicated and it is often approximated by transformations to the normal probability distribution. Methods using the simulation technique are promising for the derivation of estimates from the random series generated.

3.1.4.2 Distribution Involved in Analysis of Variance

The distributions involved in analysis of variance are used in the estimation of population parameters based on random samples from this population.

They include:

– Probability distribution chi-square; this distribution is used, for example, for

Table 3.13 Estimators of a sample from a population with normal probability distribution

Population parameter – quantity	Sample estimator
mean of sample means	$\mu(\bar{x}) = \mu(x)$
variance of sample means	$\sigma^2(\bar{x}) = \dfrac{\sigma^2(x)}{n}$
standard deviation of sample means	$\sigma(\bar{x}) = \dfrac{\sigma(x)}{\sqrt{n}}$
mean of sample variances	$\mu(s_x^2) = \dfrac{n-1}{n}\,\sigma^2(x)$
variances of sample variances	$\sigma^2(s_x^2) = \dfrac{2(n-1)}{n}\,\sigma^4(x)$
interval estimator for sample mean	$\mu(x) - t_p \dfrac{\sigma(x)}{\sqrt{n}} < \bar{x} < \mu(x) + t_p \dfrac{\sigma(x)}{\sqrt{n}}$
interval estimator for sample variance	$\dfrac{\chi^2_{1-p}}{n}\,\sigma^2(x) < s_x^2 < \dfrac{\chi^2_p}{n}\,\sigma^2(x)$
interval estimator for sample standard deviation	$\sqrt{\dfrac{\chi^2_{1-p}}{n}}\,\sigma(x) < s_x < \sqrt{\dfrac{\chi^2_p}{n}}\,\sigma(x)$
interval estimator for sample variance of a large sample	$\sigma^2(x)\left(1 - t_p\sqrt{\dfrac{2}{n}}\right) < s_x^2 < \sigma^2(x)\left(1 + t_p\sqrt{\dfrac{2}{n}}\right)$

χ^2_p, χ^2_{1-p} – tabulated values of a standardized variable having the chi-square probability distribution for probability $p = 1 - \alpha/2$

t_p – tabulated values of the standard variable having the normal distribution for probability $p = 1 - \alpha/2$, where α = level of significance

the sample standard deviation of a random sample from a normally distributed population;

– Student probability distribution; the standardized value of the sample mean if the sample standard deviation is used for this standardization, and also the standardized difference between two different samples from the same population have Student distribution;

– Distribution *F* (Snedecor); this distribution is the ratio of two different variances of samples from the same population;

– The Fisher distribution is a logarithmic transformation of the Snedecor distribution.

Table 3.14 Distributions involved in analysis of variance

Quantity	Student probability distribution	Chi-square probability distribution	Snedecor probability distribution F
p.d.f.	$f_m(t) = \dfrac{\Gamma\left(\dfrac{m+1}{2}\right)}{\sqrt{m\pi}\,\Gamma\left(\dfrac{m}{2}\right)}\left(1+\dfrac{t^2}{m}\right)^{-(m+1)/2}$	$f_m(\chi^2) = \dfrac{1}{2^{m/2}\,\Gamma\left(\dfrac{m}{2}\right)}(\chi^2)^{m/2-1}\,e^{-\chi^2/2}$	$f_{m_1 m_2}(F) = \dfrac{m_1}{m_2}\,\dfrac{\Gamma\left(\dfrac{m_1+m_2}{2}\right)}{\Gamma\left(\dfrac{m_1}{2}\right)\Gamma\left(\dfrac{m_2}{2}\right)}\cdot\left(\dfrac{m_1}{m_2}F\right)^{(m_1/2)-1}\left(1+\dfrac{m_1}{m_2}F\right)^{-(m_1+m_2)/2}$
mean	$\mu(t) = 0$	$\mu(\chi^2) = m$	$\mu(F) = \dfrac{m_2}{m_2-2}$
variance	$\sigma^2(t) = \dfrac{m}{m-2}$	$\sigma^2(\chi^2) = \sqrt{2m}$	$\sigma^2(F) = \dfrac{m^2}{(m_2-2)^2}\,\dfrac{2(m_1+m_2-2)}{m_1(m_2-4)}$
coefficient of skewness	$C_s(t) = 0$	$C_s(\chi^2) = 2\sqrt{\dfrac{2}{m}}$	
coefficient of kurtosis	$E(t) = \dfrac{6}{m-4}$		
remarks	$t = \dfrac{z\sqrt{m}}{\sqrt{v}}$ m – degrees of freedom z – independent random variable with normal probability distribution v – independent random variable with chi-square probability distribution	m – degrees of freedom	$F = \dfrac{u/m_1}{v/m_2}$ u, v – independent random variables with chi-square probability distribution For $z = \frac{1}{2}\lg F$ Fisher probability distribution m_1, m_2 – degree of freedom for u and v resp.

The p.d.f. and the most important statistical moments of these distributions are given in Table 3.14.

The application of these distributions to point and interval estimation of population parameters and to testing statistical hypotheses is demonstrated in sections 3.1.4.3 and 3.1.5, respectively.

3.1.4.3. Estimators of Population Parameters with Normal Probability Distribution

The estimators are classified as point and interval estimators. In point estimation, the population parameter is estimated by a single value. In interval estimation, two statistics are determined in such a way that the estimated parameter will be inside the interval bounded by them with a given large probability. Interval estimation is, in general, more useful; the range of the interval is related to the accuracy of the estimation.

a) Point estimation

The point estimator of the unknown population mean $\mu(x)$ is the sample mean $\bar{x}$, i.e. $\hat{\mu}(x) = \bar{x}$. The point estimator of the population variance $\sigma^2(x)$ is given by

$$\hat{\sigma}^2(x) = \frac{n}{n-1} s_x^2 = \frac{1}{n-1} \sum_{i=1}^{n} (x_1 - \bar{x})^2 \tag{3.63}$$

Since the mean of the sample variances $\mu(s_x^2)$ is equal to

$$\mu(s_x^2) = \frac{n-1}{n} \sigma^2(x) \tag{3.64}$$

the point estimator of the population standard deviation is

$$\hat{\sigma}(x) = \sqrt{\frac{n}{n-1} s_x^2} = \sqrt{\frac{1}{n-1} \sum_{i=1}^{n} (x_1 - \bar{x})^2} \tag{3.65}$$

b) Interval estimation

With probability $p = 1 - \alpha/2$ (called the confidence coefficient; α will be defined in section 3.1.5 as the level of significance), the *population mean* $\mu(x)$ is inside the interval

$$\bar{x} - t_p \frac{\sigma(x)}{\sqrt{n}} < \mu(x) < \bar{x} + t_p \frac{\sigma(x)}{\sqrt{n}} \tag{3.66}$$

It is a two-sided confidence interval. The one-sided confidence interval with probability $p = 1 - \alpha$ is

$$\mu(x) < \bar{x} + t_p \frac{\sigma(x)}{\sqrt{n}} \text{ with lower boundary } -\infty \tag{3.67}$$

or

$$\mu(x) > \bar{x} - t_p \frac{\sigma(x)}{\sqrt{n}} \text{ with upper boundary } +\infty \tag{3.67'}$$

The values t_p are determined from the tables of normal distribution. If the population standard deviation $\sigma(x)$ is not known, it is replaced by its point estimator s_x, and t_p is determined from the tables of the Student distribution for $m = n - 1$ degrees of freedom, where n is the size of the sample from the population. The Student distribution converges asymptotically to the normal distribution for $n \to \infty$.

With probability $p = 1 - \alpha/2$, the population variance $\sigma^2(x)$ is inside the interval

$$\frac{n}{\chi_p^2} s_x^2 < \sigma^2(x) < \frac{n}{\chi_{1-p}^2} s_x^2 \tag{3.68}$$

where χ_p^2 is determined from the tables of the chi-square c.d.f. If the point estimator of variance is known, the following inequalities can be written

$$\frac{n-1}{\chi_p^2} \dot{\sigma}^2(x) < \sigma^2(x) < \frac{n-1}{\chi_{1-p}^2} \dot{\sigma}^2(x) \tag{3.69}$$

Often the one-sided confidence interval is used for the determination of the upper confidence limit of the unknown population variance

$$\sigma^2(x) < \frac{n}{\chi_{1-p}^2} s_x^2 \tag{3.70}$$

or

$$\sigma^2(x) < \frac{n-1}{\chi_{1-p}^2} \dot{\sigma}^2(x) \tag{3.70'}$$

For large samples the confidence interval becomes

$$\frac{s_x^2}{1 + t_p \sqrt{\frac{2}{n}}} < \sigma^2(x) < \frac{s_x^2}{1 - t_p \sqrt{\frac{2}{n}}} \tag{3.71}$$

where t_p is determined from the tables of normal distribution (standardized variable t for probability $p = 1 - \alpha/2$).

The confidence coefficient p used for the estimation of population parameters is often chosen in the interval 95 to 99% (in exceptional cases 90%). On the other hand the necessary sample size can be calculated for a given confidence coefficient.

If the acceptable relative error of the mean is given by

$$\vartheta = \frac{t_p \frac{\sigma(x)}{\sqrt{n}}}{\mu(x)} \tag{3.72}$$

then

$$n = \left(\frac{t_p\, C_v(x)}{\vartheta}\right)^2 \tag{3.73}$$

where

$$C_v(x) = \frac{\sigma(x)}{\mu(x)} \tag{3.74}$$

t_p is to be found in tables of normal c.d.f. for probability $p = 1 - \alpha/2$

Example IV

Calculate the interval estimators of the population mean and standard deviation based on the sample of size $n = 280$ with normal probability distribution, sample mean $\bar{x} = 10.37$ and variance $s_x^2 = 0.225$ from example I. The confidence coefficient is 99.5%, the level of significance is 1%.

Solution

1. The confidence interval for $\mu(x)$ is

$$\bar{x} - t_p \frac{\dot{\sigma}(x)}{\sqrt{n}} < \mu(x) < \bar{x} + t_p \frac{\dot{\sigma}(x)}{\sqrt{n}} \tag{IV.1.}$$

where t_p can be found from tables of the Student probability distribution for $m = n - 1$ degrees of freedom. In our case $\alpha = 0.01$; $m = 280 - 1 = 279$; $t_p =$ $= 2.594$ for $p = 1 - 0.01/2 = 0.995$.

$\dot{\sigma}(x)$ is the point estimator of the population standard deviation that is calculated as the sample standard deviation with denominator $(n - 1)$ instead of n, i.e.

$\dot{\sigma}(x) = s_x = 0.475$

Substituting in (IV.1) we get

$$10.37 - 2.594 \frac{0.475}{\sqrt{280}} < \mu(x) < 10.37 + 2.594 \frac{0.475}{\sqrt{280}}$$

$$10.297 < \mu(x) < 10.444$$

2. The confidence interval of the population standard deviation $\sigma(x)$ is given by inequalities

$$\dot{\sigma}(x) \sqrt{\frac{n-1}{\chi_p^2}} < \sigma(x) < \dot{\sigma}(x) \sqrt{\frac{n-1}{\chi_{1-p}^2}} \tag{IV.2.}$$

For the confidence coefficient $p = 1 - \alpha/2 = 0.995$,

$$0.475 \sqrt{\frac{279}{\chi_{0.995}^2}} < \sigma(x) < 0.475 \sqrt{\frac{279}{\chi_{0.005}^2}}$$

In tables of the chi-square distribution with $m = 279$ degrees of freedom it was found that

$$\chi^2_{0.995} = 343.6 \quad \text{and} \quad \chi^2_{0.005} = 221.9$$

giving

$$0.428 < \sigma(x) < 0.533$$

3.1.5 Tests of Significance and Tests of Hypotheses

The testing of statistical hypotheses involves:

(1) Testing the values of parameters of the probability distribution that is assumed to be known, and

(2) testing the hypothesis of the possibility to fit some probability distribution. A null hypothesis is formulated and its validity is tested with a certain high statistical confidence; i.e. at a given level of significance α, this is the probability (small) of rejecting the null hypothesis when it is true. Its value is selected arbitrarily, e.g. in hydrology and WRS problems values of 1% and 5% are mainly used.

Another, more general, possible classification distinguishes between the testing of parametric statistical hypotheses (investigating parameters of the c.d.f. and their sample estimators) and non-parametric ones (investigating other phenomena). The methods of parametric and non-parametric testing may, but in some cases need not, depend on the type of probability distribution.

3.1.5.1 Testing Parametric Hypotheses

Testing parametric hypotheses deals mainly with the problem of decision if a given sample is a random sample from a population whose c.d.f. depends on a parameter; this testing method has been thoroughly investigated for the normal probability distribution:

a) To testing the relationship of a population with mean μ and the random sample with mean $\bar{x}$.

The null hypothesis that the random sample with mean $\bar{x}$ was derived from a normally distributed population with parameters $\mu(x)$ and $\sigma(x)$ is tested. The criterion has the following form

$$t_0 = \frac{|\bar{x} - \mu(x)|}{\sigma(x)} \sqrt{n} \tag{3.75}$$

or

$$t_0 = \frac{|\bar{x} - \mu(x)|}{s_x} \sqrt{n - 1} \tag{3.75'}$$

The null hypothesis is accepted of $t_0 < t_p$; t_p (for probability p) is found in the tables of normal distribution, using the test criterion (3.75) or the criterion (3.75′) for $p = 1 - \alpha/2$, i.e. the level of significance is $\alpha = 1 - 2p$.

b) To testing the relationship of the population with variance $\sigma^2(x)$ and the random sample with variance s_x^2.

The test criterion with the probability distribution chi-square has the following form

$$\chi_0^2 = n \frac{s_x^2}{\sigma^2(x)} \tag{3.76}$$

If $\chi_0^2 > \chi_{1-p}^2$ (for $n - 1$ degrees of freedom) with $s_x^2 < \sigma^2(x)$ and $\chi_0^2 < \chi_p^2$ (for $n - 1$ degrees of freedom) with $s_x^2 > \sigma^2(x)$, the null hypothesis is accepted (i.e. the random sample with variance s_x^2 was derived from the normally distributed population with variance $\sigma^2(x)$).

c) Statistical significance of the difference between two sample means.

Two independent random samples of size n_1 and n_2 and with means $\bar{x}_1$ and $\bar{x}_2$, respectively, are from the same population if

$$t_0 = \frac{|\bar{x}_1 - \bar{x}_2|}{\sigma(x)} \sqrt{\frac{n_1 n_2}{n_1 + n_2}} < t_p \tag{3.77}$$

or

$$t_0 = \frac{|\bar{x}_1 - \bar{x}_2| \sqrt{n_1 + n_2 - 2}}{\sqrt{n_1 s_{x_1}^2 + n_2 s_{x_2}^2}} \sqrt{\frac{n_1 n_2}{n_1 + n_2}} < t_p \tag{3.78}$$

The value t_p in (3.77) is the 100 p% point of the normal distribution and in (3.78) it is the 100 p% point of the Student distribution at the level of significance $\alpha = 1 - 2p$. If the variances $s_{x_1}^2$ and $s_{x_2}^2$ are clearly different an alternative method of testing is necessary (Votruba and Nacházel, 1969).

d) Statistical significance of the ratio of two sample variances.

Two independent random samples of size n_1 and n_2 with variances $s_{x_1}^2$ and $s_{x_2}^2$ respectively, are from the same population if

$$F_0 = \frac{\dot{\sigma}^2(x_1)}{\dot{\sigma}^2(x_2)} = \frac{s_{x1}^2}{s_{x2}^2} \text{ (for } s_{x_1}^2 > s_{x_2}^2\text{)} < F_p \tag{3.79}$$

F_p is the 100 . p% point of the Snedecor distribution F and the level of significance is $\alpha = 1 - p$ with $(n_1 - 1)$ and $(n_2 - 1)$ degrees of freedom. A few tests for a population with other than normal probability distribution have been developed.

Example V

Let the hydrological time series of annual flows Q_r at point X for $N = 100$ years be the population with mean $\mu(x) = 200.4$ m^3 s^{-1}. Decide if the deviation of this value from the sample mean $\bar{x}_1 = 196$ m^3 s^{-1} of a 30-year sample from this population is statistically significant. The sample standard deviation is $s_{x_1}^2 = 65$ m^3 s^{-1}.

Solution

Test the null hypothesis that the random sample with mean $\bar{x}_1$ is from the population with mean $\mu(x)$ and standard deviation $\sigma(x)$, which is not known.

The test criterion has the following form

$$t_0 = \frac{|\bar{x}_1 - \mu(x)|}{s_{x_1}} \sqrt{n-1} = \frac{|196 - 200.4|}{65} \sqrt{30 - 1} = 0.365 \tag{V.1.}$$

For different samples with means $\bar{x}_i$ the values t_0 have the Student probability distribution. The level of significance is $\alpha = 1 - 2p = 0.05$, therefore $p = 0.975$. For this p and for $m = n - 1 = 29$ degrees of freedom of the Student probability distribution, the corresponding t_p is

$$t_p = t_{0.975} = 2.045$$

Comparing t_0 with t_p

$$t_0 = 0.365 < t_p = 2.045$$

From this comparison the following conclusion can be drawn. At the selected level of significance $\alpha = 0.05$ the null hypothesis that this sample is from the defined population with normal probability distribution can be accepted. The difference between the sample and population means is not statistically significant.

3.1.5.2 Testing Non-Parametric Hypotheses

The simple estimation of the theoretical probability distribution of a random variable X can be uncertain. The testing of the qualitative and quantitative fit of the theoretical and empirical c.d.f.s without testing their parameters can be performed by the goodness-of-fit criteria included in non-parametric statistical testing methods.

a) Chi-square (Pearson's) criterion

Let us test if the random variable X has the probability distribution of a certain type given by the sample class frequencies. The test criterion is

$$\chi_0^2 = \sum_{i=1}^{l} \frac{(n_i - nf_i)^2}{nf_i} \tag{3.80}$$

where l is the number of classes,
n – sample size,
n_i – class frequency of i-th class,
f_i – theoretical probability in the case of an accepted null hypothesis (accepted type of probability distribution). For $m = l - c - 1$ (c = number of parameters of the c.d.f. used in the null hypothesis) and level of significance $\alpha = 1 - p$, the value χ_p^2 is found from tables. If $\chi_0^2 < \chi_p^2$ the null hypothesis concerning the goodness-of-fit of the empirical and theoretical probability distributions is accepted.

b) Bernshteyn's criterion

Bernshteyn's criterion is constructed in such a way that the goodness-of-fit is accepted if

$$\frac{1}{l-1} \sum_{i=1}^{l} \frac{(n_i - nf_i)^2}{nf_i} < 1 \tag{3.81}$$

Table VI. 1 Computation of the value of the test criterion

i	Value of class mark x_i	Sample frequencies n_i	Theoretical probabilities of occurrence *) nf_i	Adjustment of the marginal intervals **) nf_i	$n_i - nf_i$	$(n_i - nf_i)^2$	$\frac{(n_i - nf_i)^2}{nf_i}$
1	9.00	1	1.19	5.55	−1.55	2.40	0.432
2	9.25	3	4.36				
3	9.50	11	12.14	12.14	−1.14	1.30	0.107
4	9.75	24	26.00	26.00	−2.00	4.00	0.154
5	10.00	42	43.00	43.00	−1.00	1.00	0.023
6	10.25	62	54.50	54.50	+7.50	56.25	1.032
7	10.50	58	54.30	54.30	+3.70	13.69	0.252
8	10.75	44	42.05	42.05	+1.95	3.80	0.090
9	11.00	19	25.85	25.85	−6.85	47.00	1.820
10	11.25	9	12.15	12.15	−3.15	9.92	0.816
11	11.50	4	3.95	5.45	−1.55	2.40	0.440
12	11.75	2	1.10				
13	12.00	1	0.40				
Σ							5.166

*) The values were transformed by linear interpolation from the values for the lower bound of the values for the lower bound of the class interval theoretical values in Table IV.2. $(nh/s)\,f(t) = nf_i$ to the class mark. As the class length is small, this interpolation will not induce a significant error

**) The marginal intervals were joined in order to reach $nf_i \geqq 5$

c) Kolmogorov–Smirnov test

This is used for testing of the goodness-of-fit of the empirical and theoretical c.d.f.s (or exceedance probability curves). The maximum difference between their values (i.e. between the c.d.f. $F(x_i)$ and the cumulative frequency $F_n(x_i)$) determines the probability

$$P\{\max |F_n(x_i) - F(x_i)| > D_\alpha(n)\} = \alpha \tag{3.82}$$

The values of the test criterion $D_\alpha(n)$ for the sample size n and level of significance α have been tabulated. The null hypothesis for goodness-of-fit of the two functions is accepted if

$$D = \max |F_n(x_1) - F(x_i)| < D_\alpha(n) \tag{3.83}$$

Example VI

The goodness-of-fit between the empirical frequencies in example I that seem to have normal distribution and the theoretical normal probability distribution is to be tested.

Solution

a) Application of the chi-square criterion

Table VI.I. contains the observed frequencies and the corresponding theoretical probabilities of occurrence in the class intervals. The input values in Table VI.I. were taken from Tables I.1 and I.2. The value of the Pearson (chi-square) criterion is

$$\chi_0^2 = \sum_{i=1}^{l} \frac{(n_i - nf_i)^2}{nf_i} = 5.166 \tag{VI.1}$$

The degree of freedom for $c = 3$ are

$$m = l - c - 1 = 10 - 3 - 1 = 6$$

For selected level of significance $\alpha = 0.05$ and one-sided confidence interval with $p = 1 - \alpha$, the following values were found from tables:

$$\chi_{0.95}^2 = 12.6$$

and comparing

$$\chi_0^2 = 5.166 < \chi_{0.95}^2 = 12.6$$

Comparison of these values produces the following conclusion: the null hypothesis concerning the goodness-of-fit of the empirical and theoretical probability distributions at the chosen level of significance can be accepted.

b) Use of Bernshteyn's criterion

Bernshteyn's criterion has the following general form

$$\frac{1}{l-1}\sum_{i=1}^{l}\frac{(n_i - nf_i)^2}{nf_i} < 1 \tag{VI.2}$$

The value of the criterion is

$$\sum_{i=1}^{l}\frac{(n_i - nf_i)^2}{nf_i} = 5.166$$

If the number of the adjusted class interval is $l = 10$, then

$$\frac{1}{10-1}\,5.116 = 0.574 < 1$$

Using this criterion for comparison the same conclusion can be made, i.e. the null hypothesis concerning the goodness-of-fit between the empirical and theoretical probability distributions at the chosen level of significance can be accepted.

3.1.5.3 Testing Hypotheses for Statistical Dependence of Random Variables

a) An infinite population with normal probability distribution has a linear statistical dependence between random variables characterized by the coefficient of linear correlation ϱ. The random samples from it have correlation coefficients r. The test criterion for checking the deviation of r from ϱ has the following form

$$t_0 = \frac{|z - \mu(z)|}{\sigma(z)} \tag{3.84}$$

where z is a function of r and the transformation has the following form

$$z = \frac{1}{2}\lg\frac{1+r}{1-r} \tag{3.85}$$

with parameters

$$\mu(z) \doteq \frac{1}{2}\lg\frac{1+\varrho}{1-\varrho}; \tag{3.86}$$

$$\sigma^2(z) = \frac{1}{n-3} \tag{3.87}$$

for sample size n.

The value t_0 is compared with the critical value t_p for $p = 1 - \alpha$ found in tables of normal probability distribution. If $t_0 < t_p$ the deviation of r from ϱ is random and and the null hypothesis (that this sample with r is from population with ϱ) is accepted.

b) Significance of the difference between two sample correlation coefficients r_1 and r_2 is tested by the criterion

$$t_0 = |z_1 - z_2| \left(\frac{1}{n_1 - 3} + \frac{1}{n_2 - 3} \right)^{-1/2} \tag{3.88}$$

where z_1 and z_2 are transformed values of r_1 and r_2 respectively, using equation (3.85), and n_1 and n_2 are the sample sizes of the correlated pairs. The value t_0 is compared with the value t_p from normal distribution for $p = 1 - \alpha/2$. If $t_0 < t_p$ the null hypothesis of statistical significance between r_1 and r_2 is rejected. Similarly, the sample regression coefficients b_1 and b_2 can be tested by a test criterion with Student probability distribution.

3.2 THEORY OF RANDOM PROCESSES

3.2.1 Basic Notions of the Random Processes Theory

The theory of random (stochastic) processes, as a discipline of probability theory, deals with the investigation of random variables as function of variable parameters, mainly time. Probability theory deals with the probability distribution of one random variable or a finite number of them. On the other hand, the theory of random processes deals with the statistical properties of different (often time related) random functions describing the same physical phenomenon.

Typical examples of random processes in WRS are the time series of river flows, precipitation, temperature, air pressure, and also the states of reservoir storage volumes, etc. Two areas of water management were particularly suitable for the application of the theory of stochastic processes, viz.:

(1) Processing of hydrological data and

(2) probability methods of reservoir design.

The theory of random processes, however, is of wider scope and application. Its methods and procedures (often referred to as stochastic methods and procedures) are used in the investigation of the properties of various random variables in many technical sectors. They are very important in the treatment of WRS problems where together with systems methods, they form the main mathematical research tool[1]).

Quantitatively, the stochastic process is described by a random function of time $X(t)$, the values of which at any time are random variables. The random functions

[1]) Based on the theory of random processes, stochastic hydrology was developed as a subject that investigates the stochastic properties of hydrological processes and constructs mathematical models of them. There is an extensive literature on stochastic hydrology (e.g. included in Loucks *et al.*, 1981; Yevjevich, 1976; Kos, 1982). The development of stochastic hydrology has been promoted at many symposiums and conferences (e.g. Warsaw, 1971, Bratislava, 1975).

are classified with respect to the continuity or discontinuity of the parameter on which they depend: if this parameter is continuous, then the function is called a random (stochastic) process; if it is discrete, then the function is called a random (stochastic) sequence. The methods of their investigation are similar.

There is an analogy between the random variable and the random process: the random variable is described by the set of all its possible values with the probability distribution, the random process $X(t)$ is determined by the set of functions obeying the laws that describe the stochastic properties of this set. The particular functions $x^{(1)}(t)$, $x^{(2)}(t)$ ... are called realization of the stochastic process $X(t)$. In practical applications in WRS the recorded observations in a given time period, i.e. chronological arrangement of random variables, are considered as these realizations.[1])

The probability properties of random processes are investigated, in general, for the whole family of realizations. An idea of such a family is shown in Fig. 3.5 where several possible realizations $x^{(1)}(t)$, $x^{(2)}(t)$, ... are sketched in one graph.

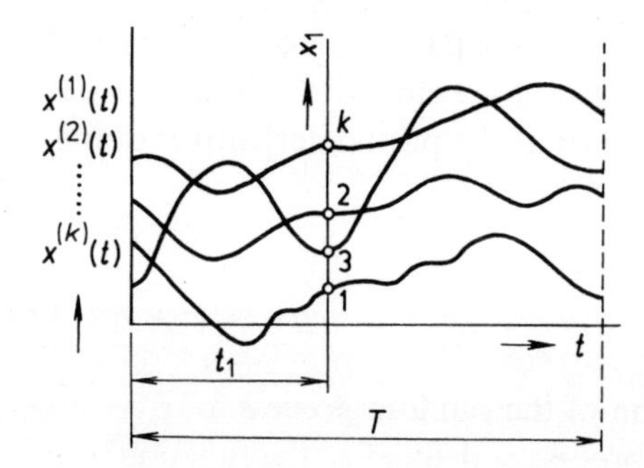

Fig. 3.5 Graph with several realizations of a stochastic process

The stochastic properties of the set (family) of realizations are described by their statistical characteristics (p.d.f., c.d.f. and numerical data, viz. k-th moments and central k-th moments), just as in probability theory these characteristics express the properties of random variables.

The principle difference is the time dependence of statistical characteristics, which is what is used for classification of stochastic processes.

In graphical representation, the straight line $t = t_1$ (perpendicular to the time axis in a fixed point $(t_1, 0)$) crosses the realizations of a set of random variables X_t. If an arbitrary value x_1 is chosen on this line we can investigate the relative frequency $P = m/k$ of occurrence of values $x^{(1)}(t_1), x^{(2)}(t_1), \ldots\ x^{(m)}(t_1)$ from the total number k of realizations $x^{(k)}(t_1)$ less than x_1. For a large number k of the values of random variable X_1, the relative frequency P converges under presumably "identical" conditions to a constant value for each value x_1 (stochastic stability of frequency P).

[1]) In sampling theory the realization of a random process can be considered as a chronological sample of the population. An arbitrary sample, however, is not a realization.

The c.d.f. of the random variable X_1 is the probability

$$F_1(x_1, t_1) = P(X_1 \leqq x_1) \tag{3.89}$$

which is a function of the selected level x_1 and the time t_1. The subscript of symbol F denotes the first-order (one-dimensional) c.d.f. of the stochastic process which apart from dependence on t_1, does not depend on the further development on the process. An idea of the dynamics of the random process is involved in the investigation of the relative frequency for two times t_1 and t_2 at the chosen levels x_1 and x_2 respectively. In that case, the second-order (two-dimensional) c.d.f. is defined by probability

$$F_2(x_1, t_1; x_2; t_2) = P(X_1 \leqq x_1; X_2 \leqq x_2) \tag{3.90}$$

which is a function of four variables x_1, x_2, t_1, t_2 and gives the probability that for time $t = t_1$ the random variable $X(t)$ is below the level x_1 and also for $t = t_2$ below the level x_2.

In the general case, an n-th order (n-dimensional) c.d.f. is defined as the probability that a random function will be at times $t_1, t_2, \ldots, t_n$ below the values $x_1, x_2, \ldots, x_n$.

The p.d.f.s of the random process are defined as the partial derivatives of the c.d.f. The first-order p.d.f. is defined as the function

$$f_1(x_1, t_1) = \frac{\partial F_1(x_1, t_1)}{\partial x_1} \tag{3.91}$$

and it is called the probability distribution of the random process.

The second-order p.d.f. of the random process is defined as the function

$$f_2(x_1, t_1; x_2, t_2) = \frac{\partial^2 F_2(x_1, t_1; x_2, t_2)}{\partial x_1 \, \partial x_2} \tag{3.92}$$

In general, the n-th order p.d.f. can be defined as the n-th partial derivative of the n-th order c.d.f. with respect to n variables.

The calculation of the c.d.f. and p.d.f. of higher orders is difficult. Consequently, it is often restricted to the calculation of statistical characteristics of the first and second orders. The treatment of many WRS problems often does not require the information provided by the p.d.f. and c.d.f. which would therefore be redundant; the numerical characteristics of the process are frequently sufficient, e.g. the time-related k-th moments of the probability distribution.

The first and second moments are the most important, i.e. the mean value of the random process,

$$m_1\{X(t)\} = \int_{-\infty}^{\infty} x \, f_1(x, t) \, \mathrm{d}x = a(t) \tag{3.93}$$

variance of the random process (second-order central moment)

$$M_2\{X(t)\} = m_2\{[X(t) - a]\}^2 = \int_{-\infty}^{\infty} (x - a)^2 f_1(x, t)\, \mathrm{d}x = \sigma^2(t) \tag{3.94}$$

and the autovariance function of the random process (second-order mixed moment)

$$m_1\{X(t_1)\, X(t_2)\} = \int_{-\infty}^{\infty}\int_{-\infty}^{\infty} x_1 x_2\, f_2(x_1, t_1; x_2, t_2)\, \mathrm{d}x_1\, \mathrm{d}x_2 = R(t_1, t_2)^{1)} \tag{3.95}$$

3.2.2 Stationarity and Ergodicity of Random Processes

The stationarity of a random process can be defined by properties of the p.d.f. and c.d.f.

A random process $X(t)$ is stationary if all its p.d.f.s of all orders do not vary with the change of the starting time from t_1 to $t_1 + \tau$. For all values of n and τ,

$$f_n(x_1, t_1; x_2, t_2; \ldots; x_n, t_n) = f_n(x_1, t_1 + \tau; x_2, t_2 + \tau; \ldots; x_n, t_n + \tau) \tag{3.96}$$

Stationarity of a random process can be defined with respect to independence of the c.d.f. of all orders on the time origin. Therefore, if

$$F_n(x_1, t_1; x_2, t_2; \ldots; x_n, t_n) = F_n(x_1, t_1 + \tau; x_2, t_2 + \tau; \ldots; x_n, t_n + \tau) \tag{3.97}$$

for any n and τ the process is stationary.

Proving the stationarity of a random process in WRS contexts using conditions (3.96) and (3.97) involves at intractable calculations. Therefore, the time independence of statistical parameters of the c.d.f. is considered satisfactory and the dependence of the autovariance (autocorrelation) function on the time lag τ only is considered as a single time factor.

Less strict requirements are used in weakly stationary random processes. Unlike strictly stationary random processes which have to satisfy conditions (3.96) or (3.97), a weakly stationary process has to meet the following conditions: (1) the mean value is a constant (time invariant), (2) the autovariance (autocorrelation) function is dependent on the time lag τ only (the probability distributions have the same covariance matrix for all τ). A strictly stationary random process with a finite covariance matrix is also weakly stationary. The converse, however, need not be true.

Random processes that do not satisfy the conditions of stationarity are non-stationary. Fluctuations of statistical characteristics during the development of the stochastic process are typical of them. Their investigation is, therefore, substantially more difficult. The theory of non-stationary processes developed later than the

---

[1]) Apart from the moments (3.93), (3.94), and (3.95), WRS problems need the coefficients of variance $C_v(t)$ and skewness $C_s(t)$ required by Pearson's probability distribution (type III) of the random process. The meaning and method of calculation of the autocorrelation function is explained in section 3.2.3.

theory of stationary processes. Therefore, an assumption of stationarity in hydrological processes was applied in WRS investigations. However, non-stationary hydrological processes, their probability properties and their effect on WRS are now being investigated (Nacházel *et al.*, 1975, Yevjevich, 1976).

The second important property of random processes that is related to stationarity is ergodicity. A random process is ergodic if each statistical characteristic calculated as the mean from set of possible realizations can be obtained (with probability converging to one) from one realization of the stochastic process using a sufficiently long time interval (i.e. a single realization yields sufficient information about the process).

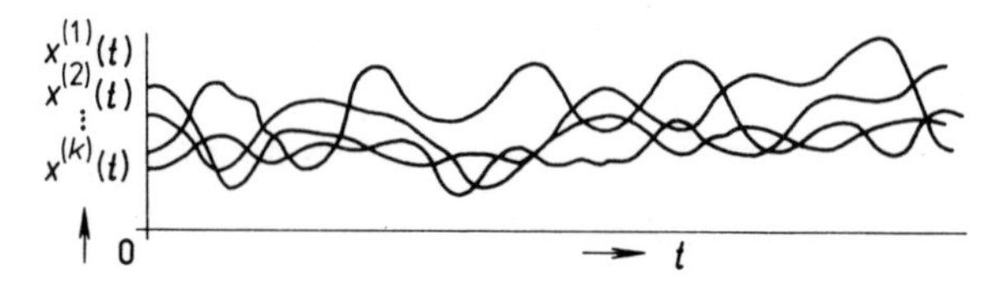

Fig. 3.6 Several realizations of a stationary ergodic random process

Ergodicity and stationarity are two different properties of random processes. Every ergodic process is stationary; however, a stationary process need not necessarily be ergodic. Figure 3.6 contains an example of a stationary ergodic random process with realizations that have a constant mean value in time. Figure 3.7 illustrates an example of a random process with realizations that have different mean values in time. Sufficient for the ergodicity of a stationary process is satisfaction of the requirement of a zero value of the time limit of the autocorrelation function for $\tau \to \infty$ (for definition of the term autocorrelation function, see section 3.2.3).

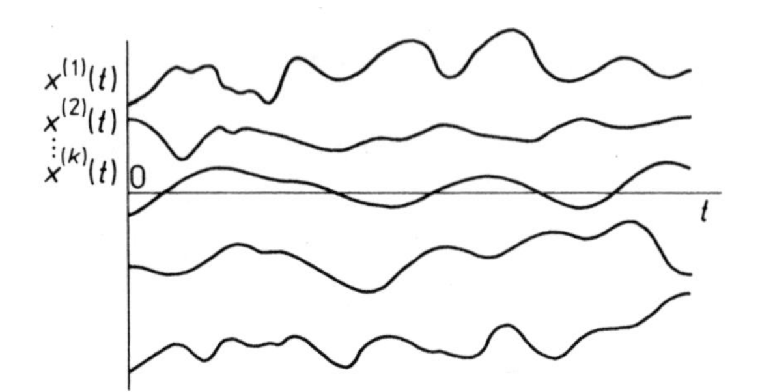

Fig. 3.7 Several realizations of a stationary nonergodic random process

The definition of ergodicity implies that each statistical characteristic of a random process can be obtained as the mean value of a family of realizations or as the mean value of a family of realizations or as the mean value of a single realization in time. The equivalence of these two methods of calculation of basic characteristics is expressed in Table 3.15.

Table 3.15 Statistical characteristics for stationary ergodic random processes

Characteristics	Mean value in time	Mean value of family of realizations
mean	$\bar{x}(t) = \lim\limits_{T\to\infty} \dfrac{1}{2T}\displaystyle\int_{-T}^{+T} x(t)\,\mathrm{d}t$	$a(t) = \displaystyle\int_{-\infty}^{\infty} x\, f_1(x, t)\,\mathrm{d}x$
variance	$\overline{x^2(t)} = \lim\limits_{T\to\infty} \dfrac{1}{2T}\displaystyle\int_{-T}^{+T} x^2(t)\,\mathrm{d}t$	$m_2[X^2(t)] = \displaystyle\int_{-\infty}^{\infty} x^2 f_1(x, t)\,\mathrm{d}x$
covariance function	$\overline{x(t)\,x(t+\tau)} = \lim\limits_{T\to\infty}\displaystyle\int_{-T}^{+T} x(t)\,x(t+\tau)\,\mathrm{d}t$	$m_1[X(t)\,X(t+\tau)] = \displaystyle\int_{-\infty}^{\infty}\int_{-\infty}^{\infty} x_1 x_2 f_2(x_1, t_1; x_2, t_2)\,\mathrm{d}x_1\,\mathrm{d}x_2$

The usefulness of ergodicity of stationary processes is related to the possibility of investigating a single realization during a long time interval instead of a large set of realizations which often are not available. This property of random processes, like that of stationarity, significantly facilitates the treatment of WRS problems. However, the direct observation of hydrological phenomena is limited in time which justifies a certain lack of confidence in the equality of the mean value determined from a set of realizations and the mean value in time.

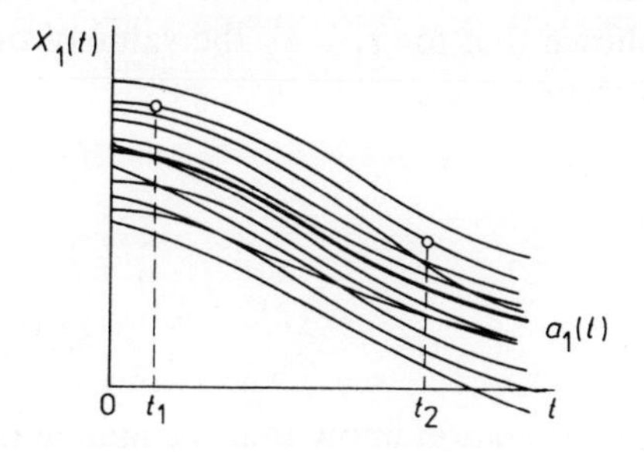

Fig. 3.8 Inner structure of realizations of a random process – regular tendency

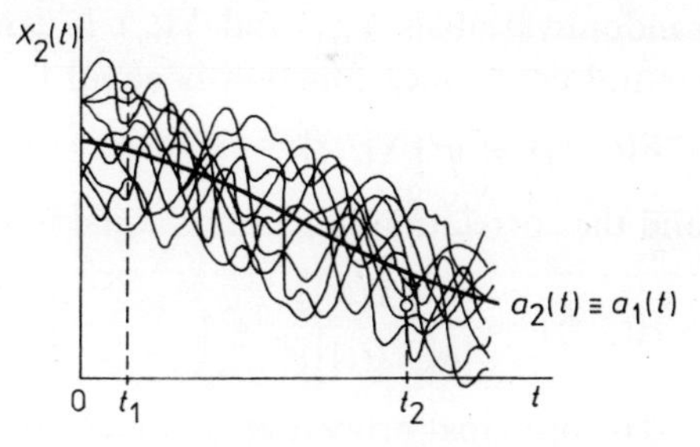

Fig. 3.9 Inner structure of realizations of a random process – irregular tendencies

3.2.3 Correlation Analysis of Random Processes

The full description of a stochastic process includes not only the p.d.f. and c.d.f. and their characteristics but also the results of correlation analysis. Figures 3.8 and 3.9 contain examples of two random processes $X_1(t)$ and $X_2(t)$, each with several realizations having the same means and variances.

The inner structure of random processes is remarkably different. For its identification the autocovariance and autocorrelation functions have been introduced to describe the correlation between the values of the process at any two times t_1 and t_2. The value of the autocorrelation function for the two given times is equal to the correlation coefficient of the corresponding random variables[1]).

The autocovariance function has been defined in eq. (3.95) and Table 3.15 shows the mean values in time and the mean values of the family of realizations. In WRS problems the covariance function of discrete random sequences is frequently calculated from the covariances of random variables at times t_1 and t_2 by the formula

$$R(t_1, t_2) = m_1\{X(t_1)\, X(t_2)\} \tag{3.98}$$

where m_1 is the mean value of the product $X(t_1)\, X(t_2)$.

The transformation from the covariance function (3.98) to the central covariance function (mixed central moments) is given by

$$\mathring{R}(t_1, t_2) = m_1[\mathring{X}(t_1)\, \mathring{X}(t_2)] = m_1\{[X(t_1) - a(t_1)]\, [X(t_2 - a(t_2)]\} \tag{3.99}$$

Dividing the values of the central covariance functions from eq. (3.99) by the product of the standard deviations $\sigma[X(t_1)]\, \sigma[X(t_2)]$ of the corresponding random variables $X(t_1)$ and $X(t_2)$, the normalized covariance function, i.e. correlation function of the random process is

$$r(t_1, t_2) = \frac{\mathring{R}(t_1, t_2)}{\sigma[X(t_1)]\, \sigma[X(t_2)]} = \frac{R(t_1, t_2) - a(t_1)\, a(t_2)}{\sigma[X(t_1)]\, \sigma[X(t_2)]} \tag{3.100}$$

The values of the correlation function are the correlation coefficients between the random variables $X(t_1)$ and $X(t_2)$. It can be shown that for $t_1 = t_2$ the value of the central covariance function is equal to the variance, i.e.

$$\mathring{R}(t_1, t_1) = \sigma^2[X(t_1)] \tag{3.101}$$

and the correlation function is equal to one

$$r(t_1, t_1) = \frac{\sigma^2[X(t_1)]}{\{\sigma[X(t_1)]\}^2} = 1 \tag{3.102}$$

The principal properties of the coefficient of correlation imply that the maximum value of the correlation function is one for $t_1 = t_2$, i.e.

$$|r(t_1, t_2)| \leqq 1 \tag{3.103}$$

The definition of the correlation coefficient implies positive and negative values of the correlation function. If the time lag between the values $X(t_1)$ and $X(t_2)$ is small

---

[1]) Apart from the autocorrelation function there are other types of correlation function which, however, have other meanings (see Beneš, 1970). When there is no danger of ambiquity, the simpler term correlation function is used.

(i.e. $t_2 - t_1 \to 0$), a close, positive correlation can be expected. For a greater time lag, the correlation function can assume different values according to the degree of dependence or independence. In any case, the values of the correlation function should be statistically tested to distinguish the statistically significant values and random fluctuations.

The correlation function is an important index for stationary and ergodic processes. The covariance and correlation functions of a stationary random process are functions of a single argument, viz. the time lag $\tau = t_2 - t_1$, i.e. for the covariance function

$$R(t_1, t_2) = R(\tau) \tag{3.104}$$

and for the correlation function

$$r(\tau) = \frac{R(\tau) - a^2}{\sigma_x^2} = \frac{\mathring{R}(\tau)}{\sigma_x^2} \tag{3.105}$$

The covariance function is an even function and hence

$$R(\tau) = R(-\tau) \tag{3.106}$$

The same relationship holds for the correlation function[1]).

If the process is stationary and ergodic its correlation function is equal to the correlation function of this realization.

A typical characteristic of a stationary random process is the decrease in interrelatedness between the random variables with increasing time lag τ, until, at the limit $\tau \to \infty$, these variables are independent. Independence can be defined, for example, in this way: for $\tau > \tau_0$ the absolute value of the correlation coefficient is less than some given number, e.g.

$$|r(\tau)| < 0.05 \tag{3.107}$$

where τ_0 is the time of the duration of the correlation.

---

[1]) These properties of covariance and correlation functions are important in practical applications. If the correlation function depends on a single argument τ, then it expresses the relationship of random variables with the time lag τ irrespective of the time origin. As the correlations function is even, the calculation can be performed with either a positive or a negative time lag.

However, the correlation function of some special random processes and functions (e.g. a stationary process with a periodic component, a non-stationary process, or periodic functions of various types) need not necessarily converge. Their correlation analysis is based on the fact that the correlation function of a periodic function is also a periodic function with the same period as the given function and independent of the phase.

In fact, there are various types of correlation functions; they are distinguished by their form, duration, convergence or divergence, periodicity or statistical significance of individual members, all of which describe the nature of the random process and the consequences for further processing.

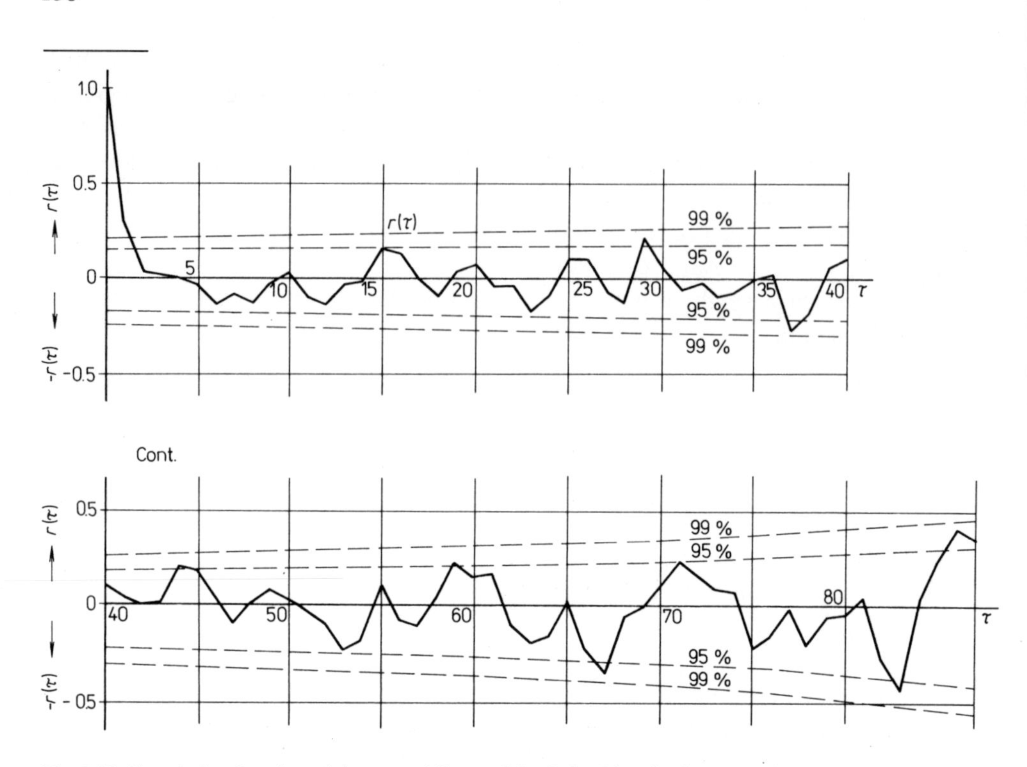

Fig. 3.10 Correlation function of the annual flows of the Labe River in the town of Děčín at time period 1851 – 1957

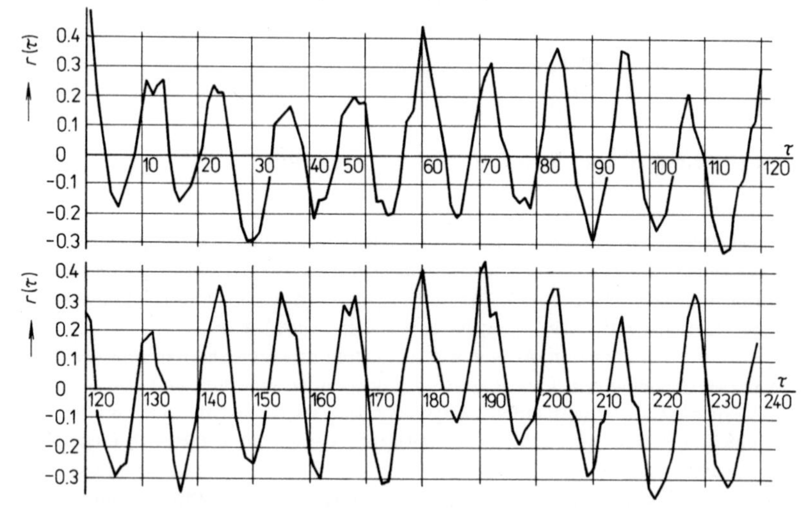

Fig. 3.11 Correlation function of the monthly flows in the Labe River in the town of Děčín at time period 1851 – 1960

If a series of discrete random variables is given in a certain observation period, the correlation function can be calculated in two ways:

(1) The requirement to use a constant number of pairs of random variables results in the calculation of the correlation function from its definition, i.e. the formula

$$r(\tau) = \frac{\sum_{i=1}^{n} (x_i - \bar{x}_i)(x_{i+\tau} - \bar{x}_{i+\tau})}{\sigma_i \, \sigma_{i+\tau}(n - 1)} \tag{3.108}$$

Figure 3.10 contains the correlation function of the annual flows of the River Labe in the town of Děčín for the period 1851 – 1957. It is periodic in form, which is typical of many hydrological patterns. In a long-term analysis, a period of 13 to 15 years can be clearly distinguished and proved by special analysis (see 3.2.4).

Figure 3.11 shows the correlation function of the monthly flows of the River Labe in the town of Děčín for the period 1851 – 1960. The correlation function is again periodic, but it has a pronounced annual cycle of flow variation.

The long-term variations in the activity of the sun are important factors in the analysis of geophysical and hydrological processes and their interrelations. Figure 3.12 contains the correlation function of the mean annual relative Wolf numbers for the period 1831 – 1894 (the calculation was carried out up to time lag $\tau = 35$). The pattern of this line is harmonic and regular, showing the well-known 11-year cycle of the sun's activity. Compared with the correlation functions of the flows or rainfall time series, in this case the correlation relationship is very close, for direct or indirect correlation. The periodicity is obvious and significant.

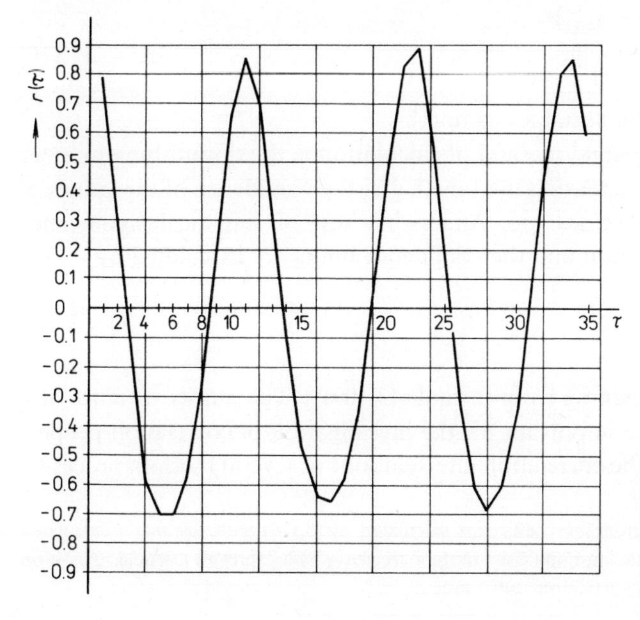

Fig. 3.12 Correlation function of the mean annual relative Wolf's numbers at time period 1831 – 1894

where x_i are random variables with values x_1 to x_n,
$x_{i+\tau}$ – random variables with values $x_{1+\tau}$ to $x_{n+\tau}$,
τ – the time lag argument of the correlation function, for $\tau = 0$ to $\tau = m - n$ where m is the total number of elements in the series (length of observation),
n – the chosen sample section of the series,
$\bar{x}_i, \bar{x}_{i+\tau}$ – the mean values of random variables x_i and $x_{i+\tau}$ respectively, and
$\sigma_i, \sigma_{i+\tau}$ – the standard deviations of these series.

The advantage of calculating the correlation function from formula (3.108) is that all the correlation coefficients have the same weight, which simplifies the test of statistical significance. The confidence limits are determined by the approximative formula (Anderson, 1942)

$$r_p = \frac{-1 \pm t_p \sqrt{n - 2}}{n - 1}$$

(for another method of calculation see (3.84) to (3.87)) where t_p is the standardized normal variable corresponding to the level of significance α and $p = 1 - \alpha$. For the usual levels of significance in hydrology viz. 1% and 5%, $t_{0.99} = 2.326$ and $t_{0.95} = 1.645$.

(2) A disadvantage attached to formula (3.108) is that the correlation coefficients do not include all the terms of the series and some information is lost. Therefore the correlation function can be calculated by

$$r(\tau) = \frac{\sum_{i=1}^{m-\tau} (x_i - \bar{x}_i)(x_{i+\tau} - \bar{x}_{i+\tau})}{\sigma_i \sigma_{i+\tau}(m - \tau - 1)} \tag{3.110}$$

with the same meaning of the variables as in (1).

The disadvantage of the second method of calculation is the diminishing number of pairs in correlation with increasing the time lag τ; the confidence of the values of the correlation function decreases too. Anderson's test of statistical significance takes this fact into consideration and the confidence limits are calculated by

$$r_p = \frac{-1 \pm t_p \sqrt{m - \tau - 2}}{m - \tau - 1} \tag{3.111}$$

where m, t_p has the same meaning as in formula (3.109)[1]. The autocovariance and autocorrelation functions are important for the investigation of correlation properties of one random process; the correlation interrelations of several random processes

[1]) The confidence limits for correlation coefficients calculated by (3.109) are constant; if calculated by (3.111) they are a function of time lag τ and their range increases, which expresses a stricter condition for the statistical significance of the correlation function.

are investigated by cross-covariance and cross-correlation functions. The simplest of these is the cross-covariance function of two random processes, which is defined in the following way:

$$R_{xy}(t_1, t_2) = m_1[X(t_1)\, Y(t_2)] \tag{3.112}$$

$$R_{yx}(t_1, t_2) = m_1[Y(t_1)\, X(t_2)] \tag{3.113}$$

It is also

$$R_{xy}(t_1, t_2) = R_{yx}(t_2, t_1) \tag{3.114}$$

The cross-covariances can also be calculated as the central and normalized moments when the cross-correlation function is obtained in the following form

$$r_{xy}(t_1, t_2) = \frac{R_{xy}(t_1, t_2) - m_x(t_1)\, m_y(t_2)}{\sigma[X(t_1)]\, \sigma[Y(t_2)]} \tag{3.115}$$

$$r_{yx}(t_1, t_2) = \frac{R_{yx}(t_1, t_2) - m_y(t_1)\, m_x(t_2)}{\sigma[Y(t_1)]\, \sigma[X(t_2)]} \tag{3.116}$$

The pattern of the cross-correlation function indicates the degree of the correlation relationship (coherency) between the realizations of the random process with some time lag.

There are many types of cross-correlation functions similar to the autocorrelation function, and they can be distinguished by the nature of the individual random processes. Their investigation is the subject of specialist literature (Beneš, 1961; Levin, 1965; Ventcel, 1964).

3.2.4 Spectral and Filter Analysis of Random Processes

The spectral distribution function and the spectral density function are further important statistical characteristics of a random process or its realization, particularly in the investigation of periodic properties. The spectral density function (s.d.f.) of a random process $X(t)$ is defined by the expression (Beneš, 1961)

$$s_{xx}(\omega) = \lim_{T \to \infty} M \left[\frac{1}{2T} \, |X_T(\mathrm{i}\omega)|^2 \right] \tag{3.117}$$

where $X_T(\mathrm{i}\omega)$ is the Fourier transform of the random process $X(t)$ in a finite interval $\langle -T, T \rangle$ (see section 3.4), and is given by

$$X_T(\mathrm{i}\omega) = \int_{-\infty}^{\infty} X(t)\, \mathrm{e}^{-\mathrm{i}\omega t} \mathrm{d}t \tag{3.118}$$

where $X(t)$ is defined by the interval $\langle -T, T \rangle$.

The spectral density function (s.d.f.) is, in fact, a limit of expectation in a set of of periodograms for $T \rightarrow \infty$; the periodogram of a realization of a random process is a Fourier transform[1]).

The s.d.f. (spectral density function) of a random process can be calculated from its autocorrelation function. Wiener and Khinchin have shown that between the autocorrelation function of a stationary process and its s.d.f. the following relationships apply:

$$\mathring{R}(\tau) = \int_0^\infty S_x(\omega) \cos \omega\tau \, d\omega \tag{3.119}$$

$$S_x(\omega) = \frac{2}{\pi} \int_0^\infty \mathring{R}(\tau) \cos \omega\tau \, d\tau \tag{3.120}$$

Equations (3.119) and (3.120), called Wiener–Khinchin equations, express fundamental and very important relationships as they facilitate the transformation from one set of statistical characteristics to a second one (the s.d.f. is a Fourier transform of the autocorrelation function and, conversely, the autocorrelation function is an inverse Fourier transform of the s.d.f.).

If, analogously to the normalized autocovariance function, the normalized s.d.f. is introduced

$$s_x(\omega) = \frac{S_x(\omega)}{\sigma_x^2} \tag{3.121}$$

then equations (3.119) and (3.120) can be arranged in the form

$$r(\tau) = \int_0^\infty s_x(\omega) \cos \omega\tau \, d\omega \tag{3.122}$$

$$s_x(\omega) = \frac{2}{\pi} \int_0^\infty r(\tau) \cos \omega\tau \, d\tau \tag{3.123}$$

In some technical applications it is mathematically easier to express the spectral and correlation functions by complex variables. Substituting in equations (3.122) and (3.123)

$$\cos \omega\tau = \frac{e^{i\omega\tau} + e^{-i\omega\tau}}{2}$$

and letting $s_x(\omega) = 2s_x^*(\omega)$ and increasing the range to $(-\infty, \infty)$, we get

$$r(\tau) = \int_{-\infty}^\infty s_x^*(\omega) \, e^{i\omega\tau} \, d\omega \tag{3.124}$$

[1]) The investigation of periodic properties of random processes using their spectra (spectral representation) is similar to harmonic analysis, which is used for the investigation of non-random processes. Spectral analysis is a generalization of this by introducing the time expectation of spectral representations obtained from realizations.

$$s_x^*(\omega) = \frac{1}{2\pi} \int_{-\infty}^{\infty} r(\tau)\, e^{-i\omega\tau}\, d\tau \tag{3.125}$$

The s.d.f. $s_x(\omega)$ differs from $s_x^*(\omega)$ by different scales on ordinates and by $s_x(\omega)$ not being defined for negative frequencies (Fig. 3.13).

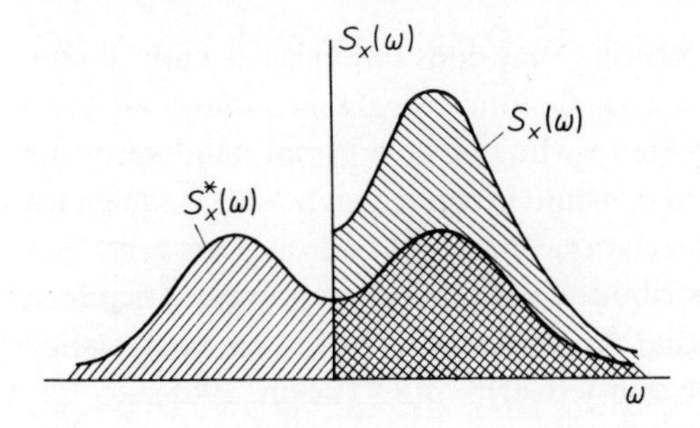

Fig. 3.13 The complex representation of the s.d.f.

Just as the pattern of correlation functions indicates the correlation tendencies in the random process, the pattern of the s.d.f. shows its periodic properties. The limiting cases with characteristic patterns of the s.d.f. are used as the basis for this analysis. For example, a periodic function with the frequency ω_0 has an infinite value of the s.d.f. for the value ω_0 (Fig. 3.14a). This limiting cases, however, is not possible in an actual stochastic process in actual physical conditions. A stochastic process with pronounced periodicity and with a periodic correlation function has the s.d.f. in a narrow frequency band concentrated around the given frequency ω_0 where the maximum occurs (Fig. 3.14b).

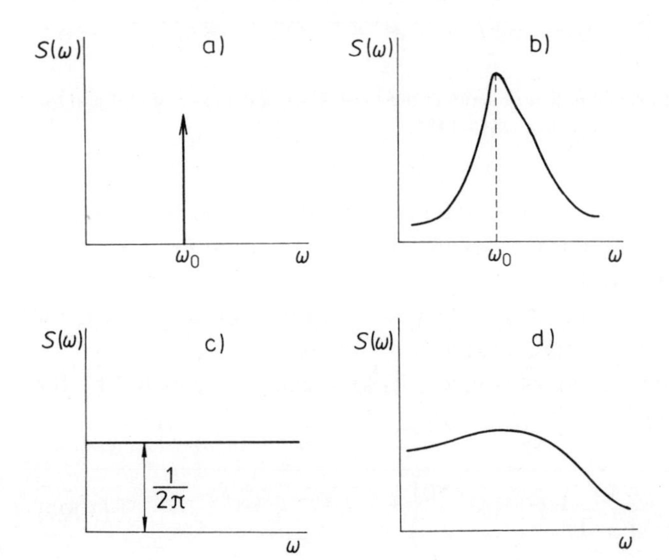

Fig. 3.14 Graphs of s.d.f.
a – periodic function, *b* – narrow band spectrum process, *c* – white noise, *d* – broad band spectrum process

The opposite case is the s.d.f. of a random process that is constant for all frequencies (Fig. 3.14c). This process is called white noise and its values are uncorrelated for any time lag τ (with the exception of $\tau = 0$). Therefore its correlation function is

$$r(\tau) = \begin{cases} 1 & \text{for } \tau = 0 \\ 0 & \text{for } \tau \neq 0 \end{cases} \tag{3.126}$$

However, white noise is also an ideal process that does not exist in natural conditions since the random functions with a small time lag are in almost all cases dependent. In real processes which converge to white noise a broad-band spectrum occurs (Fig. 3.14d), which can be considered constant in a given interval of frequencies.

White noise has a uniquely defined correlation function and s.d.f. However, they do not uniquely define the probability distribution of the random process. Therefore, white noise is an example of a group of random processes with the same correlation function and s.d.f. but with different probability distributions (Levin, 1965).

Since the correlation function and s.d.f. are two statistic characteristics related by the Fourier transformation, this relationship is often used in hydrology. Figure 3.15

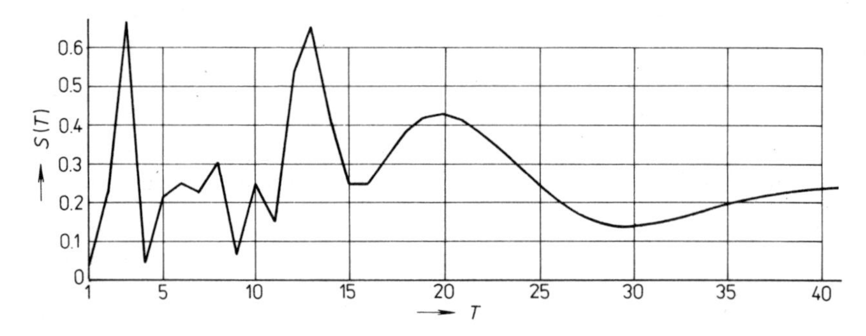

Fig. 3.15 Graph of the s.d.f. of the annual flow time series in the Labe River in the town of Děčín in the period 1851 – 1957

contains an example of the s.d.f. of the annual flows of the River Labe in the town of Děčín. In the graph, the most important extreme values are reached for $T = 3$ years (range of short-term periods) and for $T = 13$ years (range of medium-term periods). The maximum values for $T = 20$ and $T = 42$ years are less pronounced and for $T > 35$ the spectrum has a broad-band character.

If a series of discrete random values is given, its s.d.f. can be calculated by the formula (Reznikovsky, 1969)

$$S = \frac{1}{2\pi}\left[1 + 2\sum_{\tau=1}^{m}\left(1 - \frac{\tau}{m+1}\right) r(\tau) \cos\frac{2\pi\tau}{T}\right] \tag{3.127}$$

where m should be limited by $m = n/2$ (n is the length of the input series). T is the number of years related to the frequency by

$$T = \frac{2\pi}{\omega}$$

In expression (3.127) the various weights of the correlation coefficients of the correlation function are taken into account (3.110).

If the length of the input hydrological time series is limited, the periodic properties determined by spectral analysis need not be statistically significant. The s.d.f. has to be statistically tested; the numerical method of treatment is described in the references (Reznikovsky, 1969; Votruba and Nacházel, 1971).

The theory of the spectral analysis of random processes was developed for stationary stochastic processes. It can be generalized by the Fourier transformation of n-dimensional correlation functions to allow treatment of problems of non-stationary random processes (Batkov, 1959; Lebedyev, 1959; Leonov, 1956; Levin, 1965; Pugachev, 1957).

The method of stochastic filters is a useful complementary method of time series statistical analysis which involves the investigation of their periodic properties. Filters are used for the estimation of random variables, the random components being detected in a given random sequence and then separated. The underlying idea is the concept of the realization of a random sequence as the sum of deterministic variables and random variables in the form of white noise. By special algorithms (filters) the two sequences are separated, which facilitates the probability analysis of the input series.

The method of moving averages (smoothing of the sequence) can also be regarded as a method of filters. The aim is the calculation of a filtered sequence with statistical characteristics (viz. correlation function and s.d.f.) that can be more reliably used for the determination of the nature of the sequence. That is the main practical purpose of the methods of filters (Nacházel and Patera, 1974).[1])

3.2.5 Markov Processes

White noise cannot be applied in all technical problems as the random variables are often stochastically dependent. Hydrological processes are typical correlated random processes that have a great variety of properties. The investigation of random processes with a general type of correlation function is often difficult because the

[1]) In Fig. 3.16 there is an example of the s.d.f. of annual flow time series in the River Labe in the town of Děčín, filtered by binominal filters (Nacházel and Patera, 1974) in three alternatives. Their parameters are interesting in that they indicate the medium-term period T=15 years as the most significant one for all chosen degrees of the filter. This is a good fit with the estimated period of the correlation function of the observed flow series.

development and history of the process (its inner structure) has to be followed in large, theoretically infinite, time intervals. The idea of following the history of a random process in a limited time interval led to the development of random processes of special types, the theory being based on the work of the Russian mathematician A. A. Markov (1856–1922).[1])

If a Markov process is at time t_k in state[2]) $X(t_k)$, then the probability of occurrence of some future state depends only on $X(t_k)$ and not on the preceding history, i.e. not on states $X(t_i)$, $X(t_j)$ where $i < j < k$.

Markov processes can be classified similarly to general random processes using the probability distribution and continuity of the time argument. In WRS issues Markov processes with discrete time (Markov chain) are frequently applied.

Markov chains are described by the transition probabilities, i.e. by conditional probabilities $p_{i,j}$ of moving from state X_i at one time to some state X_j at the following time. If the process can assume a finite number of states $X_1, X_2, \ldots, X_n$, then for each state X_i $(i = 1, 2, \ldots, n)$ at time t_k there are n probabilities P_{ij} $(j = 1, 2, \ldots, n)$ of moving from state X_i to state X_j at time t_{k+1}. The total number of conditional transition probabilities is n^2; they are arranged in a square matrix $(n \cdot n)$, called a transition matrix, which has the following form

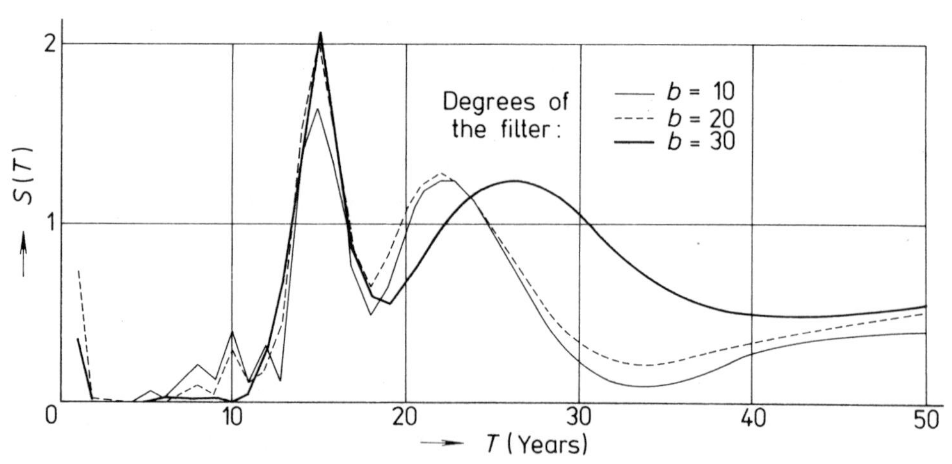

Fig. 3.16 Graph of the s.d.f. of the filtered annual flow time series in the Labe River in the town of Děčín

[1]) The theory of Markov processes is now the principal section of the theory of random processes (Dupač, 1955; Dupač and Dupačová, 1975; Dynkin, 1963). In WRS problems, Markov processes have mainly been used in mathematical models of the hydrological processes and in the probability theory of reservoir release operation. Some WRS optimization methods are likewise based on these principles.

[2]) The state is defined as the numerical value that the random variable $X(t)$ can assume.

$$\mathbf{p}_{ij} = \begin{matrix} p_{11}, p_{12}, \ldots, p_{1n} \\ p_{21}, p_{22}, \ldots, p_{2n} \\ \vdots \\ p_{n1}, p_{n2}, \ldots, p_{nn} \end{matrix} \tag{3.128}$$

Each row vector of this matrix represents the transition probabilities of moving from an initial state to all terminal states and, therefore, the sum of the values of the elements in each row has to be equal to one

$$\left(\sum_{j=1}^{n} p_{ij} = 1 \quad \text{for } i = 1, 2, \ldots, n \right).$$

Each column vector represents the probabilities of moving from different initial states to a given terminal state; the sum of the values of the elements in each column can be different.

The transition probabilities of a Markov chain can be schematically represented (Fig. 3.17) or a digraph can be used (Fig. 3.18) where the nodes of the graph represent

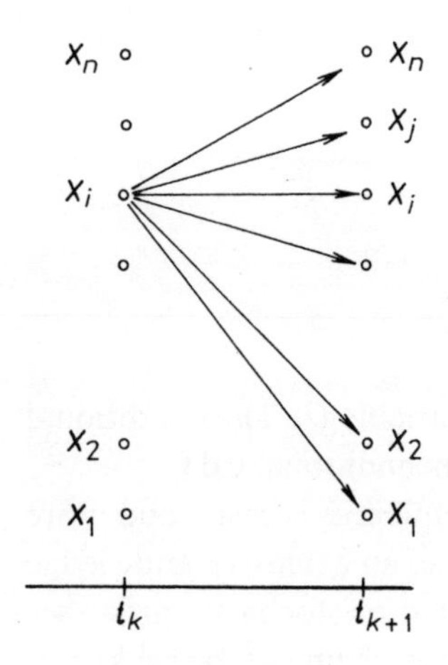

Fig. 3.17 Schematic representation of transition probabilities

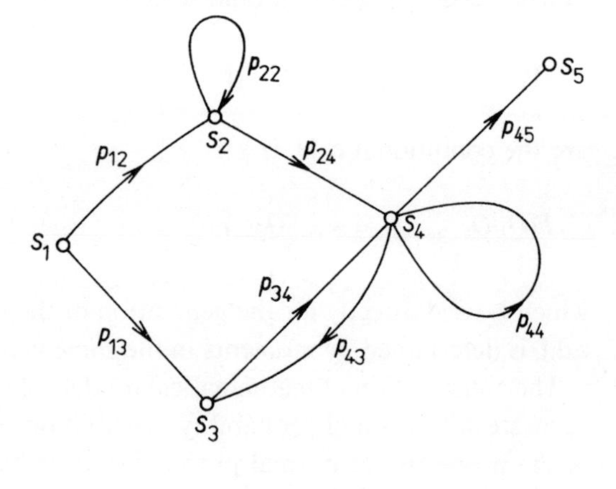

Fig. 3.18 Digraph of transition probabilities

the states of the Markov chain and the arcs of graph the transitions between the states. The missing arcs of the graph of the transition matrix $5 \cdot 5$ indicate the zero value of the corresponding element of the transition matrix (Beneš, 1970).

The Markov process is determined by the transition probabilities from time t_k to times $t_{k+1}, t_{k+2}, \ldots$ as indicated by its definition. If this relationship is confined to only two, successive times, the process is defined by a single transition matrix, i.e. it is a first-order (simple) Markov process. If n relationships are taken into consideration, compound Markov processes (or chains) or n-th order Markov processes (chains) are formed.

The form of the Markov chain is often the basis of the mathematical models of hydrological sequences. The hydrological time series is treated as a realization of a compound Markov chain, where some correlation relation occurs not only between consecutive elements of the sequence but also between the elements with time lag 2, 3, ..., τ. The conditional transition probabilities express the probability of occurrence of a random variable Q_i on condition that, at the preceding times, the random variables were equal to

$$Q_{i-1}, Q_{i-2}, \ldots, Q_{i-\tau}$$

which can be symbolically described as

$$p(Q_i/Q_{i-1}, Q_{i-2}, \ldots, Q_{i-\tau})$$

The corresponding conditional p.d.f. is

$$f(Q_i/Q_{i-1}, Q_{i-2}, \ldots, Q_{i-\tau})$$

are the conditional c.d.f. is

$$F(Q_i/Q_{i-1}, Q_{i-2}, \ldots, Q_{i-\tau})$$

which is used directly for the generation of the random variable Q_i. The conditional c.d.f. is determined by moments in the same way as the unconditional c.d.f.

The construction of mathematical models of random sequences is easier and more accurate with normal probability distribution. In this case, an extensive knowledge of the properties of normal probability distribution related to stochastic processes can be utilized and the assumption of a compound Markov chain can be related to multivariate regression analysis (Hájek and Anděl, 1969).

If the c.d.f.s are skew, some functions are sought for transformation to the normal c.d.f. or some other special methods are used. Experience has shown that these numerical methods often affect the confidence of the mathematical model, and therefore this fact has to be taken into consideration in the evaluation of the deviations of the model output statistical parameters (Nacházel *et al.*, 1975).

3.2.6 Mathematical Models of Stochastic Processes and Their Application in WRS

The importance of modelling of stochastic process in various technical tasks is a consequence of the development of the theory of stochastic processes. Their analysis was considered complete if their characteristics were expressed analytically. The analytical solution was regarded as a further stage of investigation regardless of the mathematical difficulties and degree of tractability. Research development has been influenced principally by the possibility of mathematical modelling of random process realizations of arbitrary length and of using these realizations without solution of the analytical problem.

Modelling of stochastic processes is widely used for WRS problems. The stochastic hydrological time series facilitate the investigation of many aspects of the behaviour of hydrological systems and, in addition, they form a basis for the solution of WRS problems, including hydroelectric power systems, that otherwise could only be partially solved, if at all.

This advantage, which also includes a higher degree of confidence in the calculation of water supply reliability, is a major one and therefore future research into these methods of modelling stochastic hydrological time series will centre on the theory of reservoir storage and the theory of WRS optimization[1]).

In theory, mathematical models of stochastic processes are predominantly based on the assumption of a regression relationship among the variables, on their periodic properties or on a combination of both these principal attributes of stochastic processes. Regression models have now been developed to such a degree that their principles have been successfully applied to monthly hydrological time series and the accompanying systems.

3.2.6.1 Model for Generation of Sequences of Annual Stochastic Hydrological Variables

Applying the results of section 3.2.5, construction of the mathematical model of annual values of random variables (flows, precipitation, etc.) is based on the assumption that there is a linear regression relationship between the value generated at time t and the given values at preceding times (years) $t - 1, t - 2, \ldots, t - k$. This relationship reflects the probability properties of the sequence observed (re-

[1]) That is why such attention is being paid to the development of stochastic process modelling. In Czechoslovakia, remarkable progress was achieved by, e.g., Votruba and Broža, 1966; Kos, 1969; Votruba and Nacházel, 1971; Nacházel and Bureš, 1973; Nacházel *et al.*, 1975. These references include a description of the development of these methods and a detailed explanation of them.

alization) and the number of preceding times k is determined by the order of the Markov chain. The mathematical model can be expressed by a recursive formula

$$z_t = b_1 z_{t-1} + b_2 z_{t-2} + \dots + b_k z_{t-k} + e_t, \qquad k + 1 \leqq t \leqq T \tag{3.129}$$

where T is the number of terms in the given realization (the length of the time series observed),

$b_1, b_2, \dots, b_k$ – regression coefficients related to the correlation function and

e_t – the residual random variable.

Formula 3.129 (often referred to as the linear stochastic regression model) is often used for normally distributed standardized stochastic variables[1]) which facilitates calculation and testing of regression coefficients b_i $(i = 1, 2, \dots, k)$ and the residual random variable e_t. They are derived by the method of least squares to minimize the variance of the residual random variable e_t. Regression coefficients can be expressed as follows:

$$b_1 = \frac{\mathbf{D}_1}{\mathbf{D}}, \; b_2 = \frac{\mathbf{D}_2}{\mathbf{D}}, \; \dots, \; b_k = \frac{\mathbf{D}_k}{\mathbf{D}} \tag{3.130}$$

where $\mathbf{D}$ is the determinant of the set of linear equations, including the correlation coefficients

$$\mathbf{D} = \begin{vmatrix} 1, & r_1, & r_2, & \dots, & r_{k-1} \\ r_1, & 1, & r_1, & \dots, & r_{k-2} \\ \vdots & & & & \\ r_{k-1}, & r_{k-2}, & \dots, & 1 & \end{vmatrix} \tag{3.131}$$

$\mathbf{D}_i$ is the determinant obtained by replacing the i-th column in $\mathbf{D}$ by the column vector $(r_1, r_2, \dots, r_k)$.

The standardized residual variance of the random variable e_t is defined as the difference between the total variance and its theoretically "explained" component i.e.

$$s^2(e_t) = m_1(e_t^2) = 1 - b_1 r_1 - b_2 r_2 - \dots - b_k r_k \tag{3.132}$$

Often the residual variance (3.132) is calculated for a number of alternatives until its minimum is reached. This alternative also determines the optimal number of terms in the linear expression (3.129) which is equal to the order k of the Markov chain.

---

[1]) Transformation $y = \lg(x - x_0)$ where x is the observed variable and x_0 is the lower boundary, often normalizes the probability distribution (there are also normalizing transformations for the Pearson probability distribution, type III – Kos, 1969; Votruba and Nacházel, 1971). Standardization is performed by means of the formula $z = (y - \bar{y})/s_y$ where $\bar{y}$ is the sample mean and s_y is the sample standard deviation.

Because of the limited length of the observed time series the degrees of freedom are accounted for by

$$\bar{s}^2 = \frac{m_1(e_t^2)\, T}{T - k - 1} \tag{3.133}$$

The values of random variable e_t are generated using the formula

$$e_t = \bar{s} d_t \tag{3.134}$$

where $\bar{s}$ is the standard deviation, i.e. the square root of (3.133), and d_t is the standard random variable (white noise), which can be taken from the tables of random numbers but more often is generated by computers using algorithms for the generation of standard random variables with a given probability distribution.

Having determined the regression coefficients (3.130) and the values of random variable e_t (3.134) we can generate a random sequence of desired length. If standard and transformed variables have been used, an inverse transformation of the sequence z_t is necessary.

In Soviet water resource literature (Svanidze, 1964; Reznikovsky, 1969) special linear regression models are presented, based on the assumption that the conditional c.d.f. and the unconditional c.d.f. belong to the same type and that they can be approximated by the Pearson probability distribution, type III, with $C_s = 2C_v$ (the coefficient of skewness is twice the coefficient of variation).

Stochastic sequences are generated for any order of the Markov chain by the general formula

$$Q_i = \bar{Q}_i - \sum_{j=1}^{\tau} (Q_{i-j} - \bar{Q}_{i-j}) \frac{\sigma_i}{\sigma_{i-j}} \frac{\mathbf{D}_{i(i-j)}}{\mathbf{D}_{ii}} + \Phi_i \sigma_i \sqrt{\frac{\mathbf{D}}{\mathbf{D}_{ii}}} \tag{3.135}$$

where $\mathbf{D}$ is the determinant of the correlation matrix

$$\mathbf{D} = \begin{vmatrix} r_{ii}, & r_{i(i-1)}, & r_{i(i-2)}, & \ldots, & r_{i(i-\tau)} \\ r_{i(i-1)}, & r_{ii}, & r_{i(i-1)}, & \ldots, & r_{i(i-\tau-1)} \\ \vdots & & & & \\ r_{i(i-\tau)}, & \ldots, & & & r_{ii} \end{vmatrix} \tag{3.136}$$

and $\mathbf{D}_{ii}$ and $\mathbf{D}_{i(i-j)}$ are supplements of the coefficients of correlation r_{ii} and $r_{i(i-j)}$ respectively, in the determinant $\mathbf{D}$. Φ_i are the values of standardized random variable with the Pearson c.d.f. with $C_v = 1$ and $C_s = 2C_v = 2$.

For a stationary sequence[1]),

$$\bar{Q}_i = \bar{Q}_{i-1} = \dots = \bar{Q}_{i-j} = \bar{Q}_{i-\tau} = \bar{Q}$$

$$\sigma_i = \sigma_{i-1} = \dots = \sigma_{i-j} = \sigma_{i-\tau} = \sigma$$

and the equation can be simplified as

$$Q_i = \bar{Q} - \sum_{j=1}^{\tau} (Q_{i-j} - \bar{Q}) \frac{\mathbf{D}_{i(i-j)}}{\mathbf{D}_{ii}} + \Phi_i \sigma \sqrt{\frac{\mathbf{D}}{\mathbf{D}_{ii}}} \tag{3.137}$$

The conditional standard deviation and the coefficient of skewness are determined by the expressions

$$\sigma_{\text{cond}} = \sigma \sqrt{\frac{\mathbf{D}}{\mathbf{D}_{ii}}} \tag{3.138}$$

$$C^i_{\text{s,cond}} = \frac{2\sigma \sqrt{\dfrac{\mathbf{D}}{\mathbf{D}_{ii}}}}{\bar{Q} - \sum\limits_{j=1}^{\tau} (Q_{i-j} - \bar{Q}) \dfrac{\mathbf{D}_{i(i-j)}}{\mathbf{D}_{ii}}} \tag{3.139}$$

If the stochastic sequence is calculated in relative values then (3.137) becomes

$$k_i = 1 - \sum_{j=1}^{\tau} (k_{i-j} - 1) \frac{\mathbf{D}_{i(i-j)}}{\mathbf{D}_{ii}} + \Phi_i C_v \sqrt{\frac{\mathbf{D}}{\mathbf{D}_{ii}}} \tag{3.140}$$

3.2.6.2 Model of Generation of Monthly Stochastic Variables

Generation of stochastic hydrological time series with a time interval shorter than one year is more difficult than generation of annual time series, because the model has to take into account the fluctuations of probability distributions of random variables during the year.

---

[1]) Stochastical hydrological time series have been generated for stationary processes and some attempts have been made to use non-stationary processes as well. In the Hydrological Department of the Faculty of Civil Engineering (Technical University of Prague) research into some non-stationary hydrological time series has been carried out on the basis of differences, in development of particular statistical characteristics (Nacházel *et al.* 1975).

Annual flows were generated on the basis of periodic models. Their deterministic component is given (as compared with the regression models) by a harmonic function, the form of which is optimized by spectral analysis. The random component has the same meaning as in the regression models, and it is often generated as white noise (Anděl and Balek, 1969; Balek, 1975).

There are some special types of mathematical models formed as the sum of the deterministic periodic and random regression components.

Svanidze's disaggregation method (1964) used the technique of double sampling and provided a relatively good fit between input and output annual statistical parameters. However, disaggregation using the monthly "fragments", in combination with the generated annual flows, distorted some statistical characteristics of monthly flows, and application of the method without careful analysis could result in incorrect results (Votruba and Broža, 1966).

Remarkable progress in the generation of monthly flows was achieved by Kos, 1969. From many samples of observed time series, he proved that the principle of the linear regression model can be applied to monthly flows where all possible combinations of correlation relationships are involved (e.g. between the time series of monthly flows in February and January, March and January, etc.). Expression (3.141) constructed analogously to (3.129) is the regression equation for the generation of sequences of monthly values in a chosen month m

$$z_{c,m} = b_{1,m}z_{c,m-1} + b_{2,m}z_{c,m-2} + \dots + b_{k,m}z_{c,m-k} + e_m \tag{3.141}$$

where m is 1, 2, ..., 12 (month of the year)

c – 1, 2, ..., $T/12$ (number of generation cycle – i.e. year, where T = the total number of month generated)

e_m – random deviate in month m

The monthly generating model requires twelve equations of type (3.141). The generating process uses the step equal to 12, i.e. in annual cycles: having chosen 12 random variables in the first year (e.g. by observed values or by monthly means), in the second cycle the values $z_{2,m}$ are calculated etc. In this manner a sequence of random variables in chronological order is calculated analogously to the observed time series. The unknown regression coefficients $b_{i,m}$ and random deviates e_m are determined in the same way as in the annual model. The regression coefficients are determined by

$$b_{1,m} = \frac{\mathbf{D}_{1,m}}{\mathbf{D}_m}, \; b_{2,m} = \frac{\mathbf{D}_{2,m}}{\mathbf{D}_m}, \; \dots, \; b_{k,m} = \frac{\mathbf{D}_{k,m}}{\mathbf{D}_m} \tag{3.142}$$

where the determinants $\mathbf{D}_m$ and $\mathbf{D}_{i,m}$ ($i = 1, 2, \dots, k$) have the same meaning as in the annual model but are calculated for each month m.

The residual variance in each month m is calculated by the equation

$$s^2(e_m) = m_1(e_m^2) = 1 - b_{1,m}r_{1,m} - b_{2,m}r_{2,m} - \dots - b_{k,m}r_{k,m} \tag{3.143}$$

A number of alternatives are used in order to minimize this variance and the corresponding k is the order of the Markov chain.

Because of the limited length of the series

$$\bar{s}_m^2 = \frac{m_1(e_m^2)\, C}{C - k - 1} \tag{3.144}$$

where

$$C = \frac{T}{12}.$$

The random variates e_m in each month are calculated by

$$e_m = \bar{s}_m d_t \tag{3.145}$$

The generating procedure has already been described. The last stage of calculation is the inverse transformation of the generated values if standardization has been used.

The goodness-of-fit of the linear regression model has been tested many times in water resource research and hydrological studies by statistical and other tests. The goodness-of-fit of input and output parameters was established and the deviations were not statistically significant. If the order k is less than 12 the annual parameters are not taken into account. A possible improvement of the model is being tested (Nacházel *et al.*, 1975; Kos, 1982).

In future development of monthly hydrological time series generation, the periodic models could be tested according to the obvious annual cycle.

Attempts to use the methods described for daily hydrological time series have failed; further research is desirable in this respect.

3.2.6.3 Model of Flow Generation in a System of Stations

The WRS model often requires as input values the monthly flows generated in a system of stations. In this system the elements are stochastically dependent. Therefore the stochastic time series has to be generated in a system to maintain the cross-correlations, i.e. the cross-correlations of the generated time series should correspond to those of the observed time series. The model of a system of stations should involve two subsystems of correlation relations, viz.:

(1) The inner serial correlation of individual stations (with the length given by the Markov chain) and

(2) cross-correlations among the flows in different stations.

The simplest method of modelling a system of stations is that of the central and satellite stations. Using the linear regression model (see section 3.2.6.2) the flows are generated in the central (the most important) station. Coincident flows in other satellite stations are calculated by simple linear regression

$$y_{i,j,k} = \bar{y}_{j,k} + b_{j,k}(x_{i,j} - \bar{x}_j) + u_i d_{j,k} \tag{3.146}$$

$$d_{j,k} = s_{j,k}\sqrt{1 - r_{j,k}^2} \tag{3.147}$$

where $y_{i,j,k}$ is the flow of serial number i, in month j, at point k,
$\bar{y}_{j,k}$ – mean monthly flow in month j and at point k,
$b_{j,k}$ – regression coefficient in month i for the relationship between the central station and the satellite station k,
$x_{i,j}$ – flow in the central station of serial number i in month j,
$\bar{x}_j$ – mean monthly flow in the central station in month j,
u_i – standardized random variate,
$s_{j,k}$ – standard deviation in month j at point k
$r_{j,k}$ – correlation coefficient in month j for relationship of flows between the central station and the satellite station k.

Calculation by (3.146) and (3.147) proceeds month after month. For each combination of the central station with a satellite station 12 equations (3.146) and (3.147) are necessary.

This method can be applied to synchronous or almost synchronous time patterns. If the cross-correlation is low, this method may distort the serial correlation in the satellite station.

Orthogonal transformation is a more general method of generating flows in a system of stations. The principle here is involved in transformation of linearly dependent (coincident) flows in each month in different stations to independent (orthogonal) values. It is assumed that the observed flows $x_{i,j,k}$ in the same month j but in different stations k form correlated random vectors that are transformed to independent vectors by the expression

$$\mathbf{z}_{i,j,k} = q_{j,1,k}x_{i,j,1} + q_{j,2,k}x_{i,j,2} + \ldots + q_{j,n,k}x_{i,j,n} \tag{3.148}$$

where $i = 1, 2, \ldots, T$, (number of year of observation),
$j = 1, 2, \ldots, 12$ (number of month),
$k = 1, 2, \ldots, n$ (number of station)

In each month j the coefficients $q_{j,n,k}$ are functions of the correlation matrix with the elements given by correlation coefficients between the coincident values $x_{i,j,k}$ for different k values. There are twelve such square matrices with the dimension $n \cdot n$ (according to the number of stations), each in the form

$$\begin{matrix} r_{j,1,1}, & r_{j,1,2}, & \ldots, & r_{j,1,n} \\ r_{j,2,1}, & r_{j,2,2}, & \ldots, & r_{j,2,n} \\ \vdots & & & \\ r_{j,n,1}, & r_{j,n,2}, & \ldots, & r_{j,n,n} \end{matrix} \qquad \text{where } j = 1,2, \ldots, 12 \tag{3.149}$$

The matrix (3.149) is symmetrical about the principle diagonal, all the elements of which are 1 (correlation of the flows in the same month and station).

The orthogonal transformation produces independent flows in fictitious stations.

Its advantage is that the flows in each station can be generated independent of other stations and for any conditions including the order of the Markov chain. After transformation of the observed flows in all the stations individual time series are generated for individual stations. Having finished this calculation, we proceed to the inverse transformation that produces the output correlated stochastic flows in real stations. The inverse transformation is described by the following equation

$$x_{i,j,k} = p_{j,1,k} z_{i,j,1} + p_{j,2,k} z_{i,j,2} + \dots + p_{j,n,k} z_{i,j,n} \qquad (3.150)$$

The coefficients $p_{j,n,k}$ are the elements of the inverse matrix of the original matrix composed of elements $q_{j,n,k}$. Because this is an orthogonal matrix, its inverse matrix is the transposed matrix where the rows are replaced by columns and vice versa.

Calculation of the coefficients $q_{j,n,k}$ or $p_{j,n,k}$ is relatively complicated, particularly for a large number of stations. Therefore, computer software including special algorithms (e.g. Jacobi procedure) can be found in specialized mathematical literature (Evans, 1962; Naur, 1963).

3.3 MATHEMATICAL DESCRIPTION OF SYSTEM BEHAVIOUR

3.3.1 Introduction

Assume a system with all inputs and outputs specified and described. An analysis of the system's behaviour involves investigation of interrelations between all these inputs and outputs and, in addition, the relationships among all the n elements of the system and analysis of the individual behaviour of each element. Such a complex analysis may be difficult even for a relatively limited system.

The *behaviour of the whole system* is given by the interrelations between its inputs and outputs. The input X is given by p input components $(p \geqq 1)$ $x_1, x_2, \dots, x_p$ so that it can be described by the input vector

$$\bar{x} = (x_1, x_2, \dots, x_p) \qquad (3.151)$$

The output Y is given by q output components $(q \geqq 1)$ and it can be similarly described by the output vector

$$\bar{y} = (y_1, y_2, \dots, y_q) \qquad (3.152)$$

The behaviour can be symbolically described by transformation $\bar{T}$

$$\bar{y} = \bar{T}(\bar{x}) \qquad (3.153)$$

$\bar{T}$ is often called the transformation operator.

This behaviour can be *deterministic* or *stochastic* – the latter being the general case of system behaviour. Deterministic behaviour can be classified as combinatory of sequential. In *combinatory* behaviour, the transformation $\bar{T}$ is unique, i.e. only one output vector $\bar{y}$ corresponds to a certain vector $\bar{x}$. In *sequential behaviour*, different responses correspond to the same stimulus in relation to the previous states defined or to their sequences.

In this section we shall investigate that class of *dynamic systems* which have time-variant general properties.

3.3.2 The Concept of Dynamic System Operator

A dynamic system (Fig. 3.19) with the input time-dependent function (process) $X(t)$ and output process $Y(t)$ is described by

$$Y(t) = A\{X(t)\} \tag{3.154}$$

where A is the *dynamic system operator*. It symbolizes the set of all mathematical and logical operations applied inside the system S on the input process $X(t)$ in such a way that it is transformed into the output process $Y(t)$.

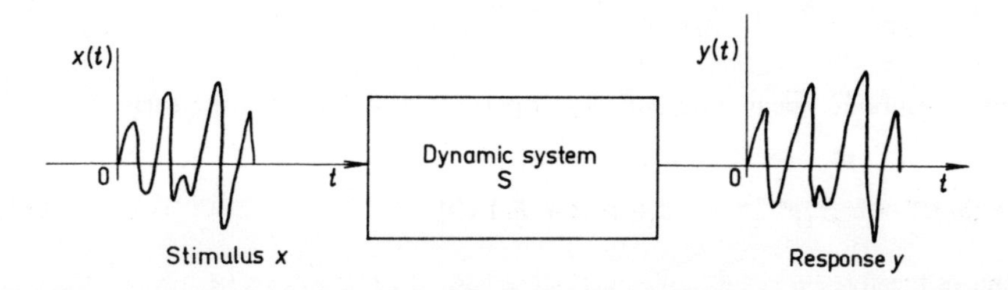

Fig. 3.19 Schematic representation of the dynamic system

The concept of the operator is a generalization of the concept of the functional (see section 3.7), just as the functional is a generalization of the concept of a function. In the case of the operator, a function (process) as the "independent variable" is related again to a function (process) as the "dependent variable".

The operator can be either linear or non-linear. *Linear operators* describing the transformation of the inputs of a linear dynamic system are characterized by satisfying the superposition principle, i.e. for any input functions $x_1(t)$, $x_2(t)$, ..., $x_n(t)$ and for any constant n, c_1, c_2, ..., c_n,

$$A\left\{\sum_{k=1}^{n} c_k\, x_k(t)\right\} = \sum_{k=1}^{n} c_k\, A\{x_k(t)\} \tag{3.155}$$

The result of the realization of the linear operator A to any linear combination of input functions is equivalent to the linear combination of the results of realization of this operator to the individual input functions.

Similarly, for an input function given by a set of differential elementary stimuli, the output function of a linear dynamic system is given by the integral of the corresponding differential responses i.e.

$$A\left\{\int_{\lambda_1}^{\lambda_2} c(\lambda)\, x(t, \lambda)\, \mathrm{d}\lambda\right\} = \int_{\lambda_1}^{\lambda_2} c(\lambda)\, A\{x(t, \lambda)\}\, \mathrm{d}\lambda \qquad (3.156)$$

A distinction can be draw between *homogeneous* (e.g. linear operator of differentiation, integration, multiplication by a function of time, etc.) and *non-homogeneous* linear operators, which are composed of a homogeneous operator and a non-random function of time.

The behaviour of the dynamic system S can be described by a linear differential equation with constant coefficients

$$a_n \frac{\mathrm{d}^n y(t)}{\mathrm{d}t^n} + a_{n-1} \frac{\mathrm{d}^{n-1} y(t)}{\mathrm{d}t^{n-1}} + \ldots + a_1 \frac{\mathrm{d}y(t)}{\mathrm{d}t} + a_0\, y(t) =$$

$$= b_m \frac{\mathrm{d}^m x(t)}{\mathrm{d}t^m} + b_{m-1} \frac{\mathrm{d}^{m-1} x(t)}{\mathrm{d}t^{m-1}} + \ldots + b_1 \frac{\mathrm{d}x(t)}{\mathrm{d}t} + b_0\, x(t) \qquad (3.157)$$

or by a symbolic linear differentiation operator $\mathrm{d}^k/\mathrm{d}t^k = p^k$ in the form

$$(a_n p^n + a_{n-1} p^{n-1} + \ldots + a_1 p + a_0)\, y(t) =$$

$$= (b_m p^m + b_{m-1} p^{m-1} + \ldots + b_1 p + b_0)\, x(t) \qquad (3.158)$$

or, more briefly,

$$A_n(p)\, y(t) = B_m(p)\, x(t) \qquad (3.159)$$

Hence

$$y(t) = \frac{B_m(p)}{A_n(p)}\, x(t) \qquad (3.160)$$

The ratio $B_m(p)/A_n(p)$ is an important characteristic of the operator of a linear dynamic system, and it is called the *transfer of the system.*

The *non-linear operator* (operator of the second and higher powers of the input function, operator of trigonometric functions, etc.) similarly describes non-linear dynamic systems.

Example VII

Assume a system S with input process $x(t)$ and output process $y(t)$. Determine its operator if

$x(t) = a \cos(t)$ and $y(t) = -a \sin(t)$

Solution

Denote the operator of this system by A. Since

$$\frac{\mathrm{d}(a \cos t)}{\mathrm{d}t} = -a \sin t \tag{VII.1}$$

the operators A equals

$$A = \frac{\mathrm{d}}{\mathrm{d}t} \tag{VII.2}$$

A is the operator of the system S and it is called the differentiation operator. This operator is linear and homogeneous if relationships (3.155) and/or (3.156) are satisfied.

3.3.3 Dynamic System with Random Inputs and Outputs

The input and output functions of a system can be random time functions. If these functions are stationary, i.e. invariant to the time origin, the system is called a *stationary dynamic system*; the opposite is a *non-stationary dynamic system* where a change in time of occurrence of the same stimulus results in a change in the response (Fig. 3.20).

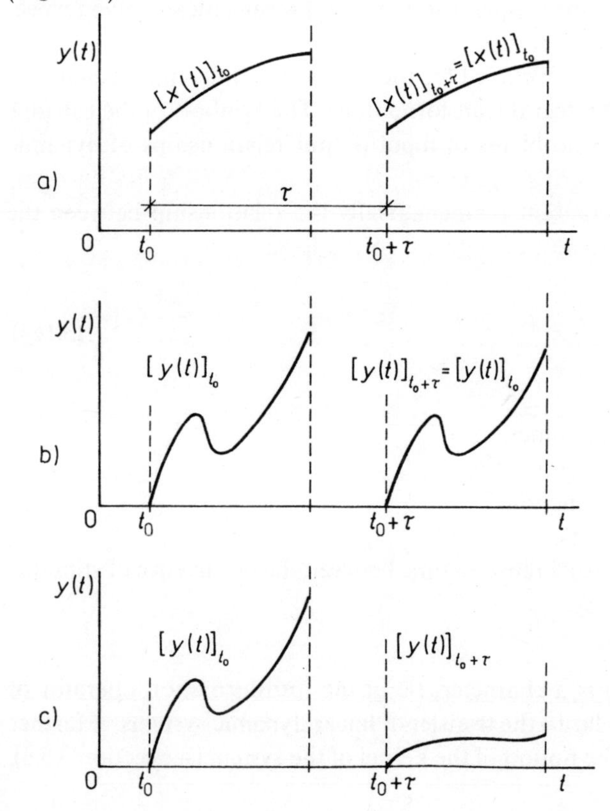

Fig. 3.20. Schematic representation of the response of stationary and non-stationary systems
a – stimulus, b – response of a stationary system, c – response of a non-stationary system

Random process occur frequently in the inputs and outputs of dynamic systems in WRS contexts. In general, it is assumed that the input and output processes are non-stationary. Assume a non-stationary dynamic system described by the linear operator A_L. The homogeneous linear operator A_0 can be separated in the transformation, i.e.

$$Y(t) = A_L\{X(t)\} = A_0\{X(t)\} + \varphi(t) \tag{3.161}$$

where $\varphi(t)$ is a non-random time function.

An input stochastic process $X(t)$ can be described, in general, by a non-stationary mean $m_x(t)$, standard deviation $s_x(t)$, autocorrelation function $r_x(t, \tau)$ and possibly by an adequate theoretical probability distribution $f_x(x, t)$. Similarly an output stochastic process $Y(t)$ is described by quantities $m_y(t)$ and $s_y(t)$ and functions $r_y(t, \tau)$ and $f_y(y, t)$, which can be derived from the input data. Then we get, for example, for $m_y(t)$

$$m_y(t) = M[A_0\{x(t)\}] + \varphi(t) \tag{3.162}$$

where A_0 is the homogeneous linear operator and M [ ] means mean value (expectation) of the expression in brackets.

The calculus of operators was developed for the solution of differential equations. It is closely related to Laplace integral transformations. The symbols of the calculus of operators can be applied in problems of input/output relationships of dynamic systems.

Sometimes it is useful to represent mathematically the relationship between the inputs and outputs by the convolution integral in the form

$$y(t) = \int_0^t x(\tau)\, h(t - \tau)\, \mathrm{d}\tau \tag{3.163}$$

where $x(\tau)$ is the input time function,
$y(\tau)$ – the output time function,
$h(t - \tau)$ – the transfer function,
t, τ – time arguments,
$(t - \tau)$ – the time lag (difference in time between the occurrence of stimulus and response).

The transfer function $h(t - \tau)$ is a characteristic of the transformation operator of dynamic systems, and it is similar to the transfer of linear dynamic systems. A further generalization the concept is the notion of the kernel of the system (see section 3.3.4).

3.3.4 Mathematical Formulation of the Behaviour of Stochastic Hydrological Systems

Dynamic systems with stochastic behaviour are frequently encountered in WRS investigation. Moreover, these stochastic systems are, in general, non-stationary and their behaviour is described by non-linear transformations.

The general form of the above-mentioned mathematical description is particularly applicable to some hydrological stochastic systems.The transformation of inputs to outputs can be given in the following form, assuming the linearity of the model (Strupczewski, 1975).

$$y(t) = H_d[x_d(t)] + H_s[x_s(t)] \tag{3.164}$$

where $x_d(t)$ is the deterministic input component (deterministic input signal),
$x_s(t)$ – stochastic input component (random input signal),
H_d – linear operator of the deterministic input signal,
H_s – linear operator of the stochastic input signal.

This concept is related to some hydrological stochastic models and their transformations (Kisiel, 1969, Yevjevich, 1976, Dooge, 1972). The behaviour of a compound linear hydrological system can be described by a convolution integral in the following general form (Huthmann, 1975)

$$y(t) = \sum_{i=1}^{N} \int_0^{\infty} h_i(\tau)\, x_i(t - \tau)\, d\tau + z(t) \tag{3.165}$$

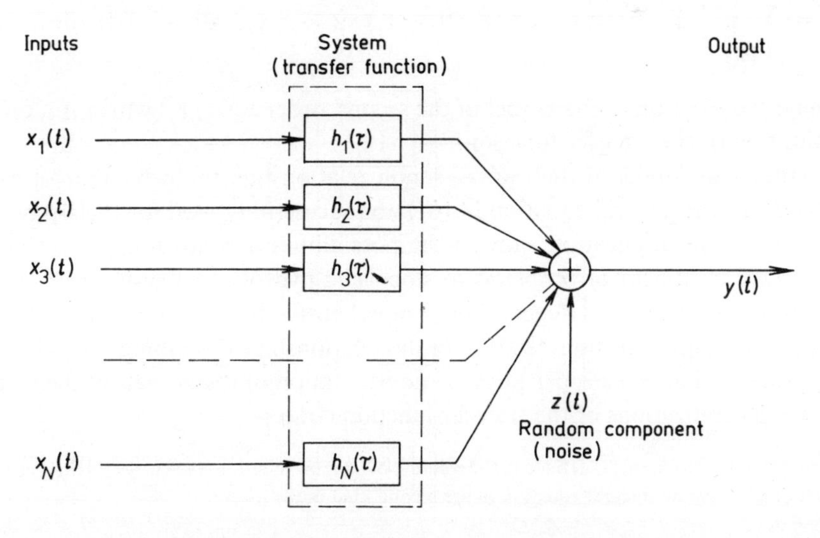

Fig. 3.21 Schematic representation of a compound linear stochastic system

where N is the number of inputs,
$y(t)$ – output,
$x_i(t)$ – i-th input,
$h_i(\tau)$ – i-th transfer function related to the i-th input,
$z(t)$ – random component (noise).

A schematic representation of this general form is given in Fig. 3.21. This formal relationship was applied by G. Huthmann in the determination of the maximum basin runoff.

A similar, more complex case with a non-linear transformation operator is given in general form in equation (3.166). The behaviour of the system is described as a multiple-dimensional convolution integral

$$y(t) = \sum_{j=1}^{\infty} \int_{-\infty}^{\infty} \dots \int_{-\infty}^{\infty} h_j(\tau_1, \tau_2, \dots, \tau_j) \prod_{i=1}^{n} x(t' - \tau_i) \, d\tau_1, \dots, d\tau_j \tag{3.166}$$

where $x(t)$ is the system input,
$y(t)$ – the system output,
$h_j(\tau_1, \tau_2, \dots, \tau_j)$ – is the kernel of the system[1])

This approach to the description of a complex, non-linear system involving the determination of the kernels of several orders was used by N. Wiener. In hydrology contexts, only the kernel of the first order, i.e. the transfer function, has a clear physical meaning. The kernels of higher orders have to be interpreted. Quimpo modified the relationship (3.166) to a finite sum of integrals in the form

$$y(t) = \sum_{j=1}^{m} \int_{-\infty}^{\infty} \int_{-\infty}^{\infty} h_j(\tau_1, \tau_2, \dots, \tau_j) \; x(t - \tau_1) \, x(t - \tau_2) \dots x(t - \tau_j) \, d\tau_1 \, d\tau_2 \dots d\tau_j \tag{3.167}$$

and found the solution of the kernel of the second order $h_2(\tau_1, \tau_2)$ which is a certain generalization of the transfer function.

The non-linear model of the rainfall-runoff relationship, the behaviour of which is described by the general equation (3.167) was successfully used for the n-day forecast of flows. This application showed the possibilities and advantages of this approach to the treatment of problems in stochastic hydrological systems. However, further applications are envisaged after comprehensive improvement of mathematical procedures aimed at discretization of the relationships describing the behaviour of the system and after a deeper physical interpretation of the kernels of the system, which are generalizations of the transfer function concept.

---

[1]) The concept of the kernel is defined in the solution of integral equations (Rektorys, 1963). The kernel is a function $K(x, s)$ in an integral equation of the second kind in the form

$$f(x) - \int_a^b K(x, s) \, f(s) \, ds = g(x)$$

3.4 FOURIER TRANSFORMATION

3.4.1 Introduction

In the investigation of WRS hydrological processes the concept of spectrum and the spectral density function (s.d.f.) are often used (Votruba and Nacházel, 1969). In the theory of spectral analysis, a transformation from the linear time argument to the frequency characteristic is performed, which facilitates and improves the investigation of complicated stochastic and deterministic processes. Various transformations are necessary in the determination of relationships between WRS inputs and outputs. The choice and analysis of such transformations in various present-day tasks of water management by means of the systems approach has not yet been developed to the stage of implementation so their practical importance is not obvious. However, they are promising elements of the theoretical basis. In inputs and outputs of dynamic systems, processes (time functions) of various types can be found. The basic classification criterion is their *deterministic* or *stochastic nature*. Deterministic (non-random) processes can be described by exact mathematical relationships, stochastic (random) processes cannot. The deterministic processes can be further subdivided into several groups (Fig. 3.22).

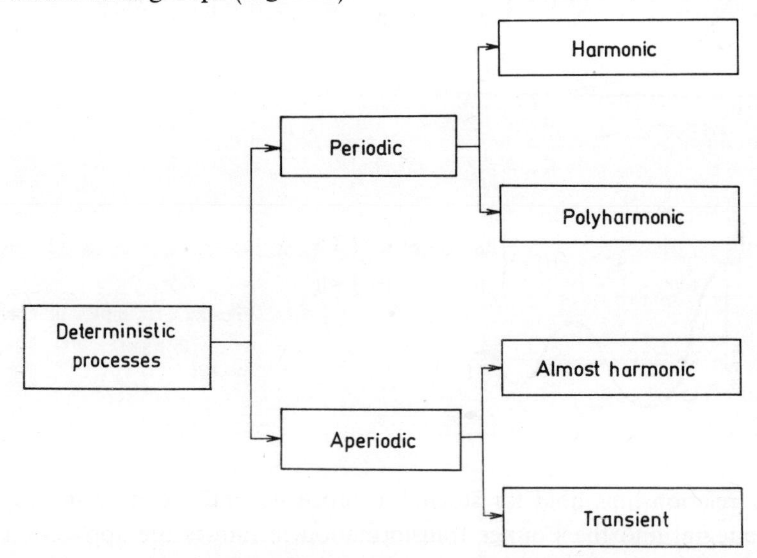

Fig. 3.22 Deterministic processes classification

Harmonic processes of a periodic function can be described in a compact and simple form by an elementary discrete spectrum with one non-zero value (see section 3.2.4 and Fig. 3.14a).

A polyharmonic process, which occurs more often in reality, is formed by the combination of two of more elementary harmonic processes, assuming that the ratio of the lengths of the periods for each pair of harmonic processes are rational numbers. In the opposite case the process in question is called almost periodic; it is composed of sine curves not only of divisible periods but with all possible periods where also an infinite period may appear. These cases occur if two or more independent harmonic processes are combined. For the polyharmonic or almost harmonic processes the spectrum can be expresses by several discrete values; in the case of almost harmonic processes some remainders are left over.

The class of transient processes includes all the other cases of periodic processes that occur in many physical contexts. The difference between harmonic, polyharmonic and almost harmonic processes, on the one hand, and transient processes, on the other, is obvious from the different form of their spectra. Transient processes cannot be expressed by a discrete spectrum. In almost all cases, however, their spectra can be described by a continuous function using the Fourier integral, which facilitates the transformation of all these types of processes to a comparable level.

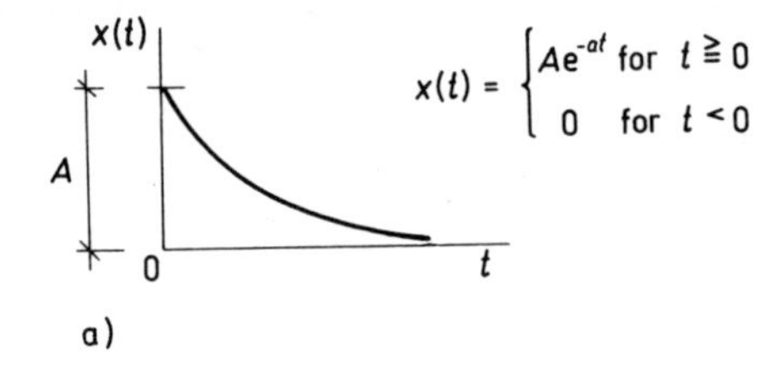

$$x(t) = \begin{cases} Ae^{-at} & \text{for } t \geqq 0 \\ 0 & \text{for } t < 0 \end{cases}$$

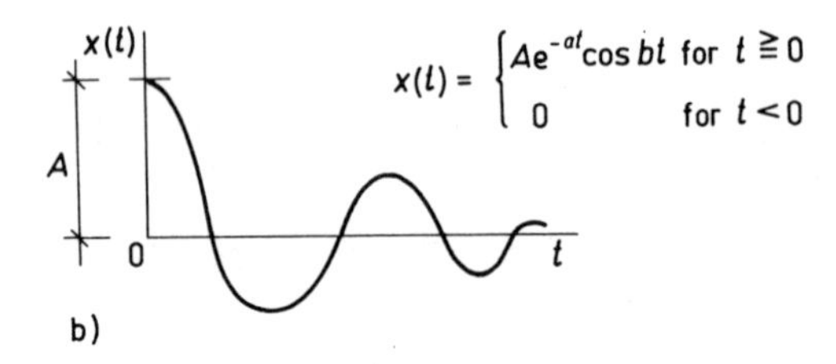

$$x(t) = \begin{cases} Ae^{-at}\cos bt & \text{for } t \geqq 0 \\ 0 & \text{for } t < 0 \end{cases}$$

Fig. 3.23 Examples of aperiodic processes and their spectra

Similar relationships hold for stochastic processes and therefore the use of the Fourier integral and the Fourier transformation facilitates the application of the unique method of investigation of processes occurring in the inputs and outputs of dynamic systems and of their relationships irrespective of the different characteristics of these processes.

Figure 3.23 shows an example of two transient processes with continuous spectra. Both these time functions can be applied in hydrology and WRS, the first case (a) being an exponential function and the second case (b) a function approximately harmonically damped. Changes taking place in them during the transition of the process through the system will be followed up. Using the argument of the spectral function, which is $\omega = 1/T$ (where T is analogous to the length of the period of periodic functions), the Fourier integral may be expressed in the form:

$$X(\omega) = \int_{-\infty}^{\infty} x(t)\, e^{-2\pi i\omega t}\, dt \tag{3.168}$$

where i is the imaginary unit $\sqrt{-1}$ (in the general case, spectrum is a complex function) and $x(t)$ is the input process with the time argument t, the spectrum of which is to be found.

Using the modulus (mean value) of the spectrum $|X(\omega)|$ and the argument $\Theta(\omega)$, the complex form may be written

$$|X(\omega)| = X(\omega) \cdot e^{i\Theta(\omega)} \tag{3.169}$$

The spectrum of function (a) has a broad-band character (Fig. 3.14d), whereas that of function (b) has a narrow-band character (Fig. 3.14b).

3.4.2 Fourier Series

The periodic function $x(t)$ with period T can be represented as the sum of a finite or infinite number of simple harmonic components (sine curves). If the periodic function analysed, $x(t)$, is bounded, piecewise continuous, and if it has inside the period T a finite number of extrema, it can be written in the form

$$x(t) = A_0 + \sum_{k=1}^{\infty} A_k \sin(k\omega t + \alpha_k) \tag{3.170}$$

where $A_0, A_1, \ldots, A_k; \alpha_1, \alpha_2, \ldots, \alpha_k$ are constants corresponding to the particular harmonic components and ω, the frequency, equals $2\pi/T$.

This equation is the general expansion for a periodic function $x(t)$ into a trigonometric series. If a new time argument is introduced

$$\tau = \omega t = \frac{2\pi t}{T} \tag{3.171}$$

so that

$$t = \frac{\tau}{\omega} \tag{3.172}$$

its function $f(\tau)$ can be written in the form

$$f(\tau) = a_0 + \sum_{k=1}^{\infty} \left(a_k \cos 2\pi k \frac{t}{T} + b_k \sin 2\pi k \frac{t}{T} \right) \tag{3.173}$$

where $a_0 = A_0$,
$a_k = A_k \sin \alpha_k$,
$b_k = A_k \cos \alpha_k$,
$(k = 1, 2, 3, \ldots)$.

The expansion of the function $x(t)$ and $f(\tau)$ in the form (3.173) is called the Fourier series in the real form. In complex (exponential) form it is

$$f(\tau) \approx \sum_{k=-\infty}^{\infty} c_k \, \mathrm{e}^{\mathrm{i}k\tau} \tag{3.174}$$

where

$$c_0 = \frac{a_0}{2},$$

$$c_k = \frac{a_k - \mathrm{i}b_k}{2} \quad \text{if} \quad k > 0,$$

$$c_k = \frac{a_k + \mathrm{i}b_k}{2} \quad \text{if} \quad k < 0, \tag{3.175}$$

i is the imaginary unit
or in a complex representation by one equation for c_k

$$c_k = \frac{1}{2\pi} \int_{-\pi}^{\pi} f(\tau) \, \mathrm{e}^{-\mathrm{i}k\tau} \, \mathrm{d}\tau \tag{3.176}$$

where $k = 0, \pm 1, \pm 2, \ldots$.

The expansion of a periodic function into a Fourier series simplifies the mathematical expression of very complicated functions that are important for technical applications. The known spectrum of a periodic function facilitates analysis of the behaviour of a system if the periodic functions are applied in the system input. This holds for both deterministic and stochastic periodic functions, assuming that the spectrum can be represented in this form.

3.4.3 Fourier Integral

The Fourier integral is a generalization of the expansion of functions, including aperiodic functions. In the Fourier series expansion the number of harmonic functions could be infinite, but the number of the frequencies was finite. If the sum of an infinite number of simple harmonic functions is expressed with frequencies that differ by an increment approaching zero, a Fourier integral results. However, it exists and it can be calculated under certain conditions called *conditions of convergence*.

Figure 3.24 shows a schematic representation of the spectrum of an aperiodic function given by a continuous or piecewise continuous curve, and the discrete (point) spectrum of a polyharmonic function with three components T_1, T_2, T_3 with frequencies ω_1, ω_2, ω_3, respectively. The spectrum of an aperiodic function can be described by a continuous, often bell-shaped curve, with the ranges of frequencies (ω_a, ω_b) and differences between successive frequencies equal to the differential d_ω.

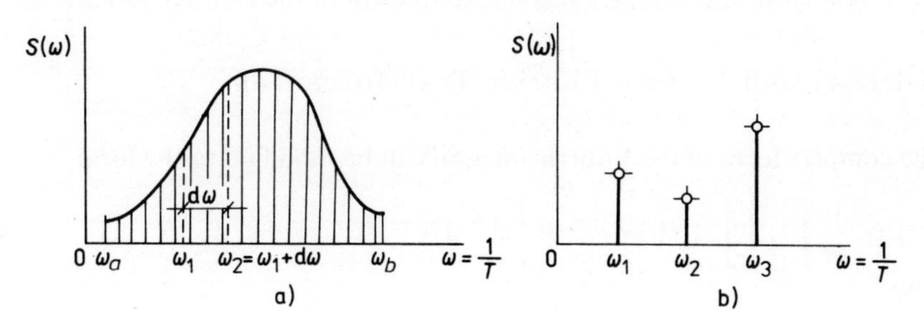

Fig. 3.24 Schematic representation of spectra of aperiodic and periodic functions

A sufficient condition of the Fourier integral convergence for the function $f(\tau)$ is the existence of the improper integral

$$I_n = \int_{-\infty}^{\infty} |f(\tau)| \, d\tau \tag{3.177}$$

the function $f(\tau)$ is assumed to be piecewise continuous with a finite number of discontinuities and local extrema in each finite interval $\Delta\tau$. At each point τ_0 inside the finite interval $(-\lambda; \lambda)$ (and $|\lambda| > \tau_0$; the function $f(\tau)$ is differentiable),

$$f(\tau_0) = \frac{a_0}{2} + \sum_{k=1}^{\infty} \left(a_k \cos \frac{k\pi\tau_0}{\lambda} + b_k \sin \frac{k\pi\tau_0}{\lambda} \right) \tag{3.178}$$

If the coefficients a_0, a_k, b_k of this expansion are expressed in an integral form, we get[1]):

$$f(\tau_0) = \frac{1}{\pi} \int_0^{\infty} d\omega \int_{-\infty}^{\infty} f(\tau) \cos \omega (\tau - \tau_0) \, d\tau \tag{3.179}$$

For formal reasons we replace τ_0 by τ and τ by t:

$$f(\tau) = \frac{1}{\pi} \int_0^{\infty} d\omega \int_{-\infty}^{\infty} f(t) \cos \omega (t - \tau) \, dt \tag{3.180}$$

[1]) A detailed mathematical description can be found in the mathematical literature (Lapa, 1971; Raven, 1966).

If the function $f(t)$ is integrable in the interval $(-\infty; \infty)$, and if it possesses in any finite interval Δt a finite number of discontinuities, then the expression (3.180) holds for every point where it can be differentiated, and it is the Fourier integral. This integral (rather, two-dimensional integral) can be written in a complex form

$$f(\tau) = \frac{1}{2\pi} \int_{-\infty}^{\infty} \mathrm{d}\omega \int_{-\infty}^{\infty} f(t)\, \mathrm{e}^{\mathrm{i}\omega(\tau - t)}\, \mathrm{d}\tau \tag{3.181}$$

which is related to the complex (exponential) form of the Fourier series.

3.4.4 Direct and Inverse Fourier Transformation

The complex form of the Fourier integral can be modified to the form

$$f(\tau) = \frac{1}{2\pi} \int_{-\infty}^{\infty} \left[\int_{-\infty}^{\infty} f(t)\, \mathrm{e}^{-\mathrm{i}\omega t}\, \mathrm{d}t \right] \mathrm{e}^{\mathrm{i}\omega\tau}\, \mathrm{d}\omega \tag{3.182}$$

where

$$\int_{-\infty}^{\infty} f(t)\, \mathrm{e}^{-\mathrm{i}\omega t}\, \mathrm{d}t = F(\omega) \tag{3.183}$$

This relationship between the complex s.d.f., $F(\omega)$, and the real function of time $f(\tau)$ is called the *Fourier transformation* (direct Fourier transformation) (Fig. 3.25), and for the time argument defined by (3.171) is has the following form

$$F(\omega) = \int_{-\infty}^{\infty} f(\tau)\, \mathrm{e}^{-\mathrm{i}\omega t}\, \mathrm{d}\tau \tag{3.184}$$

Expressing the function $f(\tau)$ by its s.d.f. significantly facilitates the investigation of dynamic system behaviour as a response to this function.

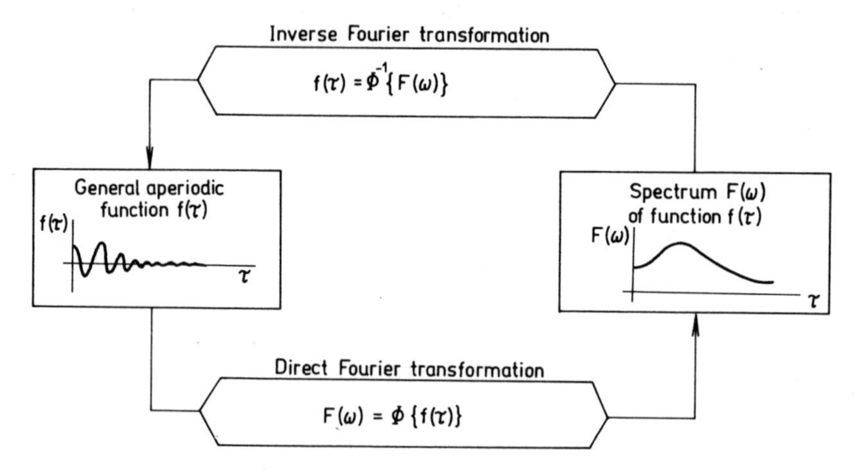

Fig. 3.25 Schematic representation of the direct and inverse Fourier transformation

The reverse of the direct Fourier transformation in the form (3.184) is the *inverse Fourier transformation* in the form

$$f(\tau) = \frac{1}{2\pi}\int_{-\infty}^{\infty} F(\omega)\, e^{i\omega\tau}\, d\omega \tag{3.185}$$

Instead of the complex form (3.185) the real form of the inverse Fourier transformation can be used

$$f(\tau) = \frac{1}{\pi}\int_{0}^{\infty} [A(\omega)\cos\omega\tau - B(\omega)\sin\omega\tau]\, d\omega \tag{3.186}$$

where

$$A(\omega) + i\, B(\omega) = F(\omega) \tag{3.187}$$

where $A(\omega)$ is an event function and $B(\omega)$ is an odd function.

3.4.5 Spectrum of a Unit Impulse

In the theory of dynamic systems and their models the unit impulse is often applied to their inputs. It can be described by the Dirac delta function (Dirac unit impulse)

$$f(t) = 0 \quad \text{for} \quad \tau \in \left(-\infty;\ -\frac{\varepsilon}{2}\right)\ \tau \in \left(\frac{\varepsilon}{2};\ \infty\right)$$

$$f(\tau) = \frac{1}{\varepsilon} \quad \text{for} \quad \tau \in \left(-\frac{\varepsilon}{2};\ \frac{\varepsilon}{2}\right) \tag{3.188}$$

where ε is a positive number that converges to zero.

The *Dirac function* is aperiodic and if eq. (3.187) is used for it, we get the following Fourier transform

$$F(p) = \int_{-\infty}^{\infty} f(\tau)\, e^{-p\tau}\, d\tau = \int_{-\varepsilon/2}^{\varepsilon/2} \frac{1}{\varepsilon}\, e^{-p\tau}\, d\tau = \frac{1}{\varepsilon_p}\, (e^{\varepsilon_p/2} - e^{-\varepsilon_p/2}) \tag{3.189}$$

The limit of (3.189) for the duration of unit impulse $\varepsilon \to 0$ is

$$\lim_{\varepsilon \to 0} \left[\frac{1}{\varepsilon_p}\, (e^{\varepsilon_p/2} - e^{-\varepsilon_p/2})\right] = 1 \tag{3.190}$$

The Dirac unit impulse of the length ε_p approaching zero and its spectrum including all possible frequencies from zero to infinity with a constant amplitude are shown in Fig. 3.26.

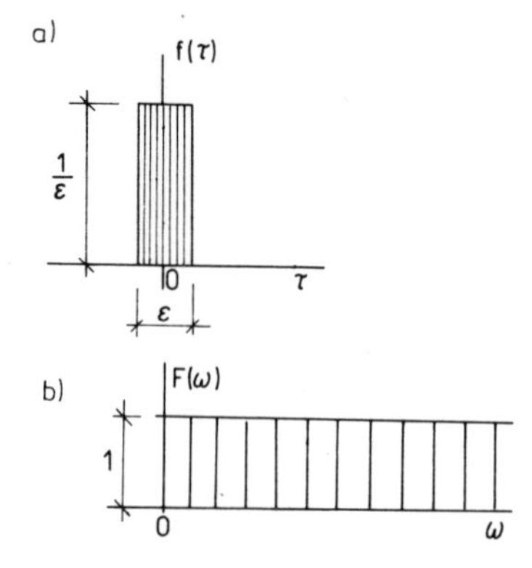

Fig. 3.26 Schematic representation of the Dirac impulse and its spectrum
a – Dirac unit impulse, *b* – spectrum of the Dirac unit impulse

3.4.6 Characteristic Function of Random Variable

The random variable X can be described by its mean value $M[X]$, p.d.f., $f(x)$ and other parameters, but, in addition, it can be handled in some random variable operations, by the mean value of the complex exponential function

$$M[e^{itX}] = g(t) \tag{3.191}$$

which is called the *characteristic function* of this random variable. This representation was introduced by Lyapunov (Ventcel, 1962).

The relationship between the p.d.f. and its characteristic function is, in fact, the direct Fourier transformation, which can be written for a continuous random variable in the form

$$g(t) = \int_{-\infty}^{\infty} e^{itx} f(x)\,dx \tag{3.192}$$

and, for the discrete random variable with probabilities p_k corresponding to random variable values x_k, in the form

$$g(t) = \sum_{k=1}^{n} e^{itx_k} p_k \tag{3.193}$$

The inverse Fourier transform is

$$f(x) = \frac{1}{\sqrt{2\pi}} \int_{-\infty}^{\infty} e^{-itx} g(t)\,dt \tag{3.194}$$

The two basic properties of characteristic functions are:

(1) If X and Y are random variables related by the linear transformation $Y = aX$ where a is a non-random coefficient, then

$$g_y(t) = g_x(at) \tag{3.195}$$

(2) Assume n random variables $X_1, X_2, \ldots, X_n$ and their characteristic functions $g_{x_1}(t), g_{x_2}(t), \ldots, g_{x_n}(t)$, then the characteristic function of the random variable

$$Y = \sum_{k=1}^{n} X_k$$

has the following form

$$g_y(t) = g_{x_1}(t)\, g_{x_2}(t) \ldots g_{x_k}(t) \ldots g_{x_n}(t) = \prod_{k=1}^{n} g_{x_k}(t) \tag{3.196}$$

An example from hydrology and WRS problems could be to determine the p.d.f. of a random variable Z which is the sum of two independent random variables X and Y with known p.d.f.s.

Using the second basic property of the characteristic function we get from $g_x(t)$ and $g_y(t)$

$$g_z(t) = g_x(t)\, g_y(t) \tag{3.197}$$

hence

$$f_y(z) = \frac{1}{2\pi} \int_{-\infty}^{\infty} \mathrm{e}^{-itz}\, g_z(t)\, \mathrm{d}t \tag{3.198}$$

However, the same problem can be solved by other methods, e.g. using the convolution principle (see section 3.5.7).

3.5 LAPLACE TRANSFORMATION

3.5.1 Introduction

The Laplace Fourier integral transformations are the basis for the description of dynamic system behaviour. Just as for the Fourier transformation, the principal idea underlying the Laplace transformation is the transformation of the time argument to the frequency one; the investigation of the input real time function is replaced by an analysis of its spectrum, which is generally a complex function. In many applications the conditions of convergence of the Fourier integral transformation (3.181) are not satisfied and its calculation is not possible. In this sense, the Laplace transformation is more useful, particularly for technical problems.

The use of integral transformations in research of systems behaviour in hydrology, theory of storage and WRS is not yet extensive enough to allow an evaluation of the potential advantages of each method; however, they form an important theoretical basis.

The Laplace transformation and its modifications replace aperiodic functions by the sum of an infinite number of damped or exponentially increasing harmonic components, whereas the Fourier transformation applies the simple (not damped) harmonic oscillations. The existence and applicability of the Laplace transformation is not universal; for the condition of its convergence see section 3.5.2.

3.5.2 Direct and Inverse Laplace Transformation

The direct Laplace transformation (or simply Laplace transformation) is given by the integral

$$L[f(t)] = F(p) = \int_{-0}^{\infty} f(t)\, e^{-pt}\, dt \tag{3.199}$$

where p is the complex variable $p = a + if$,
f – frequency,
c – limit of convergence, i.e. the real component of the complex number $Re(p) = a \geqq c$.

The function $f(t)$ is assumed to equal zero for $t < 0$.

The symbol L [ ] is formally used for the Laplace transformation; the argument of this expression is called the *original* and by its transformation the *transform* is reached. The original $f(t)$ and its transform $F(p)$ form a unique transformation pair (Fig. 3.27).

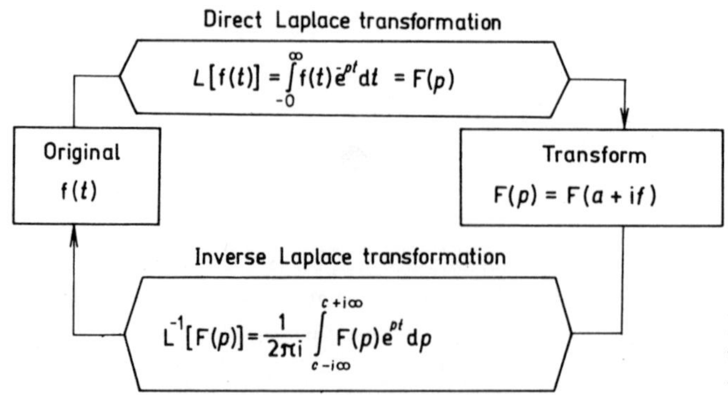

Fig. 3.27 Schematic representation of the direct and inverse Laplace transformation

The inverse Laplace transformation is an integral expression converting a function $F(p)$ back to the original time function $f(t)$ from which it came. The symbol L^{-1} [ ] is used for it

$$\mathrm{L}^{-1}[F(p)] = f(t) = \frac{1}{2\pi \mathrm{i}} \int_{c-\mathrm{i}\infty}^{c+\mathrm{i}\infty} F(p)\, \mathrm{e}^{pt}\, \mathrm{d}p \tag{3.200}$$

A function $f(t)$ is Laplace-transformable if:

(1) the function is piecewise continuous and unique in any time interval Δt (time functions are investigated),

(2) is has to be of exponential order, i.e. for each $\alpha > 0$ the following inequality has to be satisfied (for $M < \infty$)

$$|\lim_{t \to \infty} \mathrm{e}^{-\alpha t} f(t)| \leqq M \tag{3.201}$$

(3) the real component of the complex number a in (3.199) has to be greater or equal to the limit of convergence c.

3.5.3 Laplace–Wagner Transformation

The direct Laplace–Wagner transformation is defined by the integral

$$\mathrm{L}[f(t)] = p \int_0^{\infty} f(t)\, \mathrm{e}^{-pt}\, \mathrm{d}t = F(p) \tag{3.202}$$

and it gives p times greater transforms than the Laplace transformation. It has similar conditions of convergence to the Laplace transformation, viz. $Re(p) = a \geqq c$ where c is the limit of convergence.

Its application to technical problems is helpful as it retains the dimension of the original function. In the Laplace transformation the dimension of transformation is equal to the product of the dimension of the original function and time. The Laplace–Wagner transformation facilitates the checking of the dimension during the calculations.

The inverse Laplace–Wagner transformation is given by the expression

$$\mathrm{L}^{-1}[F(p)] = \frac{1}{2\pi \mathrm{i}} \int_{c-\mathrm{i}\infty}^{c+\mathrm{i}\infty} \frac{F(p)}{p}\, \mathrm{e}^{pt}\, \mathrm{d}p = f(t) \tag{3.203}$$

The Laplace–Wagner transformation is formally identical with the calculus of operations which has been traditionally used in electrical engineering.

Example VIII

a) Calculate the Laplace transformation of the unit step function in the form

$$f(t) = \begin{cases} 1 & \text{if } t \geqq 0 \\ 0 & \text{if } t < 0 \end{cases} \tag{VIII.1}$$

Solution

Using (3.199) and assuming that $f(t) = 0$ for $t < 0$ we can write for $f(t) = 1$

$$L[1] = F(p) = \int_0^\infty e^{-pt}\, dt = \int_0^\infty e^{-(a+i\omega)t}\, dt = \frac{1}{a + i\omega} = \frac{1}{p} \tag{VIII.2}$$

The value $1/p$ is the Laplace transform of the unit step function[1]).

b) Compute the Laplace–Wagner transformation for the same function

Solution

Using (3.202) we can write

$$L[1] = F(p) = p \int_0^\infty e^{-pt}\, dt = p\, \frac{1}{p} = 1 \tag{VIII.3}$$

The Laplace–Wagner transformation of the unit step function is equal to one.

3.5.4 Basic Properties of the Laplace–Wagner Transformation

The basic properties of the Laplace–Wagner transformation for L-functions, i.e. Laplace transformable functions (see section 3.5.2) can be stated in a few points. Each of these properties is, in fact, a theorem, the proof of which can be found in specialized mathematical literature.

a) Transform of a sum of functions

For n functions $f_1(t)$, $f_2(t)$, ..., $f_n(t)$ and their Laplace–Wagner transforms $F_1(p)$, $F_2(p)$, ..., $F_n(p)$, the transform of the sum of these functions is equal to the sum of the transforms of each individual function, i.e. the Laplace–Wagner transformation is

$$L[f_1(t) + f_2(t) + \ldots + f_n(t)] = F_1(p) + F_2(p) + \ldots + F_n(p) \tag{3.204}$$

This property is called the *theorem of linearity of transformation* and it is a consequence of its integral nature.

b) Transform of the function multiplied by a constant

For any constant the function $f_1(t) = \alpha f(t)$ possesses the transform $F_1(p) = \alpha\, F(p)$. Combining these two theorems, we get for the two functions $f(t)$ and $g(t)$ and the constants a, b

$$L[a\, f(t) + b\, g(t)] = a\, F(p) + b\, G(p) \tag{3.205}$$

[1]) The unit step function does not possess e.g. Fourier transformation.

c) Transform of a function with change of time scale

For any $\beta > 0$,

$$L[f(\beta t)] = F\left(\frac{p}{\beta}\right) \tag{3.206}$$

This property is called the *similarity theorem of transformation.*

d) Transformation of delayed function

For each real $\delta > 0$ the function $f_1(t) = f(t - \delta)$ delayed a time δ is shifted to the right on the time axis and $f_1(t) = 0$ for $t \leqq \delta$. Its transform is

$$F_1(p) = L[f(t - \delta)] = e^{-p\delta} F(p) \tag{3.207}$$

For the function $f_2(t) = f(t + \delta)$ shifted to the left on the time axis the transform is

$$F_2(p) = L[f(t + \delta)] = e^{p\delta} F(p) \tag{3.208}$$

e) Transform of derivatives

Assume that $f(t)$ is continuous and differentiable in each interval Δt. Then the Laplace–Wagner transform of its first derivative with respect to time is

$$F^{(1)}(p) = L\left[\frac{d}{dt} f(t)\right] = p[F(p) - f(0)] \tag{3.209}$$

If all the derivatives of the function $f(t)$ at point $t = 0$ take on zero values then the transform of the n-th derivative is

$$F^{(n)}(p) = L\left[\frac{d^n}{dt^n} f(t)\right] = p^n F(p) \tag{3.209'}$$

f) Transform of integrals

Assume, as in the preceding case, that $f(t)$ is continuous and differentiable in each time interval Δt. Then the transform of its integral $\int_0^t f(u)\, du$ is

$${}^{(1)}F(p) = L\left[\int_0^t f(u)\, du\right] = \frac{1}{p} F(p) \tag{3.210}$$

Assuming that the n-dimensional integral exists for every n, that it is continuous and is of exponential order in the sense of the Laplace transformation convergence (section 3.5.2), then its transform is

$${}^{(n)}F(p) = \frac{1}{p^n} F(p) \tag{3.211}$$

g) Convolution of transforms

The transform of the product of two functions $f_1(t)$ and $f_2(t)$ with transforms $F_1(p)$ and $F_2(p)$, respectively, has the form

$$F(p) = L[f_1(t)\, f_2(t)] = \frac{p}{2\pi i} \int_{c-i\infty}^{c+i\infty} \frac{1}{q} F_2(q) \frac{1}{p-q} F_1(p-q)\, dq \tag{3.212}$$

which can be written in a symbolic form

$$F(p) = F_1(p) * F_2(p) \tag{3.212'}$$

and is called *convolution of transforms.*

These properties can be effectively used for the computation of transforms of complicated functions without tedious calculation of the integrals that identify the transformations.

3.5.5 Use of Dictionary of Laplace Transforms for Inverse Transformation

In practical calculation of systems theory problems where the Laplace or Laplace–Wagner transformation is used, the most difficult part is the determination of the original function to the resulting transform defined by the inverse Laplace–Wagner transformation given by the general equation (3.203). Numerical calculation of this integral may be so difficult that the use of the Laplace–Wagner transformation would not be effective or advantageous. The simplest method of dealing with this problem is the application of the dictionary of Laplace–Wagner transforms, i.e. tables of the most frequent functions $f(t)$ and their transforms $F(p)$; they include also the basic properties and theorems of the Laplace–Wagner transformation. They can of course, be used for the determination of the inverse transformations.

The detailed dictionaries of integral transforms are extensive, and they are published separately (e.g. Erdélyi *et al.*, 1954) or are included in the specialized literature (e.g. Ditkin and Kuznyecov, 1954; Šalomoun, 1957). Table 3.16 provides an example of the headings in this dictionary and some of its functions and operations.

The first part includes the transform pairs for

- unit function,
- unit Dirac impulse,
- exponential function and its supplement to one,
- sine function.

In the second part the basic properties and theorems are listed for:

- the direct Laplace–Wagner transformation,
- the inverse Laplace–Wagner transformation,
- the Laplace–Wagner transformation of constants,
- linearity of Laplace–Wagner transformation,
- multiplication of the original function by a constant,
- similarity theorem for the Laplace–Wagner transformation.

Table 3.16 Sample of the dictionary of Laplace-Wagner transformations

I. Laplace–Wagner transformations	
original $f(t)$ $(t \geqq 0)$	transform $F(p)$
1	1
$\delta(t)$	p
$e^{\pm at}$	$p/(p \mp a)$
$1 - e^{\pm at}$	$a/(p \mp a)$
$\sin ft$	$fp/(p^2 + f^2)$

II. Basic transform properties and theorems of the Laplace–Wagner transformations	
original $f(t)$ $(t \geqq 0)$	transform $F(p)$
$f(t)$	$p \int_0^{\infty} f(t)\, e^{-pt}\, dt$
$\dfrac{1}{2\pi i} \int_{c-i\infty}^{c+i\infty} \dfrac{F(p)}{p}\, e^{pt}\, dt$	$F(p)$
c	c
$f_1(t) + f_2(t)$	$F_1(p) + F_2(p)$
$cf(t)$	$cF(p)$
$f(a/t)$	$F(p/a)$

3.5.6 Inverse Transformation by Heaviside Expansion

The Laplace–Wagner transformation often results in transforms in the form of a fraction of two polynomials of the complex argument p. The corresponding original function can be derived, in principle, by four methods; here the method of the residue theorem of the complex variable is used, which results in the Heaviside expansion.

Other methods e.g. partial fraction expansion with terms that are simple transforms, graphic solution of transforms and model technique seem to be more effetive only for special types of problems.

For example, different queue disciplines can be assumed, i.e. the rules determining the order in which customers entering the system are eventually served. Frequent types of queue discipline are FIFO (first in, first out), LIFO (Last in, first out; last come, first served) when the last customer is served when the server has just finished processing the previous one, random serving of customers in the queue, or *priority queue discipline* when the customers are classified into several types and priorities are assigned to these types in decreasing order of importance.

The theory of Markovian processes can be applied to systems $M/M/n$ with a finite queue. A loss of customers is assumed, i.e. the supposition that all customers will be served is not valid.

In solving problems of Markovian systems no major computational difficulties occur. The tasks can be reduced to the solution of a set of linear equations. This computational simplicity rests on the following assumptions:

– steady-state systems (i.e. we are not interested in the initial stages of the system before its stabilization),

– homogeneous Poisson input (exponential arrivals of customers),

– exponentially distributed service time.

These assumptions can be accepted in many practical cases and they are a good approximation of the real situation; however, there are many cases when they are not. Then a non-Markovian queue system with more general properties must be used.

8.2.2 Other Queuing Systems

In Kendall's notation in section 8.1 only the processes in the first row are Markovian. The applicability of other processes is more general, but their solution is more complicated. Often, a reduction to the Markovian case is attempted to allow solution by simpler methods (Zítek, 1969). A well-known procedure is called the *method of the imbedded Markov chain*[1]).

---

[1]) The principle of the imbedded Markov chain is as follows: Assume that the process being investigated is not Markovian, i.e. the condition

$$P\left\{X(t) = \frac{k}{X(s_1)} = j_1,\ X(s_2) = j_2 \ldots,\ X(s_n) = j_n\right\} = P\left\{X(t) = \frac{k}{X(s_1)} = j_1\right\} \qquad \text{(I)}$$

is not met for some choice of numbers n, $t > s_1 > s_2 \ldots > s_n \geqq 0$ and $k, j_1, j_2, \ldots, j_n$. Then a certain sequence of times

$$0 \leqq t_1^* < t_2^* \ldots < t_m^* \ldots,\ \lim_{m \to \infty} t_m^* = \infty \qquad \text{(II)}$$

can be formed in such a way that condition(I) is met if the numbers $t, s_1, \ldots, s_n$ are chosen from this time sequence (II). Therefore, a time sequence is sought where, for any $t = t_M^*$ and for natural n for any n-tuple (vector with n components) $s_i = t^*_{m_i}$ $(i = 1, 2, \ldots, n)$, $0 \leqq s_n < s_{n-1} \ldots < s_1 < t$, this condition is met.

Another method for transformation to Markovian-type processes is based on an approximate, *broader definition of the process.* Apart from reduction to the Markovian process, certain processes can be investigated directly, e.g. by the theory of *semi-Markovian stochastic processes.* Often the *Monte Carlo methods* are used for queue processes that are not Markovian. They are based on a simulation of the investigated systems, and they exemplify some of the experimental methods of probability theory. They can be used for any type of process but they require a good deal of computer time and capacity.

In the literature (Zítek, 1969) the following types of non-Markovian queue systems are listed.

System $M/D/1$

In system $M/M/n$ customer arrivals and service time are random. These assumptions may be an acceptable simplification of reality. In some queue systems model $M/D/1$ with constant service time may be more appropriate: each customer spends a fixed time interval in the serving stage.

System $M/E_n/1$

The advantage of this system as compared with the $M/M/1$ system is the Erlangian distribution of service time. This distribution is determined by two parameters, and therefore it can fit the observed frequency better than the exponential distribution determined by one parameter only. Some other modifications of this system are possible, e.g. the service can be composed of several independent, successive stages; it can include the assumption that more than one customer enters the system at the same time in the form of bulk arrivals.

System $M/G/1$

This system has been developed in a number of alternative forms that try to make the rules for customer arrivals and service time more flexible in order to produce a more general model. In computation the imbedded Markov chain is often used.

$$P\left\{X(t_M^*) = \frac{k}{X(t_{m_i}^*)} = j_i, i = 1, 2, \dots, n\right\} = P\left\{X(t_M^*) = \frac{k}{X(t_{m_1}^*)} = j_1\right\}$$

for all integer non-negative numbers $j_1, j_2, \dots, j_n, k$. The time sequence (II) is determined in such a way that the sequence of random numbers $X(t_m^*)$ for $m = 1, 2, \dots$ forms a Markov chain. If such a sequence (II) has been found, the investigation is restricted to this chain $\{X(t_m^*)\}$ and used only the theory of Markov chains. The resulting properties of this process form the basis for the estimation of properties of the original process. This imbedded chain cannot reflect all the properties of the original process. It gives information concerning the momentary states of the process at times t_m^*. This is often sufficient, e.g. in investigation of the limiting behaviour of the process for $t \to \infty$.

3.5.7 Inverse Transformation by Convolution

The preceding examples illustrated the problem of the Laplace–Wagner transformation of a function, which can be expanded into the sum of simpler functions. In the control of complex systems there are functions with the transform in the form of a product of two or more transforms of simpler functions. In this situation the convolution principle can be applied.

Assume that functions $f_1(t)$ and $f_2(t)$ are Laplace-transformable L-functions and functions $F_1(p)$ and $F_2(p)$, respectively, are their transforms. Then it can be shown that

$$f(t) = \mathrm{L}^{-1}\left[\frac{F_1(p)\,F_2(p)}{p}\right] = \int_0^t f_1(t-u)\,f_2(u)\,\mathrm{d}u = f_1(t) * f_2(t) \tag{3.217}$$

where the notation $f_1(t) * f_2(t)$ means convolution of the original functions $f_1(t)$ and $f_2(t)$ and the integral

$$f(t) = \int_0^t f_1(t-u)\,f_2(u)\,\mathrm{d}u \tag{3.218}$$

is the convolution integral. The inverse transform of $F_1(p) \cdot F_2(p)$ is the function $g(t)$ that can be expressed as a derivation of the convolution integral with respect to t, i.e.

$$g(t) = \mathrm{L}^{-1}[F_1(p)\,F_2(p)] = \frac{\mathrm{d}}{\mathrm{d}t}\int_0^t f_1(t-u)\,f_2(u)\,\mathrm{d}u \tag{3.219}$$

The integral (3.218) is generally solved by the method per partes and there are four alternatives of its application to the function $g(t)$

$$\begin{aligned}
g(t) &= f_1(t)\,f_2(0) + \int_0^t f_1(u)\,\frac{\mathrm{d}f_2(t-u)}{\mathrm{d}(t-u)}\,\mathrm{d}u\,. \\
g(t) &= f_1(t)\,f_2(0) + \int_0^t f_1(t-u)\,\frac{\mathrm{d}f_2(u)}{\mathrm{d}u}\,\mathrm{d}u \\
g(t) &= f_1(0)\,f_2(t) + \int_0^t f_2(t-u)\,\frac{\mathrm{d}f_1(u)}{\mathrm{d}u}\,\mathrm{d}u \\
g(t) &= f_1(0)\,f_2(t) + \int_0^t f_2(u)\,\frac{\mathrm{d}f_1(t-u)}{\mathrm{d}(t-u)}\,\mathrm{d}u
\end{aligned} \tag{3.220}$$

The integrals (3.220) are called *Duhamel integrals*. The integral exists if the derivatives of $f_1(t)$ and $f_2(t)$ are continuous for $t > 0$. The values $f_1(0)$ and $f_2(0)$ are the values of functions $f_1(t)$ and $f_2(t)$ at time $t = 0+$.

3.6 Z-TRANSFORMATION

While the Fourier and Laplace transformations were developed for continuous time functions in system inputs and outputs, the z-transformation is a method intended for discrete sequences often used in hydrology and WRS problems (Delleur and Rao, 1971).

The z-transformation is defined by the transformation pairs:

$$X(z) = \sum_{n=0}^{\infty} x(n\,\Delta t)\, z^{-n} \tag{3.221}$$

and

$$x(n\,\Delta t) = \frac{1}{2\pi i}\oint X(z)\, z^{n-1}\, dz \tag{3.222}$$

where z is a complex variable, i is the imaginary unit, and $x(n\,\Delta t) = x(\Delta t), x(2\,\Delta t)\ldots$ are the terms of the transformed sequence.
The line integral $\oint X(z)\, z^{n-1}\, dz$ is around the unit circle.

The z-transformation is sometimes referred to as an extension of the Laplace transformation, which is related to the possibility of its representation as a discretization of the Laplace integral by summation, which is formally identical with (3.221) with substitution of $z = e^{p\,\Delta t}$.

The transformation relationship between inputs and outputs is simpler for z-transformation than for Fourier or Laplace transformation. The z-transformation is considered in the form

$$Y(z) = H(z)\, X(z) \tag{3.223}$$

or

$$H(z) = \frac{Y(z)}{X(z)} \tag{3.224}$$

where $X(z)$ is the z-transformation of inputs $x(n\,\Delta t)$,
$Y(z)$ – the z-transformation of outputs $y(n\,\Delta t)$,
$H(z)$ – the transfer function.

3.7 BASIC NOTIONS OF CALCULUS OF VARIATIONS

3.7.1 Application of Calculus of Variations in WRS

Mathematical representation of reservoir release control and WRS problems can take the form of the *functional equation* or the *calculus of variations*. That is frequently the case with optimization of control based on Bellman's or Pontryagin's principle of optimality (Partl, 1968).

By these methods a simple queue model can be handled that uses a Poisson exponential distribution or Erlangian distribution. Erlangian distribution[1]) of k-th order is defined by the probability function (with parameters $\lambda > 0$ and $k \geqq 1$ integer):

$$f(x) = \lambda \, e^{-\lambda x}(\lambda x)^{k-1}/(k-1)! \quad \text{for } x \geqq 0$$
$$f(x) = 0 \quad \text{for } x < 0$$

It can be shown that exponential distribution is a special case of the Erlangian distribution for $k = 1$.

In the queuing theory, more complicated models occur and special methods have to be used (Dráb, 1973; Ventcel, 1966; Walter-Lauber, 1975).

The models of queuing systems can be applied for the solution of problems of various technological processes, the rational organization of the health service, traffic problems and of problems of water resources systems (see section 8.6).

8.4 UNRELIABLE SYSTEMS

So far, the queuing systems have been assumed to be reliable. This assumption is frequently unrealistic in practice. Therefore, much work has been devoted to the investigation of unreliable systems (Zítek, 1969; Gnedenko-Kovalenko, 1966; Klimov, 1966).

In models of queuing systems, unreliability is caused by the intermittent failure of servers that cannot, at some stage, serve the customers.

Often, the unreliable systems are modelled as *systems with a priority queue discipline*, as in the following example: from time to time a customer with absolute priority arrives (= failure) and the service of other customers continues only after this customer has left the system (i.e., after restarting the service).

In principle, the following systems with priority queue discipline are possible (characterized by the treatment of the customer that is being served at the moment when a customer with higher priority arrives, i.e., at the moment when the failure occurs):

a) the service of the customer is normally finished (a system with a weak priority),

b) the service is immediately interrupted (a system with strong or absolute priority),

ba) the customer whose service was interrupted immediately leaves the system (without being served)

---

[1]) The Erlangian cumulative distribution function is defined by

$$F(x) = 1 - e^{-\lambda x} \sum_{j=0}^{k-1} (\lambda x)^j / j! \quad \text{for } x \geqq 0$$
$$F(x) = 0 \quad \text{for } x < 0$$

bb) the customer with interrupted service returns to the queue and is then served

bba) the service continues at the point of interruption,

bbb) the service is restarted from the beginning.

Unreliable systems can similarly be classified by these types of failures.

Certain differences exist between unreliable systems and systems with priority discipline:

– in unreliable systems the server stops in the middle of carrying out the repair and no further failure is possible; it corresponds to the assumption that there is only one customer with absolute priority;

– the failure can occur at any time, even if the server is not busy and when no customer is being served. However, there are systems where failure is possible only when the server is busy or when the server is idle, or the service is in any case finished (this last case corresponds to the system with weak priority, the failure is modelled for a customer with weak priority).

In all unreliable systems, new data are added to the basic data describing the organization and operation of the system. These new data characterize the probability distribution of failure occurrence and the probability distribution of repair time. The probability of failure can depend on different factors, e.g., the time from starting the system after the last failure, the time spent by the customer in being served, the number of customers that have been served after the last failure, the total service time, the time that was necessary for the last repair, etc. Using these data, the probability of service without failure, the probability distribution of the possible failure, etc., can be determined. The problems are far more complicated in cases of unreliable systems with a larger number of servers.

8.5 QUEUING SYSTEMS SIMULATION

The principles of the simulation of queuing systems have been explained in the references (Buslenko and Shreyder, 1961; Zítek, 1969; Saaty, 1961; Votruba *et al.*, 1974; Kaufman and Cruon, 1961, etc.). Only the basic ideas of this approach are mentioned in this section.

Although some outstanding results have been achieved in the queuing theory, this theory is not (and probably never will be) able to solve the problems that arise in practical life. The models oversimplify the complex problems of reality, or if the models are a good approximation of reality they are mathematically intractable as there are no effective analytical methods for solution. If a rough approximation is not acceptable, simulation methods and mainly the Monte Carlo simulation models have to be used for this more complicated case. The main advantage of these methods is their universality; they can be used for any process and any model. However, they need a lot of computation and computer time and great memory capacity.

the following holds:

$$|\boldsymbol{I}[y(x)] - \boldsymbol{I}[y_0(x)]| < \varepsilon \tag{3.228}$$

b) Linearity (homogeneity) of functional

A functional is linear (homogeneous) if it satisfies the conditions:

$$\boldsymbol{I}[cy(x)] = c\boldsymbol{I}[y(x)] \tag{3.229}$$

(c is an arbitrary constant) and

$$\boldsymbol{I}[y_1(x) + y_2(x)] = \boldsymbol{I}[y_1(x)] + \boldsymbol{I}[y_2(x)] \tag{3.230}$$

c) Variation of functional

The concept of variation of a functional is based on the idea of variation of an argument (function). The increment of the argument is

$$\delta y = y(x) - y_1(x) \tag{3.231}$$

and the corresponding increment of the functional is

$$\delta \boldsymbol{I} = \boldsymbol{I}[y(x) + \delta y] - \boldsymbol{I}[y(x)] \tag{3.232}$$

which can be written in the form

$$\delta \boldsymbol{I} = \boldsymbol{I}[y(x); \delta y] + \beta[y(x); \delta y] \max |\delta y| \tag{3.233}$$

The first term $\boldsymbol{I}[y(x); \delta y]$ is a linear (homogeneous) functional in δy and is called the *variation of the functional* $\boldsymbol{I}[y(x)]$ or the main component of the increment of the functional.

The second part of the expression for the increment of the functional given, the functions $\beta[y(x); \delta y]$ converges to zero if the distance $\max |\delta y| \to 0$. It is obvious that the variation of the functional is analogous to the differential of functions.

Define function $\varphi(\alpha)$ of one variable (parameter) α:

$$\varphi(\alpha) = \boldsymbol{I}[y(x) + \alpha \delta y] \tag{3.234}$$

where $y(z)$ and δy are constants.

The variation of the functional $\boldsymbol{I}[y(x)]$ can be expressed as the derivation of the function (3.234) with respect to α at the point $\alpha = 0$, i.e.

$$\frac{\partial}{\partial \alpha} \boldsymbol{I}[y(x) + \alpha \delta y]_{\alpha=0} \tag{3.234'}$$

d) Extrema of functional

The functional assumes an extremal value for a function $y = y_0(x)$ if the corresponding variation of functional $\delta \boldsymbol{I}$ is equal to zero. The notion of the extrema of the functional can be specified and made precise by the order of the distance of curves $y(x)$ and $y_0(x)$.

The functional $\boldsymbol{I}[y(x)]$ takes on its *maximum value* if its values are not higher on any curve $y = y(x)$ in the neighbourhood of the curve $y = y_0(x)$ than on this curve, i.e.

$$\delta \boldsymbol{I} = \boldsymbol{I}[y(x)] - \boldsymbol{I}[y_0(x)] \leqq 0 \tag{3.235}$$

If the increment of functional $\delta \boldsymbol{I}$ is equal to zero for the functions $y = y_0(x)$ only, then this maximum is called a strict maximum.

On the other hand, the functional $\boldsymbol{I}[y(x)]$ takes on its *minimum value* for the function $y_0(x)$, if $\delta \boldsymbol{I} \geqq 0$. The extrema of functionals can be classified further into weak and strong.

The functional possesses a *weak extremum* if it occurs on a curve $y = y_0(x)$, which has a small distance of the first order for all curves $y = y(x)$.

The functional possesses a *strong extremum* if it occurs on a curve $y = y_0(x)$ which has a small distance of zero order for all the curves $y = y(x)$. It can be concluded that a necessary condition for the existence of the extremum of a functional $\boldsymbol{I}[y(x)]$ of the function $y = y_0(x)$ is satisfaction of the relationship

$$\frac{\partial}{\partial \alpha} \boldsymbol{I}[y_0(x) + \alpha \delta y] \big|_{\alpha=0} = 0 \tag{3.236}$$

Hence

$$\frac{\partial}{\partial \alpha} \boldsymbol{I}[y(x, \alpha)] \big|_{\alpha=0} = 0 \tag{3.237}$$

where $y(x, \alpha)$ is a family of curves starting with the curve $y = y_0(x)$ for $\alpha = 0$ and ending with the curve $y = y_0(x) + \delta y$ for $\alpha = 1$.

e) Functionals of several functions and multivariate functions

These types of functional have the form

$$\boldsymbol{I} = \boldsymbol{I}[y_1(x), y_2(x), \ldots, y_n(x)] \tag{3.238}$$

and

$$\boldsymbol{I} = \boldsymbol{I}[z_1(x_1, x_2, \ldots, x_n); z_2(x_1, x_2, \ldots, x_n); \ldots; z_m(x_1, x_2, \ldots, x_n)] \tag{3.239}$$

For example, for the functional $\boldsymbol{I} = \boldsymbol{I}[z(x, y)]$ the extremum can be found using the condition

$$\frac{\partial}{\partial \alpha} \boldsymbol{I}[z(x, y) + \delta z] = 0 \tag{3.240}$$

where $\delta(z)$ is the increment of the function $z = z(x, y)$.

3.7.5 Two-Fixed-Point Boundary-Value Variation Problem

The simplest type of functional is

$$\boldsymbol{I}[y(x)] = \int_{x_0}^{x_1} F\left[x, y(x), \frac{dy(x)}{dx}\right] dx \tag{3.241}$$

The queuing theory can be used for the similar problem of the operating policy for release, where continuous variables are used and the input is a negative exponential distribution.

Let us investigate the possibility of the direct solution of a continuous alternative of the set of equations (8.10), i.e., the equation (7.22) of section 7.4.2. For an infinite V the set of equations (8.9) resp. the equations (7.22) become

$$g(x) = f(x) \int_0^M g(x)\, dt + \int_M^{M+x} f(M + x - t)\, g(t)\, dt \tag{8.11}$$

where $g(x)$ is the probability density function of variable Y_t, $f(x)$ is the probability density function of X_t. For this equation the Laplace transform can be used for the cumulative probability function of variable X_t and $V_t = X_t + Z_t$

$$F(y) = \int_0^y f(t)\, dt, \qquad G(y) = \int_0^y g(t)\, dt \tag{8.12}$$

and in simple form

$$G(y) = \int_0^y G(M + y - t)\, dF(t) \tag{8.13}$$

This type of integral equation is well-known from the queuing theory. Lindley, 1952, derived the solution and proved that the steady-state solution exists if the mean input value is less than M, and if the initial state is zero. The distribution of variable Y_t converges on the stationary state of the process.

Smith, 1953, has shown that another useful analogy between infinite reservoirs and the queuing theory is possible, and it can be seen in the following case of the queuing theory. A queue is assumed with one server only and with regular interarrival time M. In Lindley's general theory these intervals are random, in this application it is possible to use fixed intervals. If no queue has been formed, the customer is served immediately, otherwise he must wait till all the customers in front of him are served.

The service time, denoted X_i, is a random variable with the cumulative distribution function $F(x)$, X_i being mutually independent and independent of queue characteristics. The time interval necessary for serving the whole queue that had formed immediately before the arrival time t_i, is denoted Z_i. This is the waiting time of a new customer in the queue between the time of his arrival and the beginning of his service. $Z_i + X_i$ is then the total time that a new customer spends in the system, i.e. the time interval between the arrival of a customer and the end of his service. The whole time interval necessary for serving the queue immediately before the time t_{i+1} is Z_{i+1}. It is equal to

$$\begin{aligned} Z_{i+1} &= Z_i + X_i - M \quad \text{if } Z_i + X_i - M > 0 \\ Z_{i+1} &= 0 \quad\quad\quad\quad\quad\;\; \text{if } Z_i + X_i - M \leqq 0 \end{aligned}$$

This queuing problem can be applied for an infinite reservoir, assuming that X_i represents the random inflow X_t into reservoir and reservoir storage just before the withdrawal is represented by the waiting time of the last customer. The withdrawal M from reservoir is represented by the reduction in waiting time of the new customer during the time interval (t_i, t_{i+1}). In this model (Smith, 1953) the release is not realized at once, but in regular steps. The assumption of regular withdrawals does not change the underlying theory, as the equation (8.13) is valid for this example.

It is apparent that the way the problems of reservoir operation and queuing correspond to each other is quite different from the previous case, where the length of the queue taken as a discrete variable correspond to storage of the reservoir, and bulk service was assumed. In this case, the service of a single customer is assumed and storage of reservoir is modelled by the waiting time of a new customer in the queue.

The numerical solution of this case is that of equation (8.13). It was published by Lindley, 1952, for cases where the input variable has a Pearson type III probability distribution. The resulting probability distribution (Moran, 1959) is the distribution in this case. From a mathematical point of view, this latter case of the application of the queuing theory for WRS is a continuous analogue of the previous case, often called Bailey's queuing theory with bulk service. However, the correspondence between the WRS model and queuing theory is different in each case.

Gani and Prabhu, 1957, derived the solution of equations for the infinite reservoir for the case when storage in the reservoir has a negative exponential distribution.

Moran, 1959, Kendal, 1957, and others described a number of possible approaches, the latter used the operation of reservoirs with continuous time. Input is a continuous inflow, withdrawal is also continuous, and takes place until the reservoir is empty. Both finite and infinite reservoirs were investigated. In addition, Kendall, 1957, investigated a non-stationary case of distribution of storage: the initial state in time $t = 0$ is given by storage y and the probability distribution of time interval T in which the reservoir is emptied for the first time is looked for. This method approaches the theory of common risk. All these studies are stimulative and use sophisticated mathematical theory, but the stage of development is not sufficient for its practical application. In addition, most models use an unrealistic assumption that the inflow is an additive process with a Pearson type III distribution.

The numerical calculation in these models is similar to that in the inventory theory, illustrated by a simple example in section 7.4.1.

Another approach to the application of the queuing theory for reservoir operation was described by Chorafas, 1965. Binomial probability distribution is assumed for random inflow to the reservoir in time-sequenced stages. The release from the reservoir, in each stage, is predetermined, where there is an empty reservoir, or where spillage has occurred. These models yield relationships among the following quantities: mean values and variances of reservoir inflows, reservoir storage, the chosen values of reservoir release, probability of deficits and spillage.

Solution

Functional (X.1) is of the type (3.241). Therefore the following partial derivatives are calculated

$$F_y = \frac{\partial}{\partial y} F(x, y, y') = -2y$$

$$F'_y = \frac{\partial}{\partial y'} F(x, y, y') = 2y' \tag{X.3}$$

and the Euler equation is written in the form

$$F_y - \frac{\mathrm{d}}{\mathrm{d}x} F_{y'} = 0 \tag{X.4}$$

$$-2y - \frac{\mathrm{d}}{\mathrm{d}x}(2y') = 0$$

hence

$$y'' + y = 0 \tag{X.5}$$

The solution of the differential equation (X.5) is the function

$$y = C_1 \cos x + C_2 \sin x \tag{X.6}$$

Using the boundary values (X.2) we get

$$C_1 = 0; \quad C_2 = 1 \tag{X.7}$$

and the extremal function of the function to be found is

$$y = \sin x \tag{X.8}$$

3.7.6 Principle of Two-Free-Point Boundary-Value Variation Problem

In these problems it is assumed that one or both terminal points $[x_0, y_0]$; $[x_1, y_1]$ in a two-dimensional problem are free. The class of feasible functions including the extremal function is thus enlarged. The function to be found is again the solution of the Euler equation, which contains two free constants since the differential equation is of the second order. In a two-free-point problem one or both boundary-value conditions (3.250) are not applicable. The values of the constants are calculated using condition that variation of the functional has to be equal to zero, i.e. $\delta \boldsymbol{I} = 0$.

Assume a fixed point $[x_0, y_0]$. Then all extremal functions will pass through this point and form a bunch of curves (Fig. 3.31a).

Assume that the point $[x_1, y_1]$ moves on the curve

$$y_1 = \varphi(x_1) \qquad \text{(see Fig. 3.31b)} \tag{3.251}$$

In such a case, an additional condition is applied in the form of approximate equality of the variation of the boundary values δy_1 and the differential of the function $\varphi(x_1)$, i.e.

$$\delta y_1 \approx \varphi'(x_1)\, \delta(x_1) \tag{3.252}$$

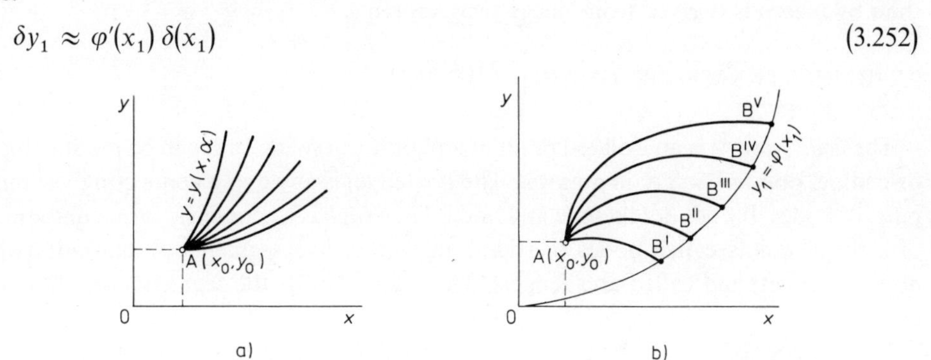

Fig. 3.31 Bunch of extremal curves in the two-free-points boundary-value problem
a – fixed terminal point $[x_0, y_0]$, b – fixed terminal point $[x_0, y_0]$ and the point $[x_1, y_1]$ on a curve $y_1 = \varphi(x_1)$

3.7.7 Application of Calculus of Variations for Stochastic Optimization Problems

Use of the calculus of variations for optimization problems and the differences in its use for deterministic and stochastic optimization problems are illustrated by a case study of water resources and hydroelectric systems.

Kartvelishvili, 1970, described a case study of the long-term operating policy of a single-purpose reservoir used for hydroelectric power generation in a system of thermal and water power plants.

The continuity equation relating reservoir inflow $Q_1(t)$, release $Q_2(t)$, quantity of water in reservoir W and any change in that quantity in the form of a derivative $\mathrm{d}W/\mathrm{d}t$ can be written in the form

$$Q_2(t) = Q_1(t) - \frac{\mathrm{d}W}{\mathrm{d}t} \tag{3.253}$$

If no spills occur (all the release passes through the turbines) the average power production of the water power plant in a particular time interval is the function

$$N(t) = N[Q_2(t); W] = N\left[Q_1(t) - \frac{\mathrm{d}W}{\mathrm{d}t};\ W\right] \tag{3.254}$$

modification called PERT/COST is used for planning the costs of the realization of these programmes and projects.

Together with these two basic methods, the method of *Resource Allocation and Multi-Project Scheduling* (RAMPS) is also used. The allocation of limited resources in space and time in different branches (WRS included) is, however, designed by different methods, e.g., by dynamic programming with relatively more possibilities than by methods derived from linear programming.

9.2 BASIC TERMS OF GRAPH THEORY

The *node* (vertex) is an isolated point in a plane or in space, or it can be the starting, or ending, point of an arc in a graph. The *arc* (edge) is any curve connecting various pairs of nodes. It does not include any node. The *graph* is a geometric form consisting of a set of points called nodes (vertices) and set of line segments joining any two nodes (vertices) and called arcs (edges). These arcs can be the segments of a line or of a curve.

In the theory of graphs, the definition of a graph G is possible without its geometrical representation as a diagram (as used in this chapter) by the set of nodes N and a set of arcs $G \equiv (N, \mathrm{A})$.

In the graph theory different kinds and types of graphs are investigated and only a part of them, under certain conditions, are useful for the schematic representation of WRS or critical path analysis of complex processes.

The *null graph* consists of a finite number of nodes that are isolated, i.e., they are not connected by arcs.

The *simple graph* is a graph that has neither self-loops (i.e., arcs having the same node as both their end nodes) nor parallel arcs (i.e., more than one arc associated with a given pair of nodes). The graph that is not simple is called a general graph (multi-graph).

The *complete graph* is a simple graph in which every pair of nodes is connected by an arc.

The *directed graph* (*digraph*, orientated graph) is a graph where every arc is represented by a line segment between the node n_i and n_j with an arrow directed from n_i to n_j.

Two graphs G_1 and G_2 are called *isomorphic* if there is a one-to-one correspondence between their nodes and arcs, such that a pair of nodes in the graph G_1 are joined by arcs, if and only if, the corresponding pair of nodes in the graph G_2 are also joined. In directed graphs the orientation is also maintained.

Planar graphs have some geometric representation which can be drawn on a plane in such a manner that no two of its edges intersect (they only meet in the node). When a node n_i is an end (or starting) node of an arc a_j, n_i and a_j are said to be *incident* to each other. The number of arcs incident to a node is called the degree of this node.

The number of arcs is then equal to half the sum of the degrees of all the nodes.

The graph is termed *connected* if there is at least one chain between every pair of nodes in this graph. The *chain* (walk) is a broken line in the graph, such that no arc is traversed more than once.

A closed chain is called a *cycle*. A cycle in which no node (except the initial and final node) appears more than once is called an *elementary cycle* (circuit).

If a closed chain in a graph traverses every arc of the graph exactly once, the chain is called a *Euler line* (named after Euler, who in 1736 developed the basis of the graph theory). Each graph that contains this line is called a *Euler graph*, and it has some special properties (e.g., it is connected).

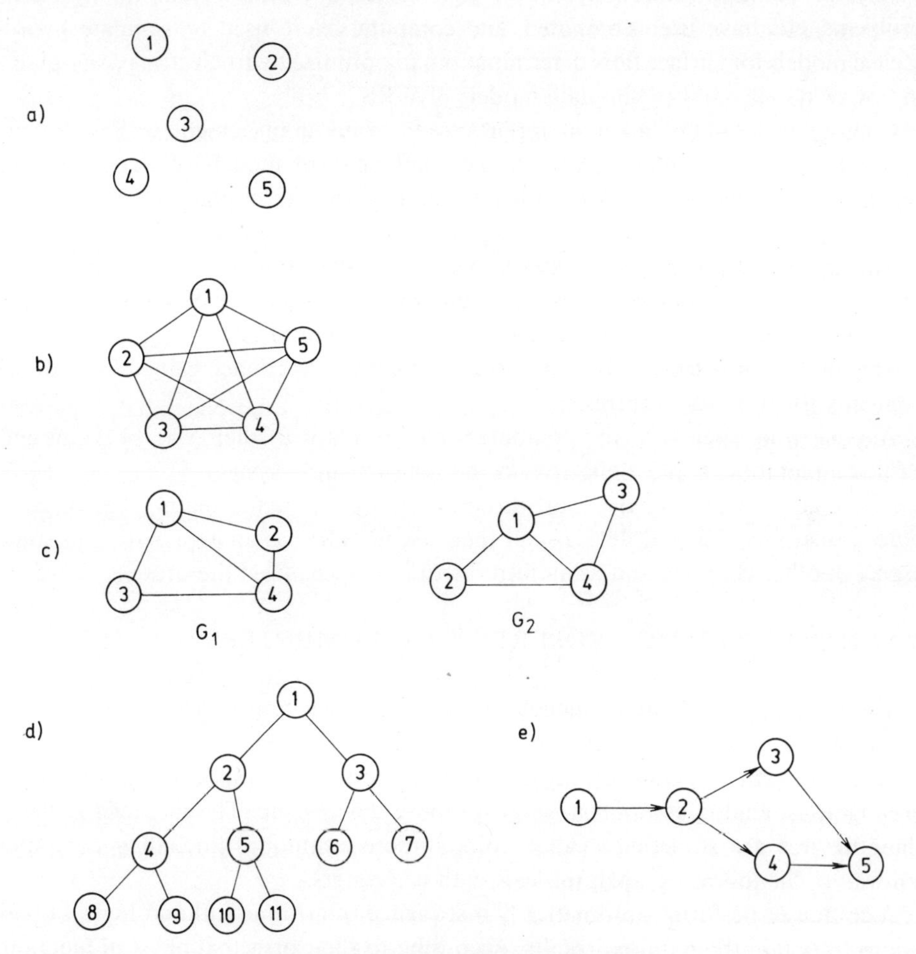

Fig. 9.1 Examples of some types of graphs
a – a null graph, *b* – a complete graph, *c* – isomorphic graphs G_1 and G_2, *d* – a tree, *e* – a directed graph (digraph)

4 APPLICATION OF COMPUTERS IN WRS

Processing of input data, and the design and operation of WRS require the use of powerful computer systems. At the beginning of the sixties data processing was the main example of the application of computers in water management (e.g. hydrological data processing, computation of statistical characteristics, data sorting, etc.). Lately, problems of reservoir storage e.g. release operations, complex hydraulic problems, etc. have been computed, and computers were used to simulate hydrological models for surface flow determination, to optimize hydroelectric power plants in power systems and to simulate models of WRS.

Computers are of fundamental importance for water management, with increasing application of the systems approach, e.g. mathematical models of water resource management in basins and, for the whole country, long-term planning, predictions, etc.

The use of computers necessitates knowledge of their function and the principles of programming, and particularly, a profound understanding of the professional problems involved in a proper analysis and formulation of tasks.

Theoretical knowledge should be accompanied by experience gained from direct contact with computer centres.

Anyone who intends to use computers has to adapt to their way of "thinking" as this adaptation is most effective in the man-machine system.

This chapter is intended as an introductory survey of the main types of computers, their facilities and devices and their use in WRS, as an explanation of some aspect of other chapters and as an introduction to specialized literature.

4.1 CHARACTERISTIC PROPERTIES OF COMPUTERS

The computation facilities include devices for easier and quicker computation, for mechanization and automation of computation and data processing. They include not only the general-purpose digital computers but also some tables, diagrams, nomographs, analog computers, single-purpose devices, etc. The methods of using these devices are sometimes called computation techniques (including computer software). The following analysis deals with computers.

A computer performs automatically a sequence of arithmetical and logical operations to obtain the required results. According to their principal physical function, computers are categorized into analogue, digital and hybrid computers. In WRS, digital computers are the most frequently used.

Analog computers

Electronic analog computers (Plander, 1969) represent the physical processes in models, and actual systems by analogous mathematical relations. The electronic computer units reflect the physical variables and numerical information by some analogous variable, e.g. by electrical force (voltage). One computing unit is necessary for each mathematical operation that takes place in an analog computer. Operational voltage is limited and this limit cannot be exceeded. Therefore the computer uses "fixed-point" methods of arithmetic.

There are two types of units:

– linear units (summator, inverter, integrator, coefficient potenciometer, etc.),

– non-linear units (multiplier, generator of functions, comparator, etc.).

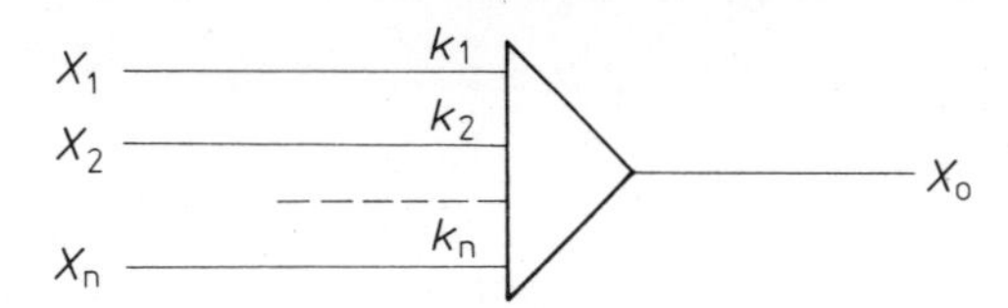

Fig. 4.1 Outline of the summator

Figure 4.1. shows a diagram of a summator, which produces a negative sum of input values multiplied by digital constants ($x_0 = -\sum_{i=1}^{n} k_i x_i$).

To set a task for an analog computer, a mathematical model of the system is first set up. The problem is formulated using physical laws so as to describe the system behaviour by mathematical relationships. By connecting the computer units, circuits are formed that carry out the computation of the problems.

Programming analog computer involves the preparation of the problem for the computer, drawing a block diagram and a scheme for the interconnection of computing units, the design of the experiments with the computer and the control of these experiments. The program, therefore, consists of formalized information on the total set of computing units and their connection, and on the dependent and independent variables which are physical realizations of variables in the problem analysed.

Some characteristic properties of the analog computer are as follows:

– dependent variables are expressed and processed in continuous form (analog computers are particularly suitable for problems of differential and integral calculus),

– programming is relatively easy,

– the computer provides a possibility for experiments on the model by variation of coefficients by potentiometers,

– the accuracy of computation is limited by the quality of computing units (0.1 to 0.01% of the full range of the maximum voltage),

– mathematical operations are carried out directly, possibilities for logical decisions and storage are limited,

– the computation is parallel, i.e. in all computer units at the same time.

Hybrid computers

Hybrid computers represent a purposeful combination of analog and digital computers. The advantages of both types of computers are utilized by employing the digital and analog sections with analog-digital and digital-analog converters. Hybrid computers were used in some optimization problems and it should, therefore, be possible to use them for WRS problems.

Digital computers

Digital computers model the algorithms by a sequence of operations that lead to the solution of the problems of the given type. These operations involve mathematical operations with numbers with a finite quantity of digits. Using the two basic physical states, not only numbers can be expressed but also alphameric and special information, logical constants and variables, etc. The possibilities of the digital computer are greater than those of the analog computer; digital computers can process any information (also non-mathematical) if the adequate algorithm is known.

Some characteristic properties of digital computers are as follows:

– the processing of information is performed in discrete steps sequentially,

– the computers are limited to the rudimentary arithmetical operations of addition, subtraction, multiplication and division; more complicated operations are performed by methods of numerical mathematics;

– the computers have the ability to carry out the logical operations of comparison, and of data storage on a large scale;

– they can use floating-point arithmetic;

– the accuracy of the computers depends only on the number of bits used for coding numerical information.

In modelling WRS, it is mainly digital computers that are used (further only "computers").

Computers can be classified in different ways e.g. according to "generation". Computers of a certain generation are characterized by certain types of construction elements. Computers of the first generation used vacuum tubes (Ural, Eniac), the second generation used transistors (Minsk, ZPA, NCR-Elliot, etc.) and the third generation uses integrated circuits (IBM 370, Siemens 4004, etc.). The principle and basic parts of computer structure are shown in Fig. 4.2. The input device serves for the input of the program and processed data into the main storage; at the same time information is transformed into a form that can be used in the computer. The usual types of input devices are:

– the punched card reader, where the carrier of information is a punched card with 80 or 90 columns,

– the paper tape reader, where the program or data are punched in 5, 7 or 8 punching positions,

– the console printer-keyboard, used for manual control of the computer.

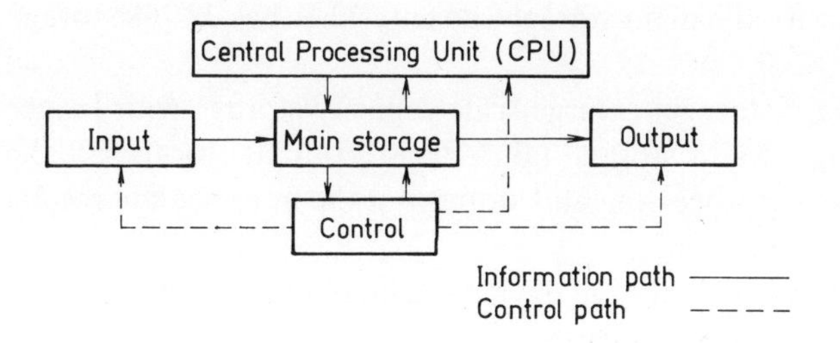

Fig. 4.2 Digital computer organization

The output devices provide the output of processed information from the computer. In general, the following output devices are used:

– a lineprinter (printer) for printing alpha characters, digits and special characters,
– a console printer-keyboard,
– a paper tape punch,
– a card punch,
– a display unit with vector graphics.

The control unit ("control") selects the instructions of the program sequentially from the main storage and controls their operation in the computer. It is the function of this equipment to interpret orders and to synchronize and direct the operation of all the units that make up the computer. Controls are required inside a computer system to (a) tell the input devices what data to enter into main storage and when to enter it, (b) tell the main storage where to place these data, (c) tell the CPU what operations to perform, where the data are to be found, and where to place the results, (d) tell them what file devices to access and what data to access, and (e) tell them what output devices the final results are to be written on.

The central processing unit performs arithmetical and logical operations in accordance with the algorithm of instructions from the control. The processor manipulations are performed one operation at a time with intermediate results being put back into main storage.

Storage is used to input, hold and output information about the program, input data, intermediate results and final results. There are two types of storage:

– main storage (primary, operational storage) provides the system with directly addressable fast-access storage data. Both data and programs must be loaded into main storage from input devices before they can be processed. The main storage consists of memory locations (cells) that can be handled directly; they can be identified by numbers called addresses. A unit of information held in a cell ("word") is formed by an ordered sequence of characters processed as a whole. The storage capacity is given in words or bytes (e.g. in the IBM system, 1 byte = 8 bits, 1 word – 32 bits = 4 bytes) that can be held in this storage. This capacity is expressed in K-1024 units, e.g. a computer (IBM) with the main storage capacity 32 K has 32 768 storage locations (bytes).

– external storage (auxiliary storage) is large-scale capacity storage with longer access time than main storage. Main storage takes from external storage all the information necessary for further processing and stores in it the intermediate results that are not necessary for the time being.

The most commonly used external storage devices include magnetic discs, magnetic tapes, magnetic drums and punched cards.

Computer devices and their construction elements are often called "hardware". The term "software" is used for programs or data not forming part a computer but used for its operation. The growing demands for capacity and the complexity of hardware require adequate software; nowadays the costs of both these parts are approximately equal. The software includes the system reference library programmes, subroutine packages (e.g. Scientific Subroutine Package), operating system and system control programs, a supervisory program, application programs, diagnostic and other service programs.

Parameters of Computer Hardware Used for WRS

For the computation of WRS problems computers of the JSEP series (Joint System of Electronic Processors) and some computers imported from western countries were used. The JSEP electronic computers are third-generation computers. They are produced in series of seven main compatible types, some of which have been used for WRS problems, i.e. EC (Electronic Computer) 1010 (made in Hungary), EC 1020 (the Soviet Union, Bulgaria), EC 1021, 1025 (Czechoslovakia), EC 1030 (the Soviet Union), EC 1040 (GDR), EC 1050 (the Soviet Union). The capacities and speed of CPU increase with higher number; the main advantages are the possibility to connect optimal devices (storage devices, printer, etc. in different computer system configuration) and compatibility of programs.

In Table 4.1. a brief survey of the basic parameters of JSEP computers is given:

For WRS problems, the most commonly used third-generation computers imported from western countries in the sixties included: IBM 360 (USA) and Siemens 4004 (Germany), ICL System 4 and ICL 1900 (both U.K.).

The basic parameters of IBM 360/40 are as follows:

CPU capacity	64 K
type of main storage	ferrit
access time	1.25 μs
external storage	magnetic tapes magnetic discs
cards read/punch	reading 1 000 cards/min punching 200 cards/min
printer	1 100 lines/min
arithmetical speed	7.5 μs (addition in fixed-point arithmetic)

Most computer programs for WRS problems in case studies of the General Water Plan used National Elliott 4120, Elliott 503 and IBM 360/40 computers.

Table 4.1 The main parameters of JSEP computers (the usual computer configuration)

Type	EC 1020	EC 1030	EC 1040	EC 1050[4])
CPU capacity KB	64 – 256	128 – 512	256 – 1024	128 – 1024
CPU cycle time μs	2	1.25	1.35	1
Punch card reader[1])	EC 6012	EC 6012	EC 6012	EC 6012
Paper tape reader[1])	EC 6022	EC 6022	EC 7902	EC 6022
Card punch[2])	EC 7010	EC 7010	EC 7010	EC 7010
Paper tape punch[2])	EC 7022	EC 7022	EC 7902	EC 7022
Printer[3])	EC 7030	EC 7030	EC 7031	EC 7030

[1]) Speed of reading EC 6012 500 punched cards/min., EC 6022 1500 characters/sec.

[2]) Speed of punching EC 7010 100 punched cards/min., EC 7022 150 characters/sec., EC 7902 100 characters/sec.

[3]) Speed of printing EC 7030 890 lines/min., EC 7031 900 lines/min., Number of symbols per line EC 7030 128; EC 7031 120.

[4]) U.S.S.R.

4.2 USE OF COMPUTERS IN MODELLING

4.2.1 Task Algorithmization

In computer operation there are some basic differences between the way a human being works and the way a computer works. A human being uses his ability and experience to change or complete the incorrect or missing rules determining data processing and computation. Therefore, only rough rules for data processing are necessary; he uses them and, if necessary, he corrects and complements them. Every detail of a task for a computer, however, has to be prepared and logically treated. Every situation that might occur during computation must be taken into account in advance. Therefore the task is divided up into relatively simple steps that can be expressed by the program used in the computer. Before programming of the computer two preliminary stages are necessary:

– formulation of the problem – i.e. a clear definition of the task and its objectives, and the form, content and type of output data,

– analysis of the problem, i.e. investigation of the logical structure of the problem; this analysis includes the exact rules governing the processing of the input data in order to obtain the output data.

The decision to investigate the behaviour of WRS by, for example, a simulation model is often based on the results of these two stages. The final stage of analysis, i.e. algorithmizing the problem, is necessary before we start programming. The processing of data is defined in sufficient detail, i.e. the algorithm of the problem is developed. There are many references (e.g. Markov, 1954) to more detailed investigation of the theory of algorithms. Markov defined an algorithm as an exact description defining the computation process that starts with the variable input data and ends with desired results.

The algorithm must be:

determinative; it must be exact, precise and comprehensive and reproducible by anybody; there should be no doubts as to the further steps of data processing,

generic; it can process variable input data, i.e. more tasks of the same type can be performed by it,

resultant; it must give the required result after a finite number of operations.

This formulation explains rather than defines the concept of an algorithm. The mathematical notion of an algorithm is sometimes taken as a principle (axiom) that cannot be transformed into simpler ideas.

In WRS analysis, the concept of an algorithm is defined as a generic, accurately determined description of procedures for a single task or for a group of analogous tasks. These procedures consist of elementary steps that should be sequentially carried out to produce the output information; the choice of the elementary steps of the process is arbitrary. There is no need for an exact knowledge of the theory of

…ally, a graphic aid is used. Graphic …and standard symbols aid the imagination, and understanding and facilitate intelligibility. They help orientation in complex problems, and sometimes they are indispensable for effective processing of information.

Graphic communication must:

– accurately and unambiguously represent complex logical problems,
– allow easy modification arising from changed conditions,
– be independent of computer type,
– provide accurately defined documentation.

The logical structure of a problem can be expressed in many ways. The most widely used form is the *flowchart*. For their own internal purposes and for dealing with customers, each of the major computer producers has, over the years, developed, adopted and published flowcharting conventions. Many committees were appointed to standardize these conventions. In the United States, this resulted in the ANSI standard, 1970 (American National Standard Institute). The 1970 revision extended the standard to match more closely that of the ISO (International Standard Organisation). The Czechoslovak standard ČSN 369 030 is similar to both these standards. The ANSI standard 1970, defines a set of graphic outlines, termed symbols. The flowchart symbols for information processing cover two major situations:

(a) representation of systems without indicating the nature of the component algorithms. It describes the system structure and relationships between activities. Other terms are procedure chart, run diagram.

are written in words or symbols or are indicated as a group ... of the outlines is not specified, and it is determined by the size of the content. The shape of the outlines should be maintained, i.e. the ratio of the width to the height and the general geometrical configuration. To indicate the sequence of operations or data, flowlines connect the outlines. To increase comprehensibility, horizontal and vertical flowlines with two preferred directions (top-to-bottom or left-to-right) are used in the positioning of outlines, and, if unambiguous, the open arrowhead can be omitted.

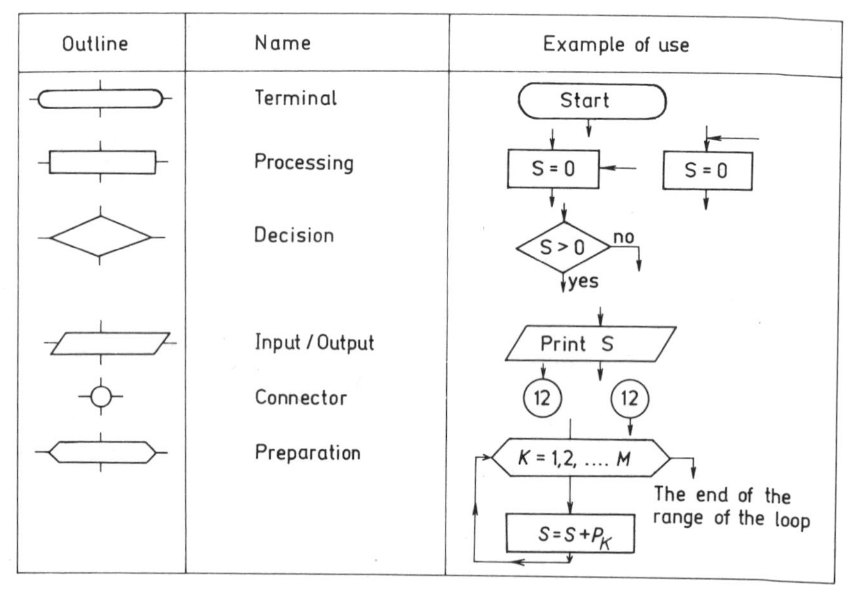

Fig. 4.3 Some outlines of the flow-charts

Figure 4.3. illustrates the outlines most frequently used in WRS modelling; they are sufficient for an understanding of the principles of flow charts:

Terminal serves to indicate a terminal point in the program – the beginning, and/or some break (start, stop, halt, delay, pause) in the usual line of flow. The terminal outlines in the first two situations are applied not only for the whole program but also for some self-contained portions, e.g. subroutines.

Processing is the general purpose outline that represents the performance of any operation (or group of operations) that is used for data transformation, data movement, or logical operations. The flowlines can enter this outline from any side, but it has only a single output.

Decision indicates comparison, decision, testing or switching operations, which determine or select among a variety of alternative flows (sequences of operations). The number of flowlines leaving a decision outline must always be greater than one (usually two or three), and these flowlines must be assigned (e.g. "yes", "no").

Input/output indicates an input or output operation, i.e. input of information for processing, recording or storing the processed information (output). It is defined for use irrespective of media format, equipment and timing. Some specialized outlines may be substituted for this outline, or the type of input/output medium or equipment may be described in words in this outline.

Connector describes the transition to another part of the flowchart (outconnector) or from it (inconnector). There are always at least two connectors, the related connectors being assigned the same symbol (letter, number, etc.).

Preparation indicates operations on the program itself. They are usually control, initialization, clean-up, or overhead operations not directly concerned with producing the output data, e.g. setting the value of a program switch.

The ANSI (and ČSN 369 030) specifies outlines for data-carrying media (document, magnetic tape, punched card, punched paper tape), for peripheral equipment (disc storage, display, manual input, off-line storage, on-line storage, communication link, etc.) for file processing (merge, sort, extract, etc.) and standard conventions for striping, cross-reference, multiple outlines, etc.

In computer processing, data are represented by *identifiers*, i.e. symbolic names. An identifier is a single alphabetical character or a string of alphameric characters; the initial character of the string must be alphabetical, e.g. A, Avarage, S10 (e.g. 2B is not identifier). Symbolic names represent the contents, i.e. concrete values at some addressed locations. For example, the statement MIN=0, written in the process outline, means that there is a zero in the location MIN. If the following process outline is B=MIN it means that the content of the location called B is identical with the content of the location called MIN and the initial content of the location MIN is maintained. The content of MIN is changed by the input of a further value into MIN. If a set of data with the same characteristics is processed an array is formed. The variables which the array comprises are called *subscripted variables.* The monthly flows for a period of 50 years can be put into the array P_I for $I = 1, 2, \ldots 600$ or into an array $P_{I,K}$ for $I = 1, 2, \ldots 12$ and $K = 1, 2, \ldots 50$.

Representation of an algorithm by a flowchart is usually performed in three stages:

– *initial operations*; these include initialization of data (e.g. by setting zeros for some variables), preparation of constants, printing of headings, etc.

– *main operations*; these perform the transformation of input data into output information. They include the input/output operations, arithmetical and logical operations, etc.

– *final operations*; arrangement of results in the desired form and their output.

The simulation model of WRS simulates the behaviour of the system at each time step – the computation is repeated in a cycle. The flowcharts are drawn for one step only with repetition drawn in the cycle (loop). Inside this loop other loops can be "nested" (i.e. they cannot exceed the range of the main loop). A group of statements in the cycle is to be executed a stated number of times while a control variable is incremented each time through the loop as long as some condition is satisfied. An example illustrating the use of a flowchart and the use of identifiers for variables is given in Fig. 4.4.

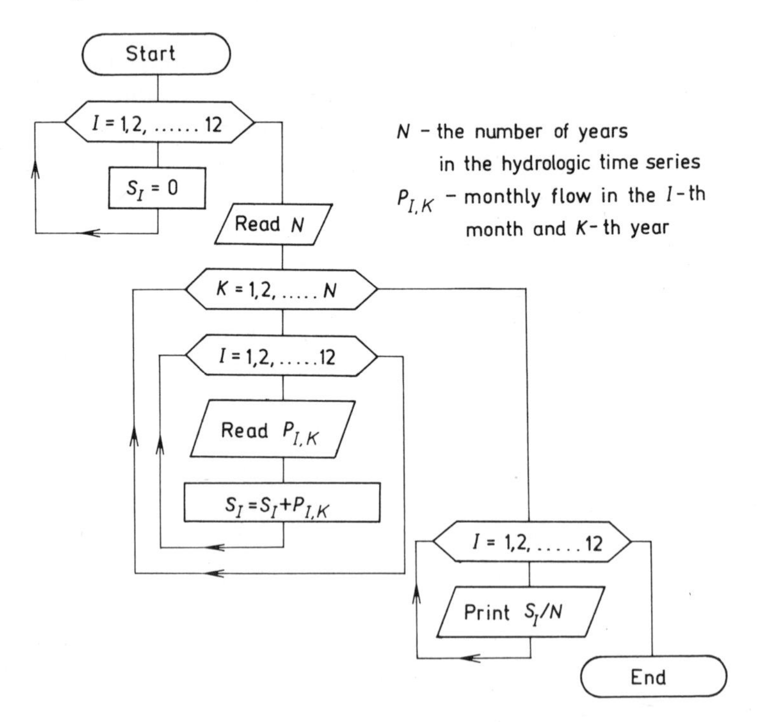

Fig. 4.4 Flow-chart of computation of average monthly flows

Two further types of graphic form of treating information-handling problems are *decision tables* and *flowcharts of operators*. The decision table is a tabular presentation of conditions (factors to consider in decision making), actions (steps to be taken when a certain combination of conditions exists) and rules (specific combinations of conditions and actions to be taken under those conditions). They are advantageous for problems with complex logical dependencies. Their use can eliminate the irrelevant combinations from a complete set of possible combinations of conditions. In flowcharting it is difficult to determine if the whole problem has been covered, if all logical elements have been defined and analysed, and if all combinations of conditions have been exhausted. The decision tables form a transition stage for converting decision alternatives into computer programs. There are three methods for this conversion: manual coding from tables, processers that convert tables into source languages for input into existing compilers, and compilers that translate tables directly into a machine language.

According to some references, programming time is reduced by as much as 60% if decision tables rather then flowcharting are used. The use of decision tables might therefore be regarded as a progressive technique for programming WRS models.

The flowcharts of operators were developed by the mathematician Lyapunov in 1962. He distinguished between computation and program flowcharts of operators. The latter were developed from the former by adding the operators specific to the computer.

4.2.3 Languages of Automatic Programming

Once the problem has been algorithmized (by a team of experts) and the algorithm has been expressed in one of the stated ways, it is necessary to describe the algorithm in a form that can be used in the computer. The transformation of a method of problem solving from a natural into a computer language is called *programming*. Programming is a precise specification of a sequence of well-defined operations that execute the chosen method of solution; the resulting specification is called a program.

The computer can process information if it is expressed in a computer code that is obtained by *machine language programming*. The computer code is composed of a set of instructions, for operations within the capability of a given computer. The machine language contains these instructions and the rules for forming the sequence to represent the algorithm. For example, the addition operation in the computer EC 1030 is 00011010.

The first stage of machine language programming in bits, was followed by representation in a decadic form. An example of an instruction in the machine language is 10 00 0201 0202. The separate parts of this instruction mean: operation code, auxiliary code, two addresses (memory location reference). The programs in this form have a great number of numbers with many digits and many errors occur in programming. Correction of errors and changes in the program are difficult. Each computer has its own code and programs are not compatible. A change of computer means learning a new machine language. Machine language programming has, however, some advantages; it (or some higher machine-oriented language) is used for programs or parts of programs if the time reduction in computation or a reduction in the main memory requirements is emphasized.

For these reasons, the most tedious part of programming was transferred to the computer. Programming languages closer to natural languages have been developed and have done away with the disadvantage of machine language programming. The first stage was the development of *symbolic machine languages*. Instead of absolute addresses, relocatable addresses are used, and for the operation code a mnemotechnic operation code is used. Translation into the computer code is performed by a *compiler*.

Further development brought *autocodes*, i.e. languages with a simple structure, the statements of which correspond to several instructions in the machine language. Their main disadvantage is a dependence on a particular type of computer.

A further advance is the system of *automatic programming* that consists of (a) the computer, (b) a source language program e.g. FORTRAN, ALGOL, PL/I, (c) a compiler that translates from the source language into the computer code, and (d) the product of translation, i.e. the program in the computer code.

The languages of automated programming do not depend on a computer type; they have the character of universal languages. Automatic programming facilitates coding of very complex algorithms and relieves the programmer of much routine work. The programs are well arranged, and easy to understand, which reduces the number of errors in their assembly. Most errors in machine language originated in coding operations. This can be done without errors by the computer itself. The main advantage, are the greater ease and shorter time of learning of the language and the absence of any tie to a particular computer type.

Survey of Programming Languages

In specialist literature more than 120 programming languages have been cited. They can be classified into several groups:

– languages for mathematical, numerical, scientific and engineering problems (ALGOL, FORTRAN, GPL, MAP, BASIC, etc.),

– automatic business data processing (COBOL),

– processing of files and strings (IPL-V, COMIT, etc.),

– formal algebraic operations (FORMAC, ALTRAN, etc.),

– general purpose languages (PL/I, FORMULA ALGOL, etc.),

– languages used in specialized disciplines (e.g. languages for programming machine tool design, civil engineering design IGES, etc.),

– design and assembling of compilers,

– simulation languages (GPPS, DYNAMO, SIMULA 67, etc.),

– languages for other purposes, e.g. query information retrieval systems (473 Query etc.).

Structure of Programming Languages

Automatic programming languages as synthetic languages are distinguished from natural languages by their simplicity. However, they have their own syntax and semantics. Syntax is thc theory of making expressions from the units of the language. It determines precise rules for the formation of sequences of characters, symbols or units that are acceptable in the language. Semantics gives the feasible symbols (character strings, etc.) a certain meaning.

The simplest elements of the programming languages are the basic symbols, i.e. letters, digits, operators (e.g. multiplication sign), delimiters (e.g. brackets), special symbols and basic word symbols. By means of the syntactic rules the more complex units of the language, i.e. words, are formed. Word consist of identifiers and numbers. Each language forms compound symbols, i.e. key words that have a special purpose and can be used only in the way the language permits. Expressions are formed from words, operators, delimiters and special symbols; the execution of an expression yields the value of the expression. Assignment and descriptive sentences (simple sentences) are formed from expressions, words and basic symbols. Sentences are grouped into compound sentences (block of statements etc.) and the compound sentences together form the subroutines ("procedures" in ALGOL). A program is the highest-level unit of the language structure.

The best-known automatic programming languages are ALGOL, FORTRAN, PL/I and COBOL.

The programming language ALGOL-60 (Algorithmic language) was developed in 1960 especially for programming algorithms of scientific and engineering problems. After it was perfected in 1968 (ALGOL-68) it became a universal language that can also be used for data processing. The following forms of it can be distinguished:

– reference ALGOL – an exact definition of the language,

– publication ALGOL – the form in publications,

– computer-oriented ALGOL – this reflects the characteristic deviations for the particular computer used.

FORTRAN (Formula Translation) was developed in 1954 for coding algorithms of scientific and engineering problems and some problems of data processing by IBM computers. Modifications FORTRAN I to FORTRAN IV have gradually been developed.

PL/I (Programming Language I) was developed as a general purpose universal language that can be used for engineering, mathematical and scientific computations and for business data processing. It uses elements of ALGOL, FORTRAN and COBOL.

COBOL (Common Business-Oriented Language) was developed in 1959 for business data processing.

Most programs for WRS problems were written in ALGOL and FORTRAN, and some in PL/I. To illustrate the basic ideas of program structure and the possibilities provided by the automatic programming languages, a simple example is presented. The program written below is a modification of ALGOL for the NCR-ELLIOTT 4120 computer[1]). In the commentary to the program lines the main elements of the ALGOL language are described.

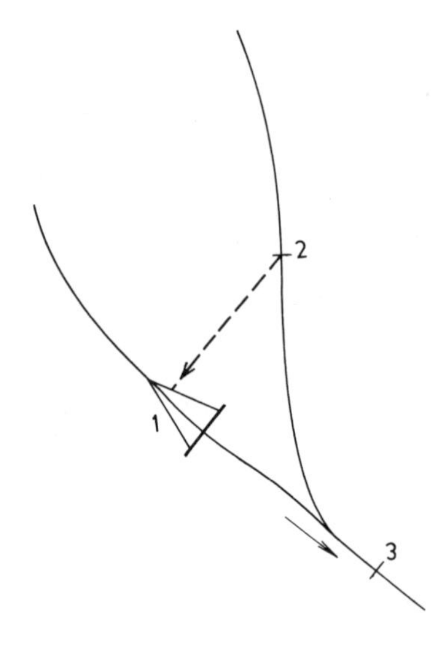

Fig. 4.5 Reservoir on tributary with diversion canal

EXAMPLE: Fig. 4.5 displays a schematic representation of a system with different elements, e.g. reservoir at site 1 and diversion channel from site 2 to site 1. The purpose of the reservoir is the river flow regulation at site 3 in order to obtain the draft *Q1* in June, July and August (e.g. reservoir operation with recreation constraints) and to obtain the draft *Q2* in other months. To keep the program relatively simple (the aim is to clarify the principles of programming of WRS) some simplifying assumptions have been introduced; e.g. all the losses of water in the reservoir were accounted for by some portion of the active storage, i.e. the calculated input value *WZA* is used (e.g. $WZA = 90\%$ of the active storage). The simulation model is applied with the time step $\Delta t = 1$ month; the hydrological data are represented by the series of monthly flows in the 40-year period.

For a chosen combination of input parameters *WZA*, *Q1* and *Q2*, the program computes the flows controlled by the reservoir operation (these flows are recorded), and the number of months when the required drafts *Q1* and *Q2* cannot be satisfied.

For a description of the program the lines have been given numbers.

---

[1]) The ALGOL language or FORTRAN is the form preferred for teaching basic programming processes. A modification for a computer was necessary to describe program inputs and outputs.

Text of the program:

```
 (1) RESERVOIR;
 (2) "COMMENT" PROGRAMMING ILLUSTRATION;
 (3) "BEGIN" "REAL" WZA, Q1,Q2, WZ, QN, WK, P;
 (4)          "INTEGER" K, I, T;
 (5)          "ARRAY" P1, P2 [1:12, 1:40];
 (6)          "PROCEDURE" HDATA (A); "ARRAY" A;
 (7)          "FOR" K:=1 "STEP" 1 "UNTIL" 40 "DO"
 (8)          "FOR" I:=1 "STEP" 1 "UNTIL" 12 "DO" "READ" A [I, K];
 (9)          PUNCH (4); SAMELINE; ALIGNED (3,3): DIGITS (3);
(10)          HDATA (P1);
(11)          HDATA (P2);
(12) L1:      "READ" WZA, Q1, Q2;
(13)          WZA: = WZA/2.6298;
(14)          WZ: = WZA; T: = 0;
(15)          "FOR" K: = 1 "STEP" 1 "UNTIL" 40 "DO"
(16)          "FOR" I: = 1 "STEP" 1 "UNTIL" 12 "DO"
(17)          "BEGIN" P: = P1 [I, K] + P2 [I, K];
(18)                  "IF" I > 7 "AND" I < 11 "THEN" QN: = Q1
                      "ELSE" QN: = Q2;
(19)                  WK: = WZ + P - QN;
(20)                  "IF" WK > WZA "THEN" WK: = WZA;
(21)                  "IF" WK < 0 "THEN"
(22)                  "BEGIN" T: = T + 1; "PRINT" - WK, K, I; WK = 0;
(23)                  "END";
(24)                  "PRINT" PUNCH (1), WZ-WK;
(25)                  WZ: = WK;
(26)          "END";
(27)          "PRINT" 'WZA=', WZA * 2.6298, 'MIL.M3',
(28)                  ''L'Q1 = ', Q1, 'M3/S',
(29)                  ''L'Q2 = ', Q2, 'M3/S',
(30)                  ''L' NUMBER OF MONTHS WITH DEFICITS=',
                      T''F'';
(31)          "GOTO" L1;
(32)          "END";
```

To explain the respective water management activities the identifiers used are listed, in the order they appear in the program (line (10) and following lines);

$P1_{I,K}$ $[m^3 s^{-1}]$ uncontrolled monthly flows at site 1, I-th month, K-th year,
$P2_{I,K}$ $[m^3 s^{-1}]$ uncontrolled monthly flows at site 2, I-th month, K-th year,

WZA [mil. m^3] calculated input value of the active storage A_z (e.g. $WZA = 0.9\ A_z$)
$Q1$ [m^3 s^{-1}] controlled flows in the months June to August
$Q2$ [m^3 s^{-1}] controlled flows in the months September to May,
WZ [m^3] the storage volume in the reservoir at the beginning of the month[1])
T current number of months with water deficits
K subscript of the annual loop
I subscript of the monthly loop
P [m^3 s^{-1}] monthly reservoir inflows
QN [m^3 s^{-1}] draft value used in computation
WK [m^3] the storage volume at the end of the month

Commentary on the program

(1) Name of the program

(2) The keyword "COMMENT" can be followed by any commentary, e.g. the goal of computation, the manner of program use, etc.; the compiler ignores the whole text between this keyword and the first semicolon.

(3) The key-word "BEGIN" marks the beginning of the program. This key-word is followed by a statement of the type, arrays and procedures (lines 3 to 8). The keyword "REAL" is followed by a list of real, floating-point variables.

(4) The keyword "INTEGER" is followed by a list of integer, fixed-point variables. Apart from the real and integer variables, logical variables ("BOOLEAN") can be used.

(5) $P1$ and $P2$ are the subscribed variables; a list of these is preceded by the key-word "ARRAY" and their identifiers are followed by the range for each dimension given by the bounds of subscripts – $P1$ and $P2$ are two-diemsional arrays.

(6) to (8) Description of the procedure for reading 40×12 values into the elements of the array $A_{I,K}$. Every procedure must have its identifier (in this case HDATA) that can be followed by a list of dummy arguments (A). These dummy arguments are used in statements of the procedure. If the procedure is used in the main program, it is called the procedure statement (identifier of this procedure is used), with dummy argument replaced by the actual parameters of the procedure. Procedures are used if the algorithm is repeated several times in different parts of the program or with different values.

(7) to (8) The "FOR" statement ("DO" statement in FORTRAN) specifies that a group of statements is to be executed a stated number of times while a control variable is incremented each time through the loop. The "FOR" statement has the following form: "FOR" ⟨variable⟩: = ⟨arithmetical expression⟩ "STEP" ⟨arithmetical expression⟩ "UNTIL" ⟨arithmetical expression⟩ "DO". The keyword "FOR" is fol-

[1]) The computation is done in units [m$^3 \cdot$ s^{-1}], the conversion to [mil. m^3] is expressed in line (13).

lowed by the control variable. The symbol := assigns the control variable its initial value. The keyword "STEP" is followed by the step (increment) and "UNTIL" is followed by the upper bound, i.e. the highest value of the sequence of controlled variable values for which the loop is executed. The initial value, the step and the upper bound of the loop can be given by a constant, by a variable or by an arithmetical expression. The keyword "DO" can be followed by any statement (or by another "FOR" statement as in our example). The statement "READ" $A[I, K]$ executes reading of input data from a punched paper tape into the memory locations with the name $A_{I,K}$. The statement in the loop is first executed for the initial value of the control variable, then it is incremented by the value of the step that is repeated until the upper bound is reached (the upper bound is included). In our case the following values are read: $A_{1,1}$, $A_{2,1}$, ..., $A_{12,1}$, $A_{2,2}$, ..., $A_{12,40}$. (Another possible form with explicitly stated values of the control variable is: "FOR" K: = 2, 4, 6, 7 "DO" "READ" $P[K]$;)

(9) The statements in this line are computer oriented and they describe the form of print: PUNCH (4) specifies output on the lineprinter, SAMELINE, print on the same line, ALIGNED (3,3) specifies the form of print of real variables (sign or blank, three digits, decimal point, three digits), DIGITS (3) specifies the form of print of integer variables, i.e. the sign and three digits.

(10), (11) Calling procedures for reading of flows.

(12) L1 is the label for labelling the statement that follows. It is the target of the "GOTO" statement, i.e. the point to which the control is transferred if the "GOTO" statement is executed. The statement following the L1 label specifies reading of values WZA, $Q1$ and $Q2$ from the paper tape.

(13), (14) Assignment statements in the form ⟨variable⟩: = ⟨expression⟩.

(15), (16) "FOR" statements.

(17) to (26) A compound statement; its general form is "BEGIN", followed by several statement ending with semicolons and "END"; the statements between these keywords are executed in the order stated. (In out case for each pair of the specified control variables of the loop K and I).

(18) The conditional statement; the general form is "IF" ⟨boolean expression⟩ "THEN" ⟨unconditional statement⟩ "ELSE" ⟨statement⟩. If the condition is satisfied (boolean expression is true), i.e. $I > 7$ and $I < 11$, then the statement that immediately follows is executed, i.e. QN: = $Q1$; if it is not satisfied, then the statement QN: = $Q2$ is executed.

(19) The assignment statement.

(20) The "IF" statement: "IF" ⟨boolean expression⟩ "THEN" ⟨unconditional statement⟩. If the condition ($WK > WZA$) is satisfied, row (20) is executed, i.e. WK: = WZA; if is not not satisfied, the following statement is executed, i.e. the "IF" statement in rows (21) to (23).

(21) to (23) The "IF" statement; if the condition is satisfied the compound statement

in row (22) is executed. The keyword "END" in row (23) ends the compound statement.

(24) The statement for punching the flows controlled by the reservoir operation into the paper tape. The number in brackets is the number of the output channel (1 – paper tape punch, 3 – console keyboard, 4 – lineprinter).

(25) The assignment statement.

(26) The end of the compound statement.

(27) to (30) The print statement to output the values on the lineprinter the statement PUNCH (4) in row (9) is global, i.e. it is valid for the whole program with the exception of row (24) where a local statement requires punching of the paper tape). The inltruction to print character strings (i.e. the sequences of characters) is given by enclosing them by single inverted commas (e.g. 'STORAGE'); printing on a new line or a specific arrangement of output data is given by a specific character in double inverted commas (e.g."L" – for printing on a new line, "F" – new page, etc.).

(31) The "GOTO" statement; the form is "GOTO" ⟨label⟩; in our case the control is transmitted to the label L1 for further reading of input data.

(32) The end of the program.

The advantages of the ALGOL programming derive from the arrangement in blocks. A block is formed by a compound statement, in which the keyword "BEGIN" is followed immediately by a declaration. The arrangement of a block is: "BEGIN", declaration, statements of the block, "END";. The variables declared in this block are found only in this block or in blocks nested in this block (they are local in this block, the variables declared in the external block are global). If a complex algorithm is coded by an automatic programming language, subsystems can be distinguished. If these subsystems are programmed in a block form, the job can be divided among several programmers, which speeds up the programming work.

4.2.4 Debugging

Debugging is an activity that helps to isolate and correct mistakes in coding (in the chosen program language) and structure of the program. Debugging prevents malfunction of the program. The overall computation tests also include the computer hardware (which is in the charge of the technical personnel of the computer centre). The probability of malfunction of the computer is negligible, but errors are possible in the functioning of the input devices, for example, if the input media (punched cards, paper tapes) are worn out.

Programming requires accuracy and precision on the part of the programmer; even a highly experienced programmer cannot avoid mistakes. The transcription of the algorithm into the programming language is usually a source of syntactic errors (clerical errors, omision, etc.). The second group of more serious mistakes can occur in the logical structuring of the program. If the definition of the problem is not complete or if it is not quite correct, some important relationships may be omitted.

Debugging is an integral part of the programmer's activities (see Fig. 4.6). These activities include main four steps: writing of the program in the source code, preparatory activities, and two debugging steps.

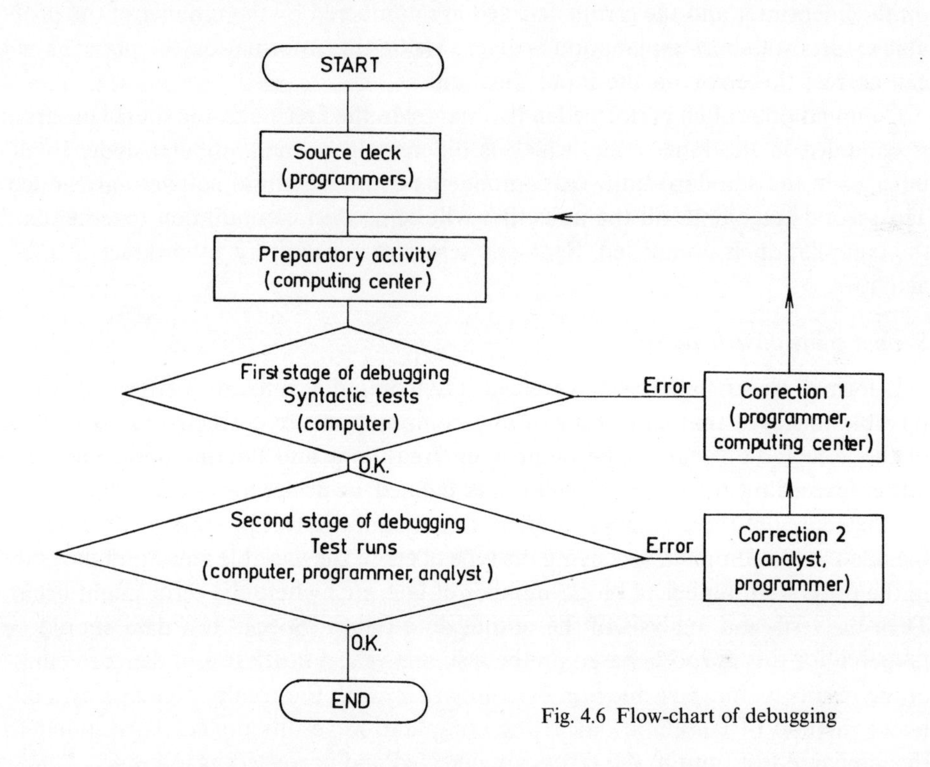

Fig. 4.6 Flow-chart of debugging

Source code

The result of the analytical modelling procedure is the flowchart (decision tables etc.) that serves the programmer in assembling the program (i.e. in producing the source deck).

Preparatory activities

The program and the input data must be transformed into a form suitable for the input device of the computer, e.g. it is punched into the punch cards or paper tape or written on the keyboard of the terminal device. Before execution of the program on the computer the operator or, in an automatic system, the supervisor (type of system program) prepares some system programs that make provisions for debugging.

First stage of debugging

The source code is transformed by the compiler into a computer code. Syntax errors in the source language are detected. In automatic programming languages the syntax tests are perfect and very detailed. Often the compiled program is printed on the lineprinter and the errors detected are numbered by the number of the probable error, or a short explanation is given. Using this information the programmer can correct the error on the input medium.

Compilation is often performed in two stages. In the first stage, the source program is compiled in the inner code, which is different from the computer code. In the inner code the standard unit, i.e. components of the standard software are added. The second stage links all the units that will be used in computation together, and the compilation is completed. Syntactic tests are a necessary by-product of these activities.

Second stage of debugging

If formal errors have been corrected, test computer runs are performed. First, possible mistakes that can occur during computation are corrected (e.g. overflow of the acceptable range of the number in fixed-point and floating-point representation, exceeding the range of subscripts defined in declaration, etc.). The search for the source of these errors is tedious, and diagnostic test programs are therefore included in the compiler, specifying the type of error, the variable name and the place in the program (number of block, number of line, etc.) where the error might occur. Then the tests and analysis of the output data follow. Special test data should be prepared for this purpose, based on the task analysis; a quick test of the correctness of the results is thus possible (e.g. by comparison with the results obtained by a different method of computation). If the computation results do not correspond to this prepared test output, the errors are detected and corrected as follows;

a) without the computer:

– test of the program by another programmer searching for the coding error in the source program,

– simulation of a computation run by a calculator using the test data and following the program instructions;

b) with the computer:

– print of the intermediate results (in the simulation model e.g. the changes in important variables at the end of the time step),

– program tracing, i.e. print out of the locations where the computation was executed,

– memory dumps, i.e. a record of memory contents when execution stops.

The method of debugging corresponds to the complexity of the program. Test data are designed to activate the execution of all the branches in the program.

Debugging of programs for computers of the second generation was done in the presence of the programmer or the analyst. Debugging of programs for computers of the third generation is often done via the terminals and the programmer can correct programs in a time-sharing and interactive mode. Not only computer time but also the programmer's time is used more effectively.

Program optimization is concerned with changes to shorten the required computer time or reduce the number of the necessary memory locations. A program must be optimized if it is used many times.

4.2.5. Documentation in Programming

The work of programming does not end with debugging, testing and obtaining the first correct results. Documentation should be an integral part of each program. Often, in some isolated tasks, documentation is omitted due to lack of time, resulting in losses of computer time and programmer's time with future use of this program. Incorrect data use and false assumptions are associated with changes and new arrangements in the program.

The first part of the documentation, which is necessary for the user of the program, should include the following information:

- the name and purpose of the program,
- different possibilities of program utilization,
- the program language used,
- the type of computer that can be used for this program,
- the necessary system configuration (necessary for computation),
- instructions for input data preparation,
- a survey of output data,
- instructions for the operators of the computer system.

The second part of the documentation is important for the analysts and programmers and should contain a detailed description of the program. The form of this description includes the flowcharts (decision tables etc.) and the program printout on the lineprinter. A list and a description of the most important identifiers are helpful.

The flowchart has to correspond to the present state of the program and should be comprehensible to programmers who will later deal with this program. Each change should be documented, otherwise the flowchart is useless.

Often, changes in the program are required. The user requires new input and output data (sometimes due to inadequate system analysis) or he has obtained some new information on the problem. The computer centre may require interruption of the computation with continuation later. The analyst may find an improvement in the logical aspects of system description or ways of speeding up computation or of obtaining more accurate results. The changes are often necessary and advantageous; however, they must be properly documented.

4.3 PRINCIPLES OF A COMPUTER CENTRE PROJECT

The computer centre costs run into millions, and the time needed for an economic return on the costs should be short-term. Therefore, a project for a computer centre and its input data should be thoroughly analysed.

If a computer centre system is to be designed, previous experience with a hired computer in a service computer centre is very important. The danger of the importance of systems analysis of the computer centre project being understimated is thus reduced. Utilization of a service computer centre before the installation of an institute's or a firm's computer centre is therefore recommended. Service computer centres can be classified into two groups:

(a) *analytical computer centre* similar to a consulting engineering service, where the customer orders the solution of his problem from the data delivered. These data and the problem are analysed (with the aid of the computer), methods of computation and data processing are designed, and programs are coded, debugged and prepared for execution;

(b) a *service computer centre*, where all the work described above is done by the customer. The computer centre is in charge of the operation of the computer, punching and testing of the transformation of input data for the computer, and the customer hires these services and the computer time.

In the design of a computer centre the following stages can be distinguished:

In the *first stage* the problem is defined and the requirements formulated. The computer tasks are specified, the method of data processing is stated and the cost-benefit analysis performed. The type of computer need not be decided upon at this stage. A comprehensive analysis of the development and future activities is all-important. The computer will be used not only for water management and scientific problems but also for business data processing, for management information systems and possibly for the operation of WRS. It is a very important, time- and labour-consuming activity and the management has to assemble a qualified and competent working group for it.

In the *second stage*, a survey of the input and output information and the system charts for the principal tasks are set up (at a rough discriminating level). By using this information in the project of the new system of data processing and computation, the requirements for the computer and the peripherals are derived (input/output devices, capacity of the main computer storage, auxiliary storage, software, etc.).

The *third stage* involves the choice and purchase of the system comprising the computer, input/output and other devices. In this choice, devices that have been purchased already and their role in the future system are taken into account. The technical parameters of the system hardware in a certain class are approximately the same, and the costs do not differ very much, but the software (compilers for the programming languages, subroutine packages, standard reference library, etc.) and the whole conception of the system (reliability of the system, prompt service by the manufacturer, flexibility in extension of the system, etc.) are often decisive. It is advantageous to obtain information from firms which deal with the same or similar problems or which own the same type of computer.

In the *fourth stage*, personnel are trained, especially the programmer and the service personnel and the programs are prepared, debugged, executed and tested in a service computer centre.

The *fifth stage* comprises the installation of the computer, tests of the system and its operation.

The activities need not always be performed in the stated order as some feedback loop may occur.

The professions represented in the computer centre include the following categories:

– *management of the centre*: director, secretary, clerical services, etc.

– *analytical and programmer's group*: the head of the group, analysts, departments for business data processing, scientific and technology tasks, econonomic problems, specialists, etc.

– *technical service group*: head of the group, service personnel,

– *operation group*: head of the group, heads of the work shifts, operators,

– *data preparation group*: personnel for punching data, typing in display stations, data control, etc.

The approximate proportions of these professions in the computer centre may be as follows: Director and heads of groups 5%, analysts 25%, programmers 37%, operators 25%, technical service personnel 8%.

In the choice of the personnel of the computer centre cooperation with psychologists is recommended. The work of the analysts is similar to that involved in management positions, but it is more diversified; a comprehensive evaluation of the analyst's psychological characteristics is therefore desirable. Programmers should possess the ability to combine information, a good insight into problems, a flexible way of thinking, etc. Operators have to be accurate and reliable. The personnel for data punching, typing, and testing should be able to concentrate, should possess good pattern and character recognition combined with a short-time but accurate memory capability. The results of the psychological tests should be supplemented by information on mental qualities, and the ability to cooperate in working groups, high motivation and a positive attitude to work, psychological stability, etc. are all important factors. The dynamic process of the transformation of these psychological qualities due to the environment and type of work needs to be taken into account.

4.4 PROSPECTS OF COMPUTER USE IN WATER RESOURCE SYSTEMS

Simulation models are often used for the design and operation of WRS. For this purpose a number of mathematical optimization models (e.g. linear or dynamic programming) have also been used. All these models use computers, not only for modelling, but also for processing hydrological input data and for the extraction of more information from the observations. If the data have been processed, corrected and transferred to a medium suitable for computer input, no further errors due to clerical oversight, reproduction failure, etc. occur.

Computers are used in hydraulic structure design computations and other fields related to water management.

At present, the development of interrogative computer systems, linked by terminals with the data base of big computers, seems to be the most progressive trend in WRS analysis and design. The computers, hardware and software can thus be used not only by programmers but by water resource engineers, planners and decision-makers and a high efficiency of work and extensive application of new ideas is possible. These users need not know in detail the algorithms of computation, programming and other activities that can be transferred to the computer system. The interrogative computer system can be used in an interactive way for the evaluation of different assumptions in project alternatives using more relevant information than could be used without this system.

The methods of business data processing by computers can be used in water balance evaluation. Some types of computers will be used for real-time control of releases from a system of reservoirs.

The information and control systems will include observation sites with remote control, data transmission, system monitoring and computer control. This system has to monitor the states of WRS, evaluate their effect on operation (even with a short-term forecast of flows) and, based on them, the system can offer a proposal for an optimal method of operation (WRS-real-time dispatching). The first stage of computer use in this system, i.e. formation of information subsystems, has been implemented in several WRS. Information retrieval systems (see Chapter 11) are a further aspect of computer use in water management with cost saving in the retrieval of information and a basic reduction in the time necessary to gain access to information about publications, reports, books, papers, etc. in water management.

5 MODELS OF OPTIMAL PROGRAMMING

The term optimal (mathematical) programming[1]) refers to a set of methods that are used for optimization of operations, e.g. the well-known linear, non-linear, and dynamic programming.

There is an extensive literature dealing with the methods of optimal programming. Therefore, references will be used and our attention will be concentrated on their application in WRS. Principles of optimal programming will be cited where necessary for computation in application models. Not all the known applications can be presented; only those directions of application that seem promising and progressive will, therefore, be given. We shall proceed from simple to more complicated models.

5.1 LINEAR PROGRAMMING

5.1.1 Models of Linear Programming

Linear programming is the most widely used and the simplest method of operations research. It is explained e.g. by Dantzig, 1963; Gass, 1969; Korda *et al.*, 1967; Walter *et al.*, 1973, and others.

Every optimization task tries to solve one of two alternative problems: either to attain a maximum (minimum) objective function with the given resources or to meet the given goal with a minimum of resources. The tasks dealt with in linear programming are expressed by:

– a set of equations and inequalities (constraints),

– the condition of non-negativity (the unrealistic condition of e.g. the negative value of production is suppressed),

– an objective (criterial) function (expressing the goal or objective that has to be met).

All the constraints and the objective function must be linear. This requirement of linearity implies simplicity but involves a limited applicability of the models of linear programming. They can be used for one-stage optimization in contrast to dynamic programming that makes a repeated optimization in a series of stages of the optimization process possible.

---

[1]) Programming used in this sense is not equivalent to programming computers, and in its broader meaning it is closer to planning.

The general mathematical model of linear programming can be formulated as follows:

$$x^{\circ} = [x_1^{\circ}, x_2^{\circ}, \ldots, x_n^{\circ}] \quad (6.1)$$

under the constraints of the task ($m < n$; the equations are linearly independent),

$$\begin{aligned} a_{11}x_1 + a_{12}x_2 + \ldots + a_{1n}x_n &= b_1 \\ a_{21}x_1 + a_{22}x_2 + \ldots + a_{2n}x_n &= b_2 \\ \ldots\ldots\ldots\ldots\ldots\ldots\ldots\ldots \\ a_{m1}x_1 + a_{m2}x_2 + \ldots + a_{mn}x_n &= b_m \end{aligned} \quad (5.2)$$

and, further, under conditions of non-negativity,

$$x_j \geqq 0 \; (j = 1, 2, \ldots, n) \quad (5.3)$$

minimizing the linear objective function,

$$z = c_1x_1 + c_2x_2 + \ldots + c_nx_n \quad (5.4)$$

where x_i are variables and a_i, b_i, c_i are known constraints.

The important fact is that there are more variables than equations. Therefore, an unlimited number of feasible solutions exists. Such cases are typical of linear programming. If the number of equations and variables were identical, a single solution would exist, and the search for an optimal solution given by an objective function would not be possible.

A simple example taken from economics can serve as an illustration of the methods of linear programming:

A factory produces two kinds of products V_1 and V_2. Two different raw materials are required, S_1 and S_2. They are available in limited quantities, 1200 kg and 1600 kg, respectively. The first product V_1 requires 3 kg of raw material S_1 and 2 kg of raw material S_2; the second product V_2 requires 1 kg of raw material S_1 and 2 kg of raw

Table 5.1 Linear programming input scheme

Raw material	Consumption of raw materials per one product (kg)		Amount of raw materials at disposal (kg)
	V_1	V_2	
S_1	3	1	1200
S_2	2	2	1600
Price of one product in $	20	10	

material S_2. The market price of product V_1 is \$ 20 and of product V_2 \$ 10. The task is to determine which product and which quantity must be produced for the factory to gain the maximum benefit. The linear programming input scheme is given in Table 5.1.

There are many feasible production schedules. Some of them are listed in Table 5.2. Which procedure can be used in search of an optimal program? It is obvious that it is not advantageous to list all the possible programs and thus try to find the best case. An enormous number of computations would be necessary even if the clearly disadvantageous alternatives are omitted.

Table 5.2 List of some possible production schedules

Number of products		Benefits	Quantity of raw materials not utilized (kg)	
V_1	V_2		S_1	S_2
400	0	8000	0	800
300	300	9000	0	400
250	450	9500	0	200
200	600	10 000	0	0
150	650	9500	100	0
100	700	9000	200	0
0	800	8000	400	0

Linear programming determines the optimal schedule as a precise and single solution in the following way:

x_1 denotes the number produced of the first product V_1; similarly, x_2 denotes the number of product V_2. Using variables x_1 and x_2, and the known quantities of the raw materials necessary for the production of each product, we can express the condition that the maximum consumption of the first raw material is 1200 kg and of the second raw material is 1600 kg. If one product V_1 requires 3 kg of raw material S_1, then x_1 products need $3x_1$ kg of raw material S_1. If the sum of the amount used for the two products is less than or equal to 1200 kg, the following inequality can be met

$$3x_1 + 1x_2 \leqq 1200$$

The same consideration for the consumption of the second raw material S_2 in the production of both products results in the inequality

$$2x_1 + 2x_2 \leqq 1600$$

If both these inequalities are considered, then couples (x_1, x_2) will be obtained expressing feasible production schedules. It is apparent that the negative solution of the inequalities has no real interpretation. Further conditions are therefore added:

$$x_1 \geqq 0; \quad x_2 \geqq 0$$

The objective of production can be formulated as the attainment of maximum benefit. This benefit denoted as z, will be expressed in terms of the price \$ 20 of product V_1 and \$ 10 of product V_2 and quantities x_1 and x_2 produced, in the following way:

$$z = 20x_1 + 10x_2$$

The production schedule, expressed by the couple (x_1, x_2) must meet the given inequalities (i.e. constraints) and maximize the expression $z = 20x_1 + 10x_2$.

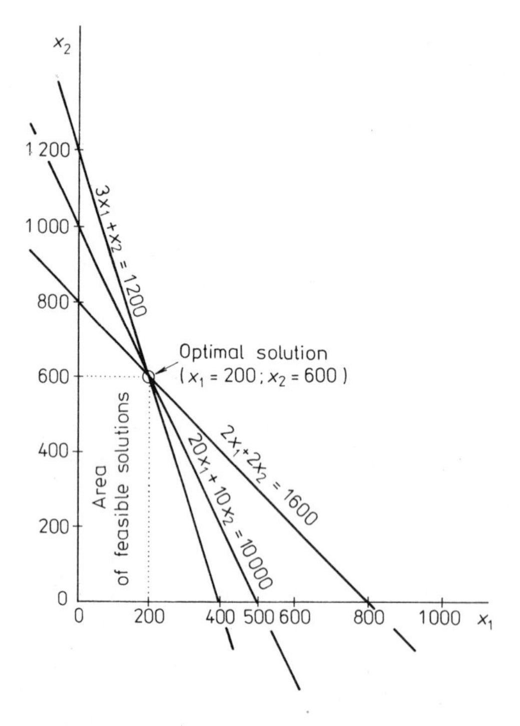

Fig. 5.1 Graphic solution of the simple example of linear programming

The task can be expressed in this way:
A couple (x_1, x_2) has to be found, meeting the inequalities

$$3x_1 + x_2 \leqq 1200 \tag{5.5}$$

$$2x_1 + 2x_2 \leqq 1600$$

$$x_1 \geqq 0; \quad x_2 \geqq 0 \tag{5.6}$$

so that the function

$$z = 20x_1 + 10x_2 \tag{5.7}$$

should reach its maximum.

The inequalities (5.5) limit or constrain the choice of program. Therefore they are called constraints of the task. The inequalities (5.6) are also constraints, and they are called non-negativity restrictions. The function (5.7) is called the objective or criterial function. All the inequalities and the objective function are linear; the problem can therefore be solved by linear programming.

Our economic task has thus been transformed to a mathematical set of inequalities (5.5) and (5.6) and the objective function (5.7).

Computation of this simple example is easy, that is, it can be performed by a graphical constructions (Fig. 5.1). In the optimal production schedule, given in the first row in Table 5.2, the numbers of products V_1 and V_2 produced will be 200 and 600, respectively. The benefits are \$ 10,000 and all the raw material is consumed. In a practical application, a set consisting of dozens of equations is solved on a computer.

Most of the tasks of linear programming are dealt with by two basic procedures:

– the simplex method that can be used generally for each model of linear programming, although in some cases it is cumbersome;

– a model of distribution type – e.g. the transportation problem, the computation of which is easier in some cases, but which requires the transported objects to be homogeneous and mutually exchangeable.

5.1.2 Application of Linear Programming in WRS

In recent years linear programming has been very often used. The final report of the Harvard Program (Maass *et al.*, 1962) includes a detailed and comprehensive analysis. Interesting applications of linear programming were published by Buras, 1972; Kos, 1972; Vedula and Rogers, 1981; Rydzewski and Rashid, 1981, and others.

These applications of linear programming have produced many positive results. The occupation is simple in principle. The software of almost every computer includes universal programs for linear programming such as the MPSX linear programming package for IBM computers, e.g. IBM 370(168) where no programming and program debugging is necessary. It is only necessary to assemble the data for the task under investigation and transform them into the form required by the program.

The main drawback is the assumption of linear relationships, particularly in the objective function. Only for a limited set of problems is this assumption acceptable without an undesirable distortion of reality. Another drawback is the one-stage optimization whereas the process of resource management is dynamic. Therefore repeated optimization is periodically required to account for the changes in the state of the system[1]). A further drawback is the deterministic nature of most linear programming models; this can be overcome by using chance-constrained models (see Chapter 10).

In general, it can be stated that the linear programming models are used mainly as "screening" models for the preliminary design of WRS at a rough discriminative level. A large number of alternatives are considered with different input values. Thus, a region with acceptable parameters is selected for further investigation by simulation models at a more detailed discriminative level. Such a procedure was used for WRS for the water supply of New York. Linear programming models are sometimes combined with dynamic programming or other models.

Linear programming for two hydrological periods

The simplest model is deterministic as it assumes known hydrological conditions. The computation is performed for one year, and only two different hydrological periods are assumed, a wet and a dry period. Both are described by the average flow in volume units. This model can be illustrated by an example:

Investigations are carried out on a WRS, the scheme of which is shown in Fig. 5.2. It includes reservoir N_1 with active storage V_1 and reservoir N_2 with active storage V_2 an irrigated area with annual water requirements V_z and a water power plant with annual production E_{prod}.

The following input values are given:

– the flows in the wet and dry periods at given points. The upper number in the scheme applies to the wet period, the lower applies to the dry period. Both are in 10^9 m^3;

– 40% of the irrigation water requirements V_z should be met in the wet period, 60% in the dry period. The return flow from the irrigation area is assumed to be 10% of V_z in the wet period and 30% of V_z in the dry period;

– power production in the water power plant must not fall below the given value E_f in any period; half of the annual production is required in the wet period and half in the dry period.

The active storage of both reservoirs, the water withdrawal for irrigation, and the power production will be determined so as to maximize the benefits from this WRS.

The constraints given in Fig. 5.2 were derived from input values and requirements.

---

[1]) The static approach of linear programming models is acceptable in, for example, optimization of the allocation of withdrawals of water from reservoirs on the River Nile for different users. Every year follows the same pattern: the reservoirs are filled at the same time every year by approximately the same volume of melt-water from the mountainous region of the upper reaches, and the withdrawals, mainly for irrigation, take place in the same season of the year. The dynamic nature of the process can be neglected without causing great distortion.

It is assumed that in the wet period the active storage of the reservoirs is filled up and in the dry period it is emptied. Therefore, the release from reservoir N_1 for example in the wet period is $0.95 - V_1$ and in the dry period $0.54 + V_1$. The other data in this scheme are clear.

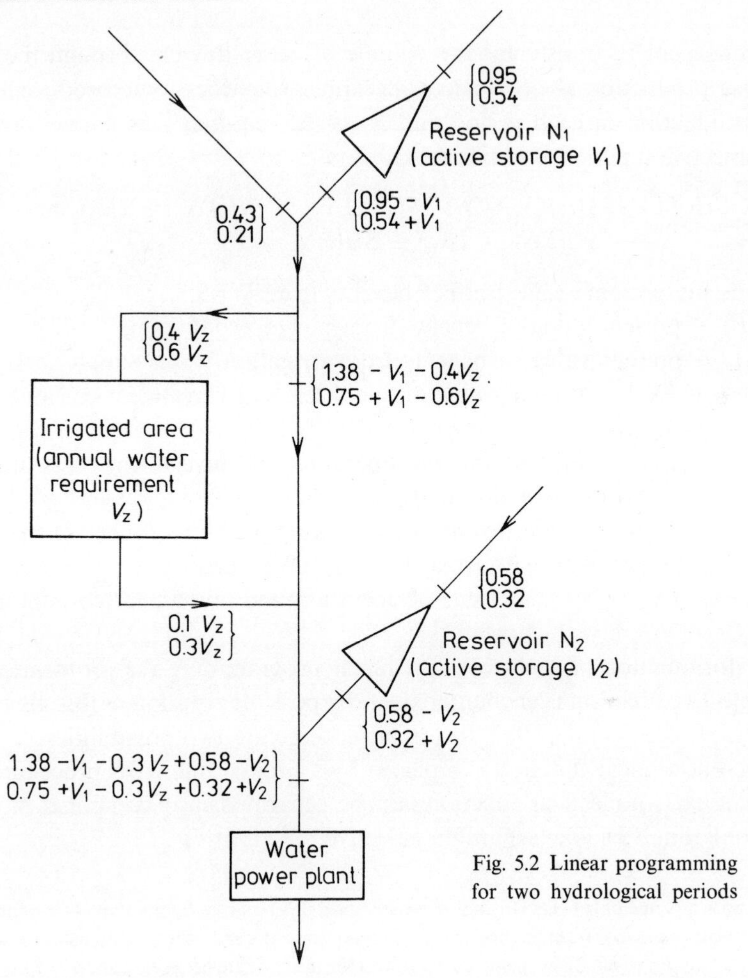

Fig. 5.2 Linear programming for two hydrological periods

The task is formulated in terms of three groups of constraints and the objective function. The constraints of the first group require that no variable should be negative (V_1, V_2, V_z, E_{prod}). The second group of constraints requires that flows at all points of the WRS should be non-negative. These constraints are

$$\begin{aligned}
&0.95 - V_1 &&\geqq 0\\
&1.38 - V_1 - 0.4V_z &&\geqq 0\\
&0.75 + V_1 - 0.6V_t &&\geqq 0\\
&0.58 - V_2 &&\geqq 0
\end{aligned} \qquad (5.8)$$

The third group of constraints requires that the flow in the power plant should be sufficient both in dry and wet periods for the production of firm energy:

$$\begin{aligned} 1.38 - V_1 - 0.3V_z + 0.58 - V_2 &\geqq 0.5k_eE_f \\ 0.75 + V_1 - 0.3V_z + 0.32 + V_2 &\geqq 0.5k_eE_f \end{aligned} \tag{5.9}$$

The coefficient k_e transforms the volume of water flowing through the turbines into power production, the coefficient 0.5 express the necessity to produce half of the annual production in each period and E_f is the required firm energy production.

The objective function is maximized:

$$\begin{aligned} \pi = SH(VE) + SH(V_v) - SH(PN_1 + IN_1) - SH(PN_2 + IN_2) - \\ - SH(PN_E + IN_E) - SH(PN_z + IN_z) \end{aligned} \tag{5.10}$$

where π is the present value[1]) of net benefits from WRS,

$SH(VE)$ – present value of benefits from power production,

$SH(V_v)$ – present value of benefits from irrigation water supply,

$SH(PN_1 + IN_1)$ – present value of operation and investment costs of reservoir N_1 with active storage V_1,

$SH(PN_2 + IN_2)$ – present value of operation and investment costs of reservoir N_2 with active storage V_2

$SH(PN_E + IN_E)$ – present value of operation and investment costs of water power plant with capacity E_{prod} (kWh/year)

$SH(PN_z + IN_z)$ – present value of operation and investment costs of irrigation system with capacity V_z.

In the formulation of such tasks for linear programming the problem of a non-linear objective function is encountered, and a possible solution of this problem was described by Maass *et al.*, 1982. In principle, there are two possibilities:

– Each non-linear function is replaced by a broken line. This procedure can be followed if each non-linear function can be separated into components, each including only one variable (separable objective function).

---

[1]) The present value of benefits (V) and operation costs (VN) is given by the sum of discounted annual values of benefits and costs from the beginning of operation to the end of the economic lifetime period of investment (economic planning horizon). If i per cent is the interest rate per period and $\alpha = [1 + (i/100)]^{-1}$ is the single period discount factor, the present value R of the sequence of R_t is

$$R = \sum_{t=1}^{N} \alpha^{t-1} R_t$$

The present value of investment (initial) costs is the sum of investment costs and interest from the beginning of construction to the beginning of operation

$$\text{Present value } I = \sum_{t=1}^{M} \beta^{M-t} I_t$$

where $\beta = 1 + i/100$ (in ČSSR $i = 6$ per cent is recommended).

– If such a separation is not possible, the linearization becomes more complicated.

If the constraints are non-linear functions of only two variables, linearization uses the analogy between a curve for a function of one variable and a surface in the case of two variables. Analogous to the line given by two points on a curve is a plane given by three points in space. Linear approximation of the function of two variables can therefore be performed by a rectangular grid so that the values of the function in all intersection nodes of the grid are determined. Linear approximation can be a weighted mean of these values. Other approaches are also possible, of course, e.g. the finite elements method.

The problem of non-linear functions being solved, further computation is not difficult in principle, and modern computers have enough memory capacity and speed for complex systems. The main drawback of the model is the assumption of a regular pattern of flow and annual control. The model described can therefore be used for a first screening.

Linear model for more than two hydrological periods

The model for two hydrological periods described in the previous section cannot be used if carry-over occurs. In that case a more comprehensive model must be applied, based on the following assumptions:

– the hydrological conditions are assumed to be known; i.e. it is a deterministic model,

– unlike the previous model, the carry-over is considered in a series of different hydrological periods.

The method is explained for four periods, so that a large series can be used. The scheme of the model is given in Fig. 5.3. To facilitate comparison with the previous model, a similar scheme was chosen with the following changes:

– reservoir N_2 was omitted,

– the total water requirement for irrigation V_z for all four periods was given,

– the values of flow were determined in 10^9 m^3 at three points for each period,

– the release from reservoir N_1 in individual periods was denoted as o_t, where $t = 1, 2, 3, 4$. The other data in the scheme are clear, and they were determined in a similar way to the previous case.

As the amount of irrigation water was given, the aim of the task is: the determination of the active storage in reservoir N_1 and the power production which maximizes benefits from WRS. The following objective function will be maximized:

$$\pi = SH(VE) - SH(PN_1 + IN_1) - SH(PN_E + IN_E) \tag{5.11}$$

where π is present value of net benefits from WRS,

$SH(VE)$ – present value of benefits from power production,

$SH(PN_1 + IN_1)$ – present value of operation and investment costs of reservoir N_1 with active storage V_1,

$SN(PN_E + IN_E)$ – present value of operation and investment costs of power plant with capacity E_{prod} (kWh/4 periods).

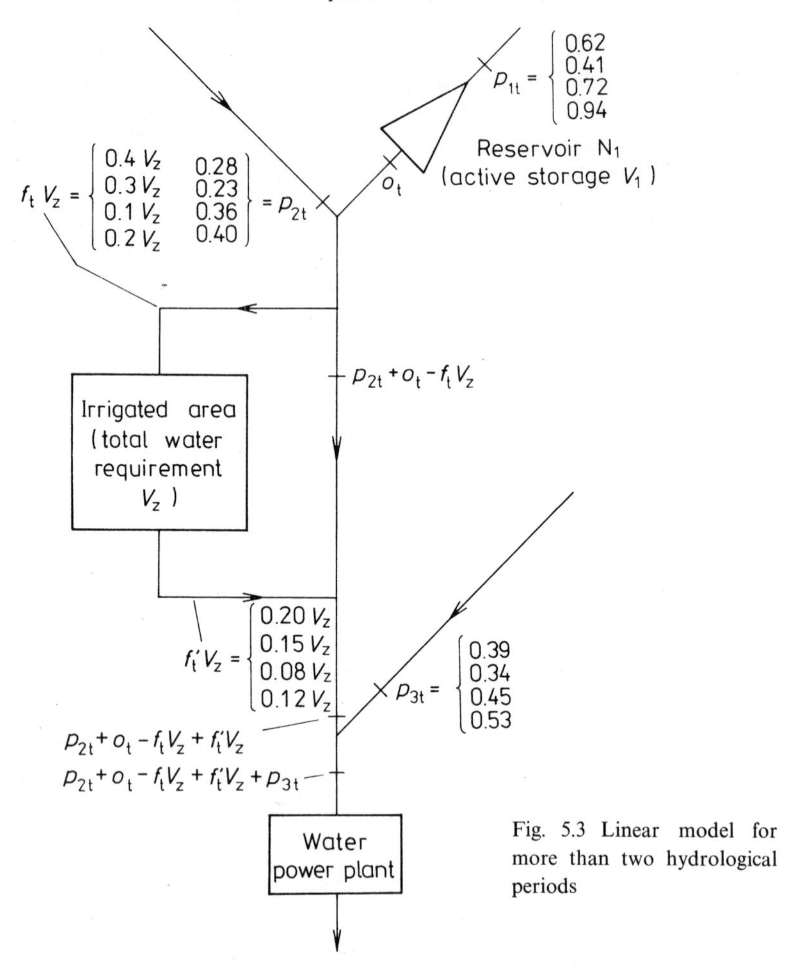

Fig. 5.3 Linear model for more than two hydrological periods

The objective function is, of course, non-linear. The methods of linear approximation are the same as in the previous example. For each hydrological season there are six constraints. The first one is apparent from the expression

$$o_t + p_{2,t} \geqq f_t V_z \tag{5.12}$$

The second constraint states that the volume of water released in each period must not exceed the sum of reservoir storage at the beginning of that period and the reservoir inflow during that period.

The reservoir storage at the beginning of period t is n_t; then

$$o_t \leqq n_t + p_{1,t} \tag{5.13}$$

The third constraint specifies that reservoir storage at the beginning of any period must not exceed the storage at the end of the previous period:

$$n_t = n_{t-1} + p_{1,t-1} - o_{t-1} \tag{5.14}$$

According to the fourth constraint, the reservoir storage at the end of any period must not exceed the active storage V_1:

$$n_t + p_{1,t} - o_t \leqq V_1 \tag{5.15}$$

The last two constraints concern power production. The fifth constraint guarantes the firm energy production

$$o_t + p_{2,t} - (f_t - f_t') V_z + p_{3,t} \geqq k_e E_{t,f} \tag{5.16}$$

where the coefficient k_e transforms the volume of water flowing through the turbines into power production.

According to the sixth, and last, constraint, power production in each period should equal, or exceed, the pre-determined proportion c_t of the total production.

$$E_{t,\mathrm{prod}} \geqq c_t E_{\mathrm{prod}} \tag{5.17}$$

The six constraints listed are valid in each period. The values of variables o_t, n_t, V_1, $E_{t,\mathrm{prod}}$, E_{prod} are determined so as to maximize the objective function. This example with 24 constraints can easily be handled by any computer. Usually, substantially more hydrological periods are considered. Then a reduction of the computation burden is achieved by the principle of decomposition of linear programs (Maass *et al.*, 1962).

Linear programming with stochastic flows

In this model the variables are not deterministic, as in the previous models, and only the expected values and the probability distributions are known. The principles can be explained by a simple example:

A WRS with the scheme given in Fig. 5.4 is considered. It includes reservoir N, an irrigated area, and a power plant. The computation is performed for three hydrological periods. In this scheme the following notation is used:

p_t – inflow to reservoir N in period t $(t = 1, 2, 3)$
o_t – release from reservoir N in period t,
z_t – withdrawal of water for irrigation in period t,
$g_t z_t$ – return flow from irrigation.

The system was simplified, compared with the previous model. The tributary inflows α, β, marked in Fig. 5.4 by a dashed line, are considered in a simple way as being deterministically related to stochastic flow p_t. In computation, the stochastic releases from reservoir N are increased by the values of the tributary inflow α. Similarly, the flows in the water power plant are increased by β, determined with reference to p_t.

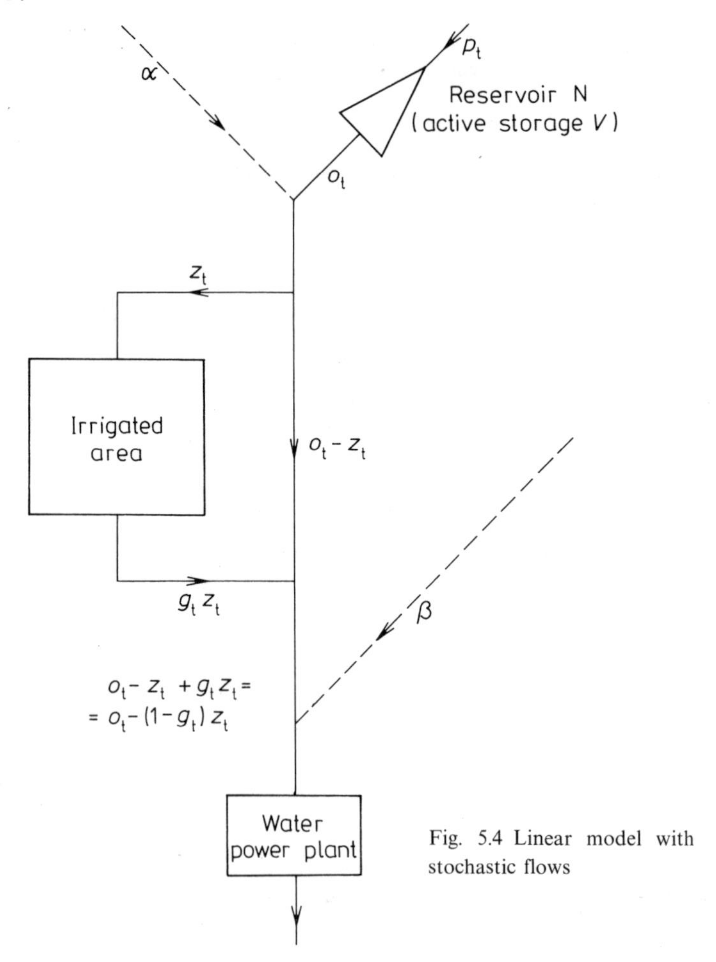

Fig. 5.4 Linear model with stochastic flows

The aim of the task is to determine the values of the following parameters that will maximize the objective function:

- active storage of reservoir N,
- irrigation water requirements V_z for all periods,
- rated generating capacity of power plant I_e,
- target power production in each period E_c.

The flows p_t are given by their probability distributions for each of the three periods involved. The flows in each period are assumed to be statistically independent.

The maximized objective function has the following form:

$$\pi = SH(VE) + SH(V_v) - SH(PN_1 + IN_1) - SH(PN_E + IN_E) + \tag{5.18}$$

+ value of dump energy − losses due to deficits in irrigation water supply and in firm energy,

where π is present value of net benefits from WRS for all periods,

$SH(VE)$ – present value from power production,

$SH(V_v)$ – the present value of benefits from the irrigation water supply,

$SH(PN_1 + IN_1)$ – the present value of operation and investment costs of reservoir N_1 with active storage V_1,

$SH(PN_E + IN_E)$ – the present value of operation and investment costs of power plant with rated generating capacity I_e.

Since the objective function is non-linear, its linearization is performed in a similar way to the first model of this section (Maass *et al.*, 1962).

Linear programming cannot be directly used for this model with a linearized objective function, since it depends on stochastic inlows p_t. Therefore the maximization of the expected value of benefits from WRS is performed and all random variables in the objective function substituted by their expectations.

As an example of some variable expectation computation, the withdrawal of irrigation water in the first period z_1 is considered. This variable can acquire four possible values $z_{1,2}$, $z_{1,3}$, $z_{1,4}$, or $z_{1,5}$ with probabilities r_2, r_3, r_4, r_5. The same probabilities r_y are assumed for flows p_1, e.g. the probability that $z_1 = z_{1,2}$ equals the probability that $p_1 = p_{1,2}$ etc. Then the expected value of irrigation withdrawals in the first period is

$$z_1 = \sum_{y=2}^{5} r_y z_{1,y} \tag{5.19}$$

where r_y is the probability that $z_1 = z_{1,y}$ or $p_1 = p_{1,y}$ (the flow in the first period p_1 will be equal to $p_{1,y}$)

The constraints for the first period are:

$$\begin{aligned}
& n_1 + p_{1,y} \geqq o_{1,y} \\
& n_1 + p_{1,y} - o_{1,y} \leqq V_1 \\
& n_1 + \sum_{y=2}^{5} r_y p_{1,y} - \sum_{y=2}^{5} r_y z_{1,y} \geqq n_2 \\
& z_{1,y} \leqq o_{1,y} \\
& z_{1,y} \leqq k_z V_z \\
& u_{1,y} \leqq o_{1,y} - (1 - g_z)\, z_{1,y} \\
& k_e u_{1,y} \leqq k_d I_e \\
& e_{1,y} \geqq E_c - k_e u_{1,y}
\end{aligned} \tag{5.20}$$

where

n_1 is the assumed reservoir storage at the beginning of the first period,
$o_{1,y}$ – reservoir releases in the first period for the given value of inflow $p_{1,y}$ when $y = 2, 3, 4, 5$,
$z_{1,y}$ – irrigation water supply in the first period for the given flow $p_{1,y}$,
$u_{1,y}$ – the flow through the turbines (power flow) for the given flow $p_{1,y}$,
$e_{1,y}$ – the value of energy added by thermal power plants in the first period for the given flow $p_{1,y}$,
n_2 – the minimum reservoir storage at the end of the first period,
k_z – the coefficient determining that proportion of the irrigation water requirement V_z that is to be supplied in the first period,
k_e – the constant that relates power flow and power output,
k_d – the duration of the period in hours multiplied by the loading factor (e.g. 5/24 if the power plant operates 5 hours per day).

In this example, there are 60 constraints for three periods, which presents no problem for computation. If the problem were more complicated, the decomposition principle mentioned above would be used. This model is also used as a screening model in the design of WRS.

Model using mixed integer linear programming

This type of model was developed in Czechoslovakia by Korsuň *et al.*, 1972, 1975, 1976, and its mathematical principles were published by Glückaufová and Trčka, 1972. Although most applications were used for irrigation systems, they can be applied to any WRS.

Mathematical principles of the model: Integer linear programming is a non-linear procedure that would be linear if it were not for the fact that some variables can acquire integer values only. These procedures may be classified into two types:

– pure integer problems of linear programming are those where all variables are integer-valued,

– mixed integer problems of linear programming are linear programs where some variables are integer-valued and others are continuous.

In practical applications, mixed integer linear programming is often used when some variables can acquire the values zero and one only. Such problems are called bivalued ("go-no-go", or zero-one) mixed integer linear programming problems. The importance of this class of problems lies in the fact that many non-linear problems, which seemingly have nothing in common with the bivalued program, can be formulated as bivalued mixed integer linear programs. The advantage of bivalued programming is the possibility of using universal programs for the optimization of large and complex problems with some non-linear and discontinuous functions. The dynamics of processes can to some degree be reflected in such models, i.e. the time dimension can be considered.

The mathematical model of WRS (Korsuň, 1976) is formed by a set of m equalities and inequalities with n variables. The equalities and inequalities are of the following type:

$$\sum_{j \in I_1} a_{ij} x_j + \sum_{k \in I_2} a_{ik} x_k \gtreqless b_i \tag{5.21}$$

where

$$\begin{aligned} & x \geqq 0 \\ & i \in \{1, 2, \ldots, m\} \\ & j \in I_1 \\ & k \in I_2 \\ & I_1 \cup I_2 = \{1, 2, \ldots, n\} \end{aligned} \tag{5.22}$$

where x_j are continuous variables used for modelling various aspects of the system investigated (e.g. production, technological, financial, economic activities, etc.),

x_k – bivalued variables, or zero-one variables; these variables can be used for modelling integer (indivisible) activities or for discontinuous or non-linear relationships,

a – technological or economic coefficients of variables,

b – the right-hand side constants in equalities and inequalities,

I_1 – the set of subscripts of continuous variables,

I_2 – the set of subscripts of bivalued variables.

The solution of the mathematical model is obtained by a computer run for the optimization criterion determined, formulated as the objective function of the model. In WRS the following objective functions are often used:

– maximization of technological parameters, e.g. the volume of water of the required quality that is produced by WRS,

– maximization of financial gross product, gross or net benefits,

– minimization of investment or operational costs of WRS,

– maximization of equivalent average return of investment costs of WRS or minimization of return period,

– a compromise objective function that minimizes the deviation of several functions from the extreme values. Different weights can be attributed to these functions.

For example, the mathematical model of the WRS on the River Kladénka and on the River Olšava in the south-eastern part of Moravia (ČSSR) was optimized by the following dynamic criterion:

$$\sum_{j \in I_1} \sum_{t=1}^{h} x_j z_{jt} v^t + \sum_{k \in I_2} \sum_{t=1}^{h} x_k z_{kt} v^t - \sum_{j \in I_1} \sum_{s=1}^{g} x_j c_{js} r^{(g-s)} -$$

$$- \sum_{k \in I_2} \sum_{s=1}^{g} x_k c_{ks} r^{(g-s)} \rightarrow \max ! \qquad (5.23)$$

where z are the incremental benefits in years of system operation for one activity unit,

c – investment costs in years of system construction for one activity unit,
r – interest rate factor $(1 + r_1)^t$; t – time in years, r_1 interest rate,
v – discount factor $(1 + r_1)^{-t}$,
g – the total construction time of the system,
s – individual years of construction of the system (index),
h – the average lifetime (planning horizon) of the system,
t – years of system operation.

This criterion expressed the requirement of obtaining the maximum difference between the total amount of discounted incremental benefits (present value of benefits) during the average lifetime of the system (starting from the end of the construction period), and the total amount of calculated investment costs (present value of investment costs) during the construction period. The computation of costs was performed to the end of the construction period.

The method of bivalued mixed integer linear programming described was successfully applied, and it is effective for large and complex systems. The model of the relatively simple WRS on the Rivers Kladénka and Olšava comprised approx. 300 equalities and inequalities with 220 variables. A further model of a rather simple WRS in the upper reaches of the River Svratka comprised 258 equalities and inequalities with 198 variables, five of which were bivalued. According to the authors of the method, modern computers can handle effectively models of up to a limit of 1000 to 1200 equalities and inequalities with 1000 – 1500 variables (with as many as 50 bivalued variables). The model for the River Dyje was developed for six existing reservoirs (completed or under construction) and a further 11 reservoirs in the design stage. The disadvantages of the method are the difficulties in handling longer hydrological time series. If monthly flows are used, optimization can be carried out for up to 3 to 5 years, since a significant increase in the length of the time series makes the dimensions of the model prohibitive. Therefore, for large systems the authors recommend the following method (involving a combination with a simulation model): The preliminary investigation is performed by a simulation model in long hydrological time series to determine one or several critical periods. Then optimization by bivalued mixed integer linear programming is carried out in only these critical periods. The resulting optimal solution is tested on a simulation model in longer time series.

The four linear models described above were illustrations of the various possibilities of applying linear programming in WRS. For a detailed description and further application of linear programming, see (e.g. Maass *et al.*, 1962; Buras, 1972; Kos, 1978, 1979; Loucks, 1981; Vedula and Rogers, 1981, etc.).

5.2 DYNAMIC PROGRAMMING

Dynamic programming is a very powerful, mathematically elegant, and well-known method of operations research. Its main advantage is the possibility of optimizing dynamic processes since the optimization is carried out in a long series of stages of the process, and the aim is to optimize control of the whole process. No restrictions are used for an objective function that need not be linear or analytical. Moreover, the stochastic nature of the processes investigated can be described easily and comprehensively.

These advantages compensate for some of its serious drawbacks. The method is complicated, and it requires a relatively broadly based investigation (it is a principle with many interpretations):

– in this case there are no standard programs for computers, unlike that of linear programming, i.e. for each task a new program must be elaborated and debugged;

– the models require the great memory capacity and high speed of computers; even with large computers a maximum of four independent parameters can be optimized by the basic method of discrete dynamic programming.

This condition determines the limits of the application of dynamic programming in WRS. For example, the maximum problem that can be optimized by dynamic programming[1]) is a WRS with four reservoirs at a rough discriminating level, where the problem is reduced to control of the releases from these reservoirs.

Reference can be made to numerous publications dealing with dynamic programming: fundamental research was done by Bellman, 1957, 1961, 1963; Bellman and Dreyfus, 1962; Bellman and Kalaba, 1965; Vencel, 1964; Ter-Manuelianc, 1966; Walter *et al.*, 1973; Howard, 1960; and others. A brief description of the principles and methods of dynamic programming can be found in each publication dealing with operations research.

[1]) In section 5.2.3 the limits of the problems are extended to 10, or even more, reservoirs by using special computational techniques.

There is an extensive literature on the applications of dynamic programming in water resource management: references given by Partl, 1968, and Buras, 1972, can be supplemented by references in Votruba, Nacházel and Patera, 1974; Kartvelishvili, 1967, papers at symposia and conferences (e.g. Karlovy Vary, 1972) and a recent publication by Yakowitz, 1982.

5.2.1 Main Principles of Dynamic Programming

The *goal of dynamic programming* is to determine a policy for the whole process that maximizes or minimizes the objective function. The process is divided into a series of stages. This division is often done with reference to time (see Fig. 5.5). The computation is often performed in a recursive way, starting from the end, i.e. the last stage. In this *last stage*, for each possible state S of the system at the beginning of the last stage (i.e. at the end of the penultimate stage) the conditional optimal policy R for the last stage is determined. It is a policy that yields the required extreme value k^n for increments of the objective function at this last stage. Next, the *penultimate stage* is investigated. The conditional optimal policy is determined for each possible state of the system at the beginning of this stage. However, this condition

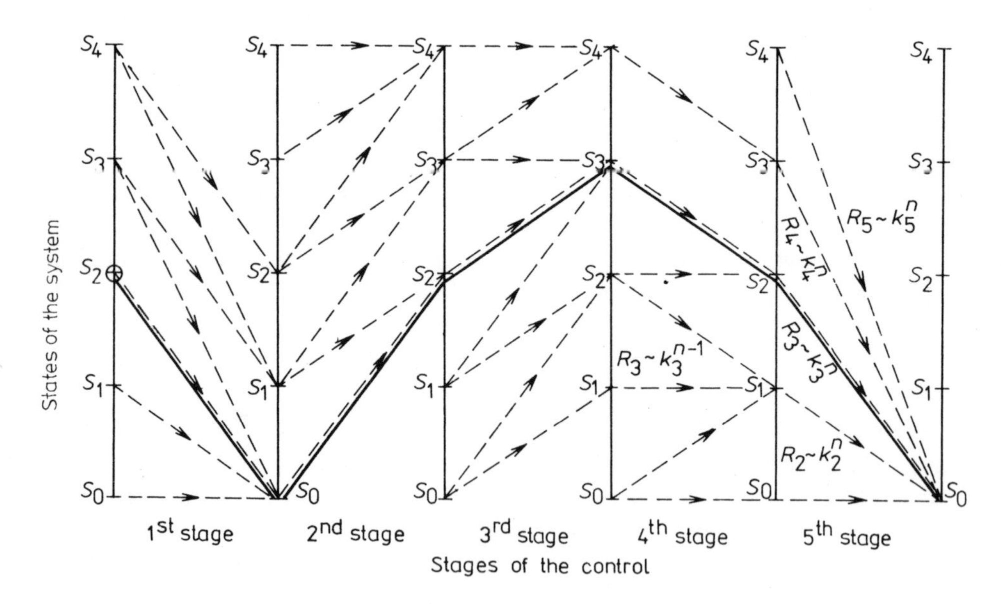

Fig. 5.5 The scheme of the method used in optimization of a five-stage process by dynamic programming

depends not only on increments of the objective function at this stage but on the sum of the last two stages. This relationship is allowed for in the following way:

The policy in the penultimate stage is sought, assuming that the state of the system at the beginning of this stage was S_2. If, for instance, a policy R_3 is used that gives in the penultimate stage the increment of the objective function k^{n-1}, the system at the end of the penultimate stage (i.e. at the beginning of the last stage) is in state S_1. For this state the conditional optimal policy in the last stage R_2 has been determined and the corresponding increment of the objective function k_2^n is known. If the sum of both increments $k_3^{n-1} + k_2^n$ is maximal, then the investigated policy R_3 in the penultimate stage is the conditional optimal policy for the state S_2 at the beginning of this stage. Similarly, the conditional optimal policy in the last two stages is determined in the penultimate stage for all possible states $S_0, \ldots, S_4$.

The method is repeated for the third, fourth, and further stages (counting from the end) until the beginning of the process is reached. At each stage, for all possible states of the system at the beginning of the stage, a conditional optimal policy is determined that gives the maximum (minimum) of the sum of the increments of the objective function at this stage and the objective function in all remaining stages of the process that were determined using the optimal policy in these remaining stages. In Fig. 5.5 the conditional optimal policy is marked with a dashed line.

Then follows the second part of the computation (in Fig. 5.5 the full broken line). Starting from the given (or chosen) state of the system, e.g. S_2, the optimal policy for the whole process is determined, based on all the computed conditional optimal policies at the individual stages (proceeding from the beginning to the end) that give the required extreme values of the objective function for the whole process. This second part of the procedure is carried out quite mechanically in Fig. 5.5. For certain initial states, e.g. S_3, there are several equivalent alternatives with the same total values of the objective function.

With this method, the determination of the optimal policy at a particular stage need not take into account all the possible consequences of this chosen policy for the rest of the process; only the chosen optimal policies in the rest of the process are considered. In this way the number of alternatives that must be compared in optimization is greatly reduced. For example, a very short process consisting of 20 stages with a system of 10 states provides 10^{20} possible alternative policies. The computation and comparison of all these alternatives is beyond the scope of even the largest computers. With the procedure described, the number of alternatives compared is reduced to $10 \cdot 20 = 200$ (i.e. the exponent is changed into a multiplier), comparison of which can be easily performed.

This procedure was developed and precisely formulated by Bellman, 1957. It can be expressed by the principle of optimality, which can be simplified to the following equation:

$$R_i^{\text{opt}} \sim \max K_i^n = \max \left[k_i + K_{i+1}^{n,\text{opt}}\right] \tag{5.24}$$

where R_i^{opt} is the optimal policy in stage i,
$\max K_i^n$ – the maximum value of the objective function from stage i to stage n (i.e. the end of the process),
k_i – the incremental objective function in stage i of the process,
$K_{i+1}^{n,opt}$ – the value of the objective function from stage $i + 1$ to stage n using the optimal policy at all these stages.

The principle of optimality can be formulated as follows: The optimal policy R_i^{opt} at stage i is a policy that maximizes the sum of the increments of the objective function at stage k_i and the objective function $K_{i+1}^{n,opt}$ in the remaining stages of this process using the optimal policy determined. Bellman's classical formulation of the principles of optimality is as follows: An optimal policy must have the property that, regardless of the route taken to enter a particular state, the remaining decisions must constitute an optimal policy for leaving that state.

5.2.2 Application of Dynamic Programming in WRS

The above-stated principle of optimality has been used for the solution of various problems in WRS. Well-known applications are those described by Little, 1955; Cvetkov, 1961; Maass *et al.*, 1962; Kartvelishvili, 1967; Buras, 1972. Recent applications are mentioned in section 5.2.3. The possibilities of this principle of optimality are apparently not yet exhausted and new approaches keep appearing.

A two-dimensional dynamic program

Firstly a simple model conforming to the scheme in Fig. 5.6 will be analysed. It includes two reservoirs and withdrawal for an industrial area. The aim is to determine the optimal river flow regulation at a point P at a distance from both reservoirs; the constraints are the given withdrawals for the industrial area. The operational policy further includes flood control and recreation which specifies flood control storage in both reservoirs, and in the recreation period, a certain minimum pool for recreation purposes is required. The runoff from an interbasin, schematized as tributary inflow, cannot be controlled. The dynamic programming approach is as follows:

– the releases from both reservoirs for the whole process are given by the optimization of river flow regulation at point P,

– the requirements of water supply for industry, flood control and recreation are used as constraints,

– as only two independent parameters are optimized (the operational policy of two reservoirs), a two-dimensional task is formulated that can be handled easily by computers.

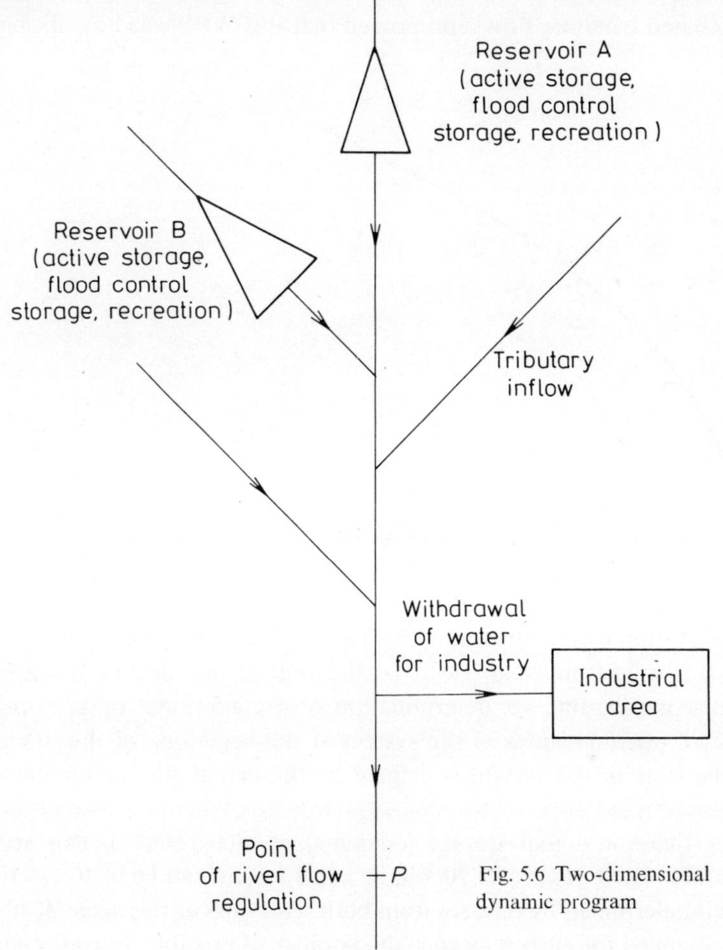

Fig. 5.6 Two-dimensional dynamic program

The choice of the form of objective function should be specified for each task individually. In this case, when optimal river flow regulation at point P was required, a parabolic function of the type $y^2 = 2x$ (Fig. 5.7) was used. This function minimized effectively both extreme values of flows, minimization of the upper values being rather more effective.

The computation was done for a period of thirty years of observed monthly flows (1931 – 1960) i.e. in monthly stages. The formulation was deterministic as the values of the flows were considered as known for the whole period in question. The optimal operational policy obtained cannot be implemented in practice; river flow regulation at point P is the optimum limited by releases from the two reservoirs A and B. Practical result: the computation showed the limited possibilities of river flow regulation at a point P at a distance from reservoirs A and B (due to the relatively large unregulated tributary flow) and proved that this WRS was not efficient.

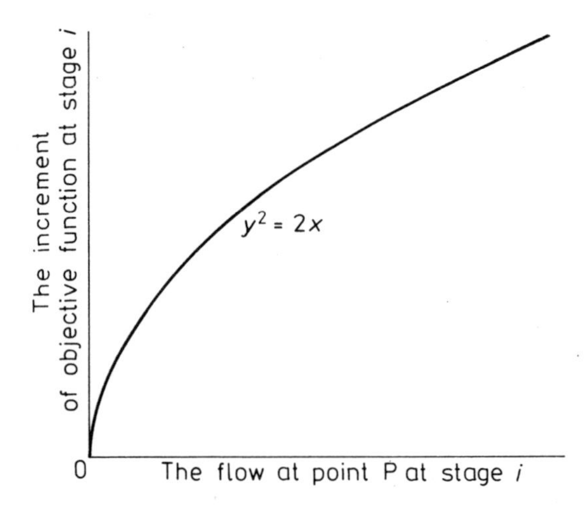

Fig. 5.7 The objective function for a two-dimensional dynamic program

The computation can be illustrated by Table 5.3 where, however, data have been simplified. It is valid from stage $n - 3$, i.e. the fourth from the end. It represents the first part of computation, i.e. determination of a conditional optimal policy (decisions) for all possible states of the system at the beginning of this stage. In this example, the state of the system is defined by the actual storage (contents) of the reservoir[1]).

Assuming that the actual storage (contents) of each reservoir can acquire ten different values (e.g. 0, 10, 20, ..., 90 mil. m^3), this system can be in $10^2 = 100$ states. The policy is determined by releases from both reservoirs at this stage. If 10 different values are assumed for each reservoir, the number of possible decisions equals 100.

In the upper part of Table 5.3 there are three numbers in each field. The first is the increment of objective function at this stage, determined by decision at this stage. The second number is the sum of the objective functions at the remaining stages until the end of the process (it has been determined by the same procedure). The third number is the sum of the first and the second. For each state the maximum sum (denoted by asterisks) determines the conditional operational policy at this stage. The free field contains the stages and states excluded by the constraints.

---

[1]) In other examples the state of the system can be characterized by other parameters, e.g. by values of flows at certain points, etc.

Table 5.3 Scheme of the computation of a conditional optimal operational policy for stage $(n - 3)$

Operational policy at the stage $(n - 3)$		State of the system at the beginning of the stage $(n - 3)$				
	0	–	–	–	–	–
	1	10 95 105	–	–	–	–
	2	19 88 107	19 95 114	–	–	–
	3	27 81 * 108	27 88 * 115	27 95 * 122		–
	4	34 73 107	34 81 * 115	34 88 * 122	34 95 * 129	–
	5	40 65 105	40 73 113	40 81 121	40 88 128	40 95 * 135
	6	–	45 65 110	45 73 118	45 81 126	45 88 133
	7	–	–	49 65 122	49 73 122	49 81 130
	8	–	–	–	52 65 117	52 73 125

* maximum sum

State at the beginning of the stage	0	1	2	3	4
Conditional optimal operational policy at the stage	3	3; 4	3; 4	4	5
Increment of objective function at the stage	108	115	122	129	135
State at the end of the stage	2	3; 2	4; 3	4	4

The lower part of the table summarizes the results. Similar tables are computed for all stages from stage n to stage 1. Then, for the known (or chosen) state at the beginning of the process, the optimal policy for the whole process is determined using the lower part of these tables.

An illustration of the results is given in Fig. 5.8; for a detailed description see Partl, 1968, 1969; Palla and Partl, 1970.

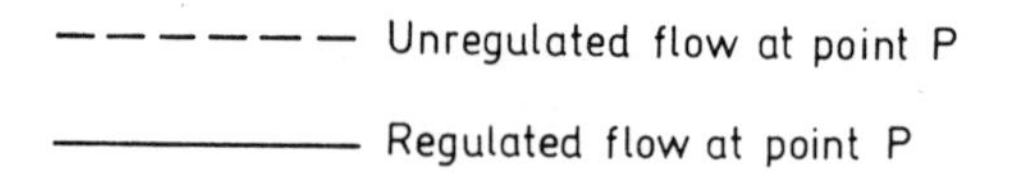

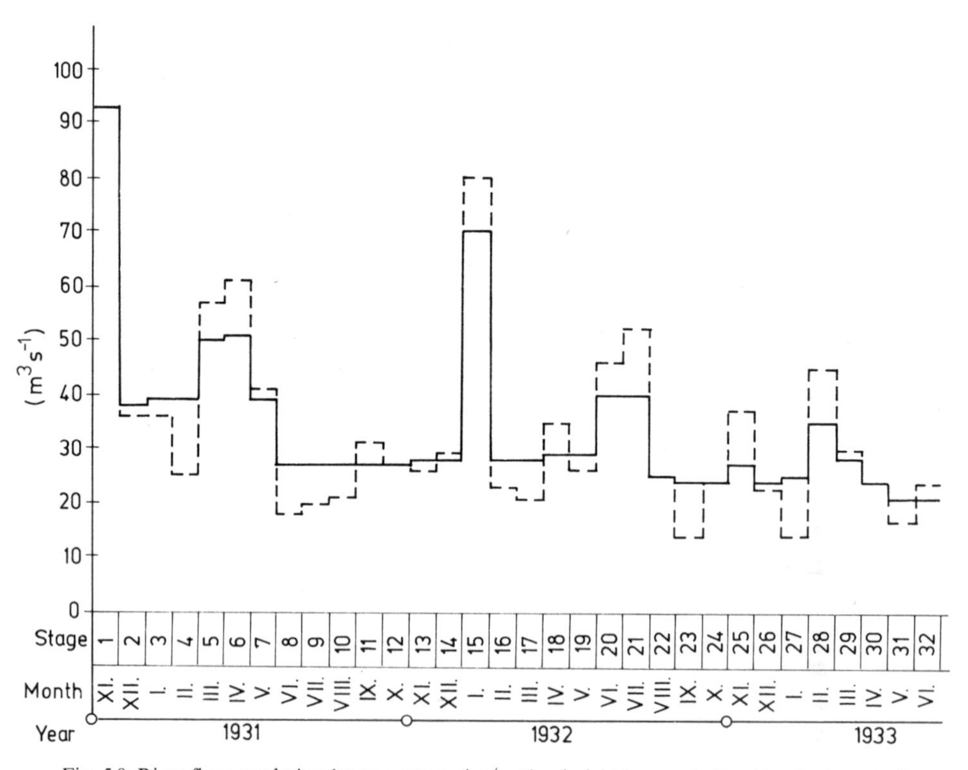

Fig. 5.8 River flow regulation by two reservoirs (optimal yield by parabolic objective function)

This example for 360 monthly stages was computed on a National Elliott 4120 computer. The computation of one stage took 15 seconds. The duration of computation is a linear function of the number of stages and an exponential function of the number of optimized parameters. The computation of a three-dimensional problem (e.g. combination of three reservoirs) took several hours and required the external memory of the computer.

A three-dimensional dynamic program

The WRS described in Fig. 5.9 has been analyzed. It contains three reservoirs. Reservoir A is used for supplying water to an industrial area, reservoir C for irrigation.

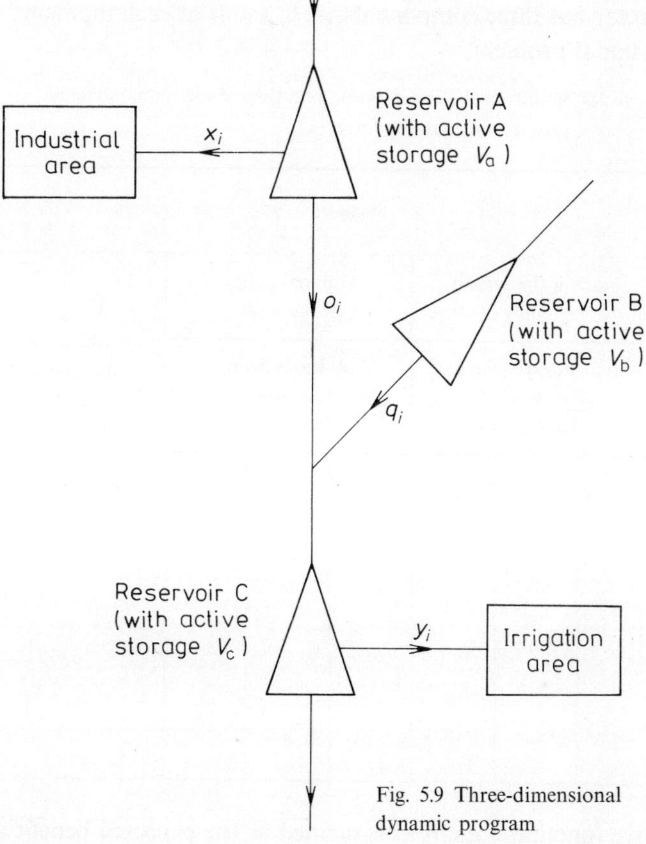

Fig. 5.9 Three-dimensional dynamic program

The economically optimal operational policy is to be determined, i.e. for each stage the releases from reservoirs A and B and withdrawal for the irrigation area that maximize the objective function for the whole period of operation. The withdrawals for industry are not optimized and are therefore not used in the objective function. They are predetermined for each stage.

The relationship between water supplies for irrigation and agricultural benefits is given. The following notation is used:

- x_i – withdrawal of water for the industrial area at stage i (it is predetermined for each stage),
- o_i – the release from reservoir A at stage i,
- q_i – the release from reservoir B at stage i,

y_i – the withdrawal for irrigation at stage i,
a_i – actual storage (contents) of reservoir A at the beginning of stage i,
b_i – actual storage (contents) of reservoirs B at the beginning of stage i,
c_i – actual storage (contents) of reservoir C at the beginning of stage i.
The state vector has three components, a_i, b_i and c_i at each moment; it constitutes a three-dimensional problem.

Table 5.4 Results of the optimization of a three-dimensional dynamic program in a tabular form

Stage $n = 9$						
State of the system at the beginning of stage			Optimal policy at the stage			Value of objective function (mil. $)
Actual storage (content) of reservoirs [mil. m^3]			Releases from reservoirs [$m^3 s^{-1}$]			
a_i	b_i	c_i	o_i	q_i	y_i	

The objective function $f_n(a, b, c)$ is defined as the expected benefit of the n-stage process using an optimal operational policy.

The stochastic nature of unregulated inflows in reservoirs is reflected in the following way: in computation of the optimal operational policy, the mean values (or quantiles, e.g. 10% quantiles) of the observed hydrological series are used. In the application of operating policy this simplification is corrected by a combination of operating policy with the actual state of the system at the beginning of the stage.

In optimization, many constraints must be complied with. These constraints are similar to those used in linear programming. In this example, the following constraints were imposed for every stage:

– the required withdrawal for industry should be met,

– the releases from all reservoirs must not decrease below the given minimum values,

– the withdrawal for irrigation must not decrease below the required minimum values,

– the actual storage (contents) of reservoirs must not be negative and cannot exceed their active storage, etc.

The result of the solution for each stage is the determination of the optimal decision vector with three components o_i, q_i, y_i.

The results for stage $n = 9$ are given in tabular form in Table 5.4. The three left-

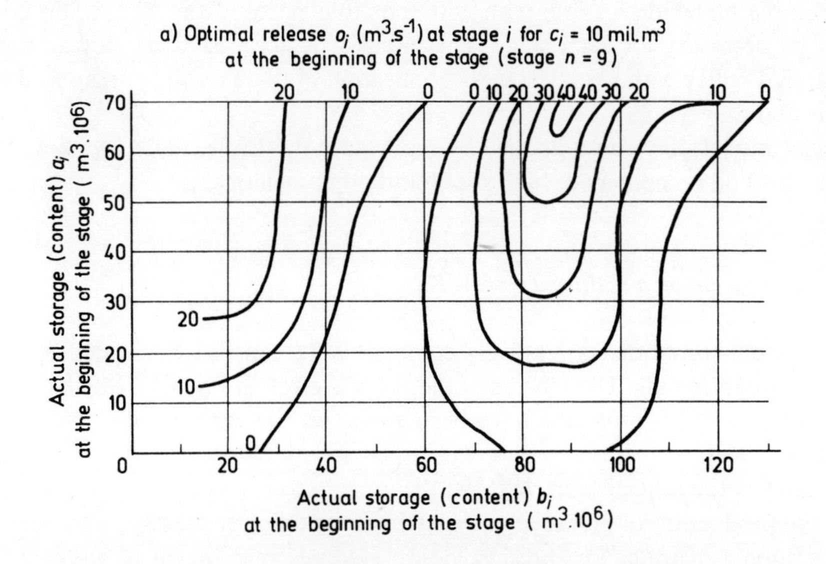

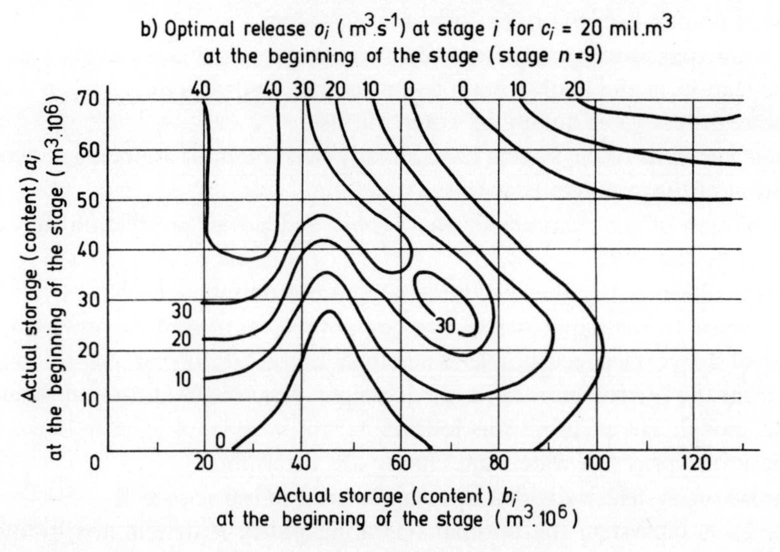

Fig. 5.10 The results of the optimization of a three-dimensional dynamic program in graphic form

hand columns of the table contain the states of the system, i.e. all possible combinations of actual reservoir storages.

If each reservoir is assumed to have ten possible states, this table has $10^3 = 1000$ rows. Sometimes the results are presented in graphical form. It is more instructive and interpolation is facilitated. Figure 5.10 provides an illustration of two graphs for stage $n = 9$. In the first graph, the actual storage (contents) of the reservoirs C at the beginning of the stage $c_i = 10$ mil. m^3; in the second graph, $c_i = 20$ mil. m^3. For the example under discussion, with 1000 possible states of the system, 30 graphs would be necessary for each stage. This number is acceptable, e.g. for an annual operational policy with weekly stages. Application of these results in graphical form is clear and easy.

Similar models have been applied by many authors (Little, 1955; Cvetkov, 1961; Buras, 1972) on computers of high speed and great memory capacity.

Dynamic program of a multi-purpose WRS

This model shows that a relatively complex WRS can be investigated by mathematical programming. The schema of the WRS is given in Fig. 5.11. Similar or even more complex WRS problems have been solved in this way (Parikh, 1966; Hall–Shephard, 1967).

The aim of the schematized WRS is:

– the production of power in three hydroelectric power plants,

– a supply of water for three irrigation systems; one system is supplied from a diversion point where the river flows are regulated,

– a municipal water supply from reservoir D,

– navigation in the main stream as far as the control point,

– the required water quality in watercourses.

As most elements of the system have already been built, or are being constructed, formulation of the problem is as follows:

a) estimation of the maximum water supply and power production that can be guaranteed by the WRS,

b) an operation policy that yields a maximum economic return.

With regard to the capacities of the computers, the use of monthly stages for a period of 10 years is acceptable. Therefore, a critical 10-year period is usually chosen from the hydrological series as a sample, which is used for computation of the WRS model. The method is as follows:

– the initial prices of water and energy are determined,

– the whole system is divided into subsystems, in our case α, β, γ, δ,

– for each subsystem the optimal operating policy is determined by dynamic programming,

– linear programming is used for determination of the operating policy of the whole system that maximizes the net benefits[1]).

– the linear program gives a set of shadow prices that are used as input values for the second run of the dynamic optimization of the four subsystems. New operational rules for releases of water from reservoirs and new optimal values of water supply and power production are thus generated in all subsystems,

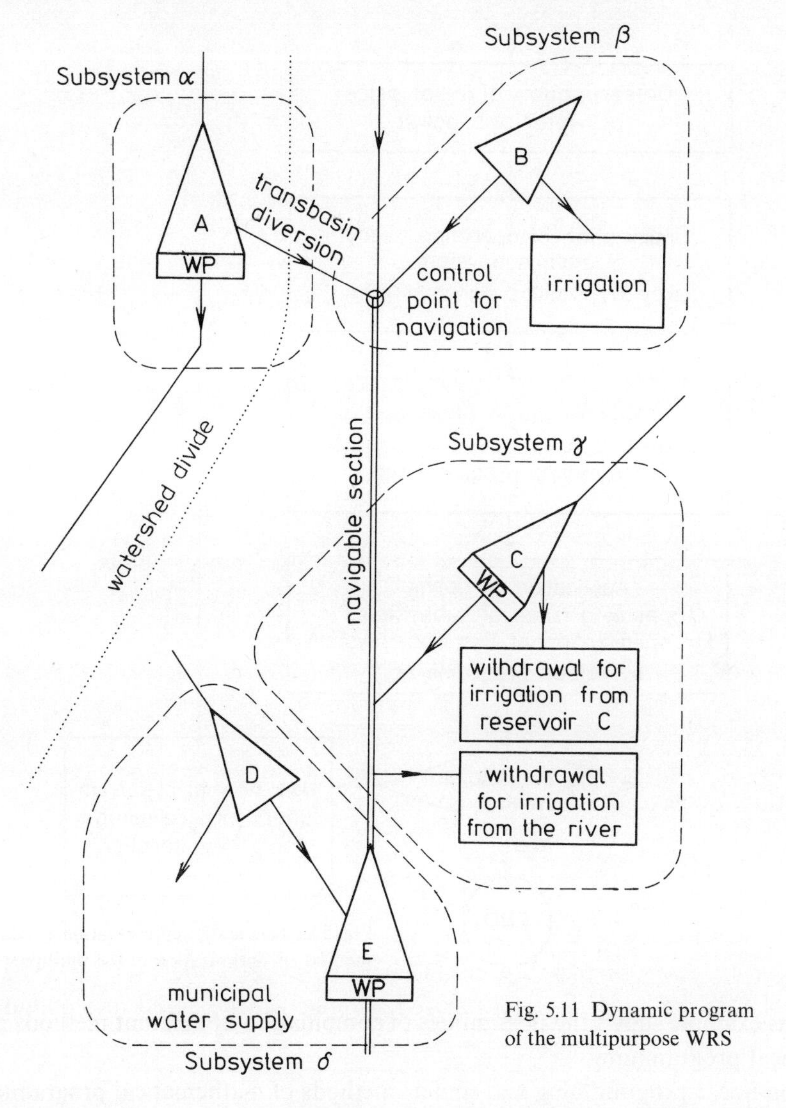

Fig. 5.11 Dynamic program of the multipurpose WRS

[1]) This linear program is often very extensive; it is therefore subdivided in accordance with the decomposition principle (Maass *et al.*, 1962; Dantzig, 1963).

– these results are used for a new run of the linear program for the whole system for the maximization of net benefits from water supply and power production.

The iterative procedure proceeds till the increment of the objective function of the linear program is lower than the given limit. The scheme of computation is given in Fig. 5.12.

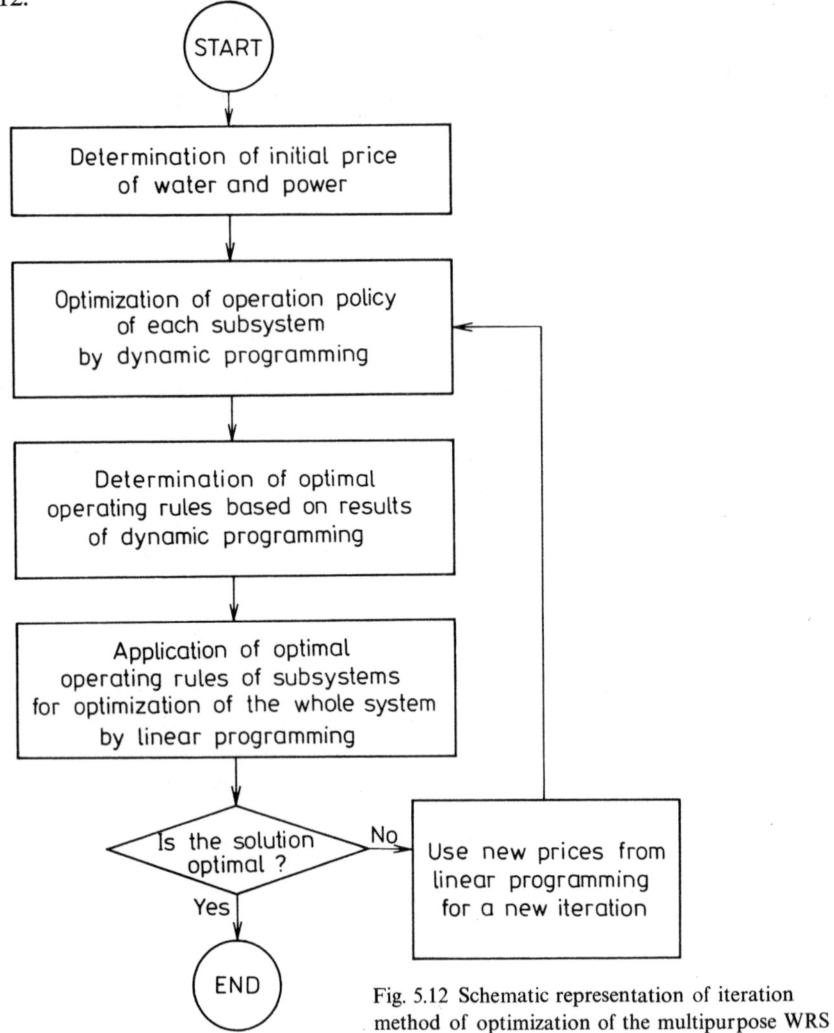

Fig. 5.12 Schematic representation of iteration method of optimization of the multipurpose WRS

This example shows the possibilities of combination of different methods of mathematical programming.

Non-linear programming and similar methods of mathematical programming are not discussed in this book as their practical application in WRS has not been reported yet.

5.2.3 The "Curse of Dimensionality" in Dynamic Programming

The computer technique of dynamic programming described so far was discrete dynamic programming. We have mentioned the exponential growth in memory and time requirements with the increase in the number of dimensions of the state vector. Three dimensions are the maximum that can be handled without difficulty by this method. Heidari *et al.*, 1971, considered a four-reservoir problem (see Fig. 5.13) which is probably the limit of this method of discrete dynamic programming because of the "curse of dimensionality".

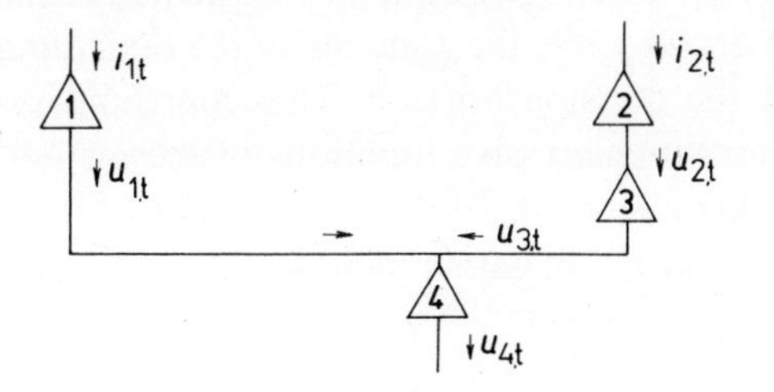

Fig. 5.13 Schematic representation of a four-reservoir problem
$i_{k,t}$ – inflow in reservoir k at stage t, $u_{k,t}$ – release of reservoir k at stage t

A new computational technique of dynamic programming called "discrete differential dynamic programming" has been proposed. It is a successive approximation algorithm using the term "corridor" in which computation is performed. This corridor substantially reduces the dimensions of the problem. However, the computation must be repeated if the conditional optimal policy reaches the limits of this corridor. The drawback of this method, and other subsequent computational techniques, is that the sequence of policies can converge to a local optimum or can diverge unless the control problem satisfies certain stringent assumptions. Chow *et al.*, 1975, used the same four-reservoir problem to text the computer time and memory requirements of discrete differential dynamic programming. In this technique the computation requirements were reduced, but the curse of dimensionality

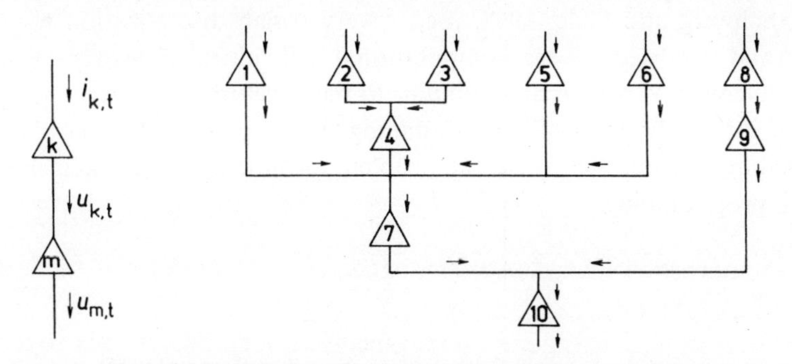

Fig. 5.14 Schematic representation of a ten-reservoir problem
$i_{k,t}$ – inflow to reservoir k at stage t, $u_{k,t}$; $u_{m,t}$ – releases from reservoirs k and m resp. at stage t

still remains. For example, a ten-reservoir problem is beyond the scope of this method.

Murray and Yakowitz, 1979, used this four-reservoir scheme for comparison with a ten-reservoir problem (see Fig. 5.14) and proposed the computational method called constrained differential programming. This method requires for its input a "good" policy (i.e. initial estimate of optimal policy) supplied by the user. The resulting algorithm describes how to compute a better policy given the initial policy.

Another method that can sidestep the curse of dimensionality is state incremental dynamic programming. Trott and Yeh, 1973, described the details of this method and used it for operations of a reservoir system. The principle of the method is the optimization of the releases of the i-th reservoir while the contents of the remaining reservoirs are kept constant throughout the decision horizon. This constraint reduces the dimensionality to one dimension. In successive iterations of the method the value of the fixed reservoir contents changes.

The main assumption of the method is that the dynamic relationship

$$x_{t+1} = f_t(x_t, u_t) \tag{5.25}$$

where x_t is the state at stage t and u_t is the policy at stage t can be inverted in u_t

$$u_t = g_t(x_t, x_{t+1}) \tag{5.26}$$

Thus, state incremental dynamic programming is essentially dynamic programming with a univariate state space.

Differential dynamic programming and state incremental dynamic programming have much greater computational possibilities than discrete dynamic programming, e.g. Yakowitz and Rutherford, 1981, solved a dynamic programming problem with as many as 40 state variables.

This remarkable progress was achieved in deterministic dynamic programming. In stochastic dynamic programming it is not possible to use these methods, and original discrete dynamic programming solution within or outside the water resource literature has the problem of having, at most, two or three state variables. For example, Schweig and Cole, 1968, used a very rough discretization for two state variables and reported severe computational difficulties. However, as water resource problems have served as a stimulus to the development of dynamic programming itself, especially stochastic programming, it is hoped that further mathematical and modelling research will solve the problem of dimensionality even in stochastic dynaming programming.

6 SIMULATION MODELS OF WATER RESOURCE SYSTEMS

The advances in computer techniques have created the possibility to construct models of large-scale and complex systems. In most cases the optimization methods of operations research cannot be used for the solution of the issues involved in these systems, and therefore the use of simulation models employing digital computers is the basic technique for modelling these systems. The problems of water resource systems are typical of large-scale and complex systems, and simulation models have found many applications in this field.

6.1 THE TERM SIMULATION MODEL

Generally, the term *simulation* refers to the technique used for the evaluation of the consequences of some decision without its being implemented in the real system concerned. This definition (Habr and Vepřek, 1973) is somewhat too general for water resource systems and applies to a broader meaning of simulation. Such a concept includes managerial games which simulate a certain economic situation when the participants have to react according to their strategies, analogue models which may be either physical models of water supply and irrigation networks etc., or analogue simulation models which may provide a method of obtaining an approximative solution to the sets of complex integral and differential equations. The simulators used for training drivers, astronauts, etc., likewise refer to this broader meaning.

In water resource system design the term simulation is used in a narrower sense as a modelling technique in which the operation of the water resource system is represented by mathematical and logical relationships in a chosen time step based on specific inputs (inflows of water into the system, demands for the supply of water and for water-derived products and services), capital costs of hydroelectric power plants, capacities of diversion tunnels, etc., and on some predetermined operational policy.

This definition can be expressed in the following concise way: a *simulation model* of a water resource system is a mathematical technique expressing by arithmetical and logical procedures (algorithms) the dynamic behaviour of the water resource system in discrete time steps.

A simulation model describes the relationships among the elements of water resource systems that can be arranged in three classes, i.e. (1) *water engineering constructions*, with one or several characteristic parameters, (2) *natural water resources*, described by time series of flows and by the variables expressing the quality of water and (3) the *requirements for water in the water resource systems*. These requirements can be constant or variable, with some trend component; they may register cyclic variations, and/or they can have a stochastic component such as the requirements for water for irrigation (Kos, 1969a, 1970, 1982).

In principle, a simulation model of a water resource system is the simulation of its operation (with a defined operational policy) on the basis of a time series of flows and/or the variables describing water quality, with predetermined parameters of water engineering structures and other facilities, and demands for water resource systems. The correctness and variables of the parameters chosen are assessed by the *objective function*, consisting of technical quantities and their reliability in different forms and/or economic quantities.

The simulation model consists of *procedures* expressed by precise relationships and symbols describing a certain system, including its elements and their interrelations, and in this way simulates a real situation. The operations are usually expressed in flow-charts, and in a procedural language that is able to describe the *algorithm* of input-data processing and their transformation into output information. The prediction of the consequences of decisions (in the form of parameters of the system) is performed in procedural steps based on input data.

Since simulation models do not use an explicit mathematical, analytical procedure for the determination of the combination of the controlled variables, it is necessary to proceed by trial and error or by the *strategy of parameter sampling*. The structure of the simulated system is so complex that its analytical expression is not possible. Otherwise, a simulation model would not be used.

Simulation models are suitable for all issues of water resource systems with interrelations among variables that can be correctly described by arithmetical and logical expressions. A practically feasible simulation model needs a required amount of *input data* including time series of appropriate duration. These data should reflect the dynamic and often stochastic properties of the simulation process.

Simulation modelling is rather expensive in terms of both model preparation, and programming (however, this part of the costs may be significantly reduced by the use of procedures and simulation languages like SIM-WRS), input-data processing and the cost of the computer time necessary for the computation of alternatives during the search for the optimum solution. The advantage of the simulation model, however, lies in its relatively accurate description of the simulated reality; the method is suitable for communication between technicians as it uses the same principles as the traditional concept of water resource system design, and the output of the simulation models complies with familiar ideas. Simulation is reproducible and therefore it is easy to check. With respect to these properties, engineering estimation and intuition can be used in processing and in the choice of alternative simulation models, and thus the number of computer runs can be reduced.

6.2 PROPERTIES OF SIMULATION MODELS

Simulation models can be classified as *static* or *dynamic*. Dynamic models correspond to the development of human society. Such models take into account the changing parameters of water engineering structures and facilities and the variations in their operation. In view of the long economic lifetime of these facilities, and depending on the difficulties in the application of dynamic changes, it is often possible to use static models with good results. The simulation model of a water resource system is considered a dynamic one if the operational policy can be dynamically adjusted and if such adjustments correspond to the system demands and related changes in system parameters.

Deterministic and *stochastic* simulation models are distinguished by the relationship of the model to the concept of probability. Real situations have a stochastic character. Therefore, Vepřek (1970) defined simulation in a strick sense with reference to the Monte Carlo method and issues of random processes. However, under certain conditions deterministic simulation models may be used. If gauged monthly flows for a period of 40 years, for example, have been used as input variables, the simulation model is called deterministic. Series of gauged flows represent a sample of the stochastic process. If such a sample is used alone and if it can be considered a correct basis for characterizing the stochastic process, the output variables are defined by given input parameters and the operational policy of the system. However, such a modelled process is not deterministic, and the future operation of the simulated system will differ from the operation obtained by the deterministic model.

If phenomena of sufficient frequency in the observed series are simulated, it may be assumed that the differences between the real values and the values determined by the deterministic model will be small so that the deterministic model can be considered a reliable approximation of reality. For phenomena of relatively low frequency (e.g. if a 97–99% reliability is required) there is some danger of underestimation of the dimensions of water engineering structures if only the historical sequence of flows is considered (Kos, 1975).

Two principal methods are used to account for stochastic properties in the simulation model:

– the synthetic flows generated by methods of stochastic hydrology are applied as input values,

– the simulation model is combined with other models that permit a stochastic solution (e.g. the chance-constrained model, Kos, 1975, etc.).

The aim of *stochastic hydrology* is to generate an unlimited number of realizations of the stochastic process on the basis of the gauged time series, which is a realization of the same process. The series generated should not be distinguishable from the series of gauged flows obtained by the methods of mathematical statistics.

Furthermore, a *degree of aggregation* can be used for the classification of simulation models. The simulation model with a detailed discriminative level (i.e. with a low degree of aggregation or high level of detail) is suitable for the investigation of the operation of existing water resource systems, particularly for systems in which gross errors in operational policy have been corrected after a sufficiently long transition period. The objective of such a model is the improvement of the system operation. Beard, 1975, published the results of a case study investigating a system consisting of 20 reservoirs, 12 hydropower plants and a system of thermal power plants. In this case a model with a rough discriminative level would not be able to improve the operation. It would tend to impair it, even if the computed values for the "optimal" solution seem to offer better results. A rough, highly aggregated model would not be able to reflect the specific heuristic policy successfully applied in the operation. A very detailed and precise algorithm had to be chosen for the description of power generation to serve power–system loads with the allocation of hydropower to "peaking" operation, taking into account the second main objective of the described system, i.e. flood control. The desired change in operation was to increase flood-control storage at times when floods are most severe and to increase active storage at other periods.

On the contrary (Jacoby and Loucks, 1972), simulation models with much more aggregated data are appropriate for the design of water resource systems as used, for example, in the investigation of the Delaware River Estuary in combination with analytical optimization models.

In Czechoslovakia, simulation models were used in case-studies of the General Water Plan (1976) and in many projects (Kos, 1970). In the General Water Plan the models encompassed entire basins and the main aim of the investigation was to plan water resource system development; therefore a medium level of aggregation was used. Demands for water supply were investigated in detail, and the optimization of operational policy was performed to the degree that was necessary for proof of design feasibility.

Aggregation of parameters and data is acceptable if it does not create significant deviations from reality. Aggregation simplifies reality. Further simplification can be achieved by *neglecting variables* that do not exercise a decisive effect on the system behaviour (this should be proved by analysis) and by a *change in the form of the variables* (variables which have only an insignificant effect on the output are considered constants). The modelling of a continuous process by a discrete model requires the assumption that the continuous changes during a defined period take place instantaneously at the end or at the beginning of the period.

Simulation models of water resource systems are *discrete* models. The modelled process however is continuous. The decision-making process is discrete. Therefore, the *model time step* is an important dimension of the model and great care must be devoted to its choice. This choice depends either on the degree of aggregation or on the time variability of the input information.

A simulation model reflects the processes in water resource systems by a series of "snapshots" in specified time steps. For water supply purposes monthly periods are generally used. These periods make it possible to reflect the seasonal variability of demand and hydrological data. On the other hand, low flows, which determine the output, change only slightly during the monthly period. The differences between the continuous reality and the monthly values in the simulation model are usually lower than the accuracy of the demand data (Kos, 1976).

In Central European conditions monthly periods are too long for flood control. Hourly gauging records are therefore applied. The reduction of the length of time steps requires an increase in their number in the simulation model. Therefore, in many projects of water resource systems, flood control simulation is investigated separately from the simulation of other objectives of the system. The flood control storage of the reservoir is determined on the basis of recorded or generated synthetic floods in one model and a second model serves for water supply and other objectives. This model takes flood control storage capacities into account. Such an approach ignores the flood control effect of active storage; the actual effect is greater than the value given by the simulation model.

6.3 DEVELOPING A SIMULATION MODEL

In developing a simulation model the following steps are necessary:

– definition of the problem,

– determination of model input and output, data requirement, availability and processing,

– description of the water resource system and its hydrological relationships, design of the model,

– definition of the parameters of existing structures, estimation of the design simulation parameters for the first model run,

– design of the operational policy of the system,

– assembly of the computer program,

– debugging of the program, model tests.

6.3.1 Defining the Problem

Defining the problem for a simulation model is a matter of prior systems analysis. This definition is not an isolated act but a continuous process of clarifying objectives and achieving precision, leading from verbal expression to a technical and quantitative specification. Many simulation models have been designed to meet the *satisfaction principle*, i.e. to satisfy some present or future demands on water resources in systems. The objective is to achieve the required results at minimum cost. With such a definition of the objective, the term optimum is reduced to a selection from a given number of alternatives (approx. 40 – 60). In view of the theoretically possible effects, this approach leads to a sub-optimum and represents a certain compromise between the requirements and possibilities created by our present state of knowledge (mainly in the economic field).

Simulation models are concerned with the quantitative aspects of the design and operation of water resource systems. An important characteristic is the ill-defined structure of a system involving difficult quantification of some objectives, such as environmental conservation, which are often expressed only verbally. Nevertheless, they influence the definition of the problem for simulation models. A heuristic approach is useful in this phase of model formulation (Kos, 1970).

A basic requirement in defining the problem, and in the formulation and running of the simulation model, is to *comply with limits*. In processes that cannot be simulated due to lack of input data or an insufficient knowledge of the process investigated (e.g. the determination of the relationship between the value of biological oxygen demand and the discharge) a minimum acceptable discharge is specified. This discharge should maintain, under certain conditions, the required water quality in the watercourse.

During the phase of model creation and definition of the problem, the question of the complexity of the model and its ability to reflect reality should be dealt with. The degree of mathematical complexity of the model aiming at a realistic representation should be considered as part of the decision-making process.

6.3.2 Input and Output Determination

There are two main types of *input data* for the simulation model: (1) the variables given by the natural conditions, e.g. the monthly flows in a system of gauging stations, and (2) parameters of water engineering structures and demands on the system, either existing or in the design stage. The acquisition of good hydrological data is not easy even with a relatively dense hydrological network and sufficiently long time series. The periods of observation often do not coincide for all the stations of the water resource system and the stations may not be located at points required by the simulation model. These discrepancies require *homogenization of the data.* If the records are interrupted and the observation periods are shorter in some stations, the records should be completed with reference to the homogenized period by methods of *hydrological analogy*. Cross-correlations between hydrological stations and meteorological and climatic elements are often applied.

In the course of the processing of observed data anomalies may occur. Sometimes peculiar physical phenomena are found (e.g. a negative contribution of a sub-basin without substantial withdrawals). Therefore, the *compatibility of data* in the system of stations is investigated and the necessary corrections are made. The final step is *space interpolation*, i.e. the transformation of the gauged values into values applying to the sites of reservoirs, diversion points, water quality observation points, etc.

In stochastic simulation models, this transformation of observed data is followed by a second phase aimed at the generation of synthetic flow series or stochastic irrigation water requirements. The methods of *synthetic flow generation* have been developed to such a degree that they are commonly used in simulation modelling (Beard, 1973; Kos *et al.*, 1974).

The generation of synthetic flows for shorter periods (daily or hourly flows) is far less reliable. Some successful approaches have been found in a combination of methods of stochastic and deterministic hydrology for a rainfall-runoff relationship, but the problem of coordinated generation in a system of stations with correct timespace relationships has not yet been solved. The stochastic simulation models of flood control and pumping systems are therefore less reliable than stochastic simulation models of water supply.

The *parameters of the system* include not only the storage capacities of reservoirs but also the acceptable minimum releases from reservoirs, the transfer of regulated flows, or of unregulated flows upstream of reservoirs for the enlargement of basins (transbasin diversion), the requirements for water quality, etc.

The input data include the demands in the demand centres and diversion points where these demands on water resource systems are summarized and covered (together with the minimum discharges) by operation of the systems. The designed capacities of reservoir transfers and diversions are variable, the demands in a static simulation model are constant. Sometimes the demands are variable due to the possibility of allocating some withdrawals to different points of the system or of changing the sites of the structures under design (e.g. thermal power plants).

Water resource balance is often the basis of the determination of demand input data. The demand data are sorted and allocated to demand centres of the water resource system, and their coverage is tested in advance. This test requires the values of the *available firm water supply* consistent with the hydrological constraints at the given point. When these data are not available, they have to be estimated by some of the common methods of water resource analysis (e.g. by Ripple's mass diagram). The accuracy of this estimation is not important, it is simply used for a reduction in the number of simulation runs; underestimation or overestimation becomes apparent from the output of the simulation model.

Sometimes some of the required input data are not available because their acquisition is either not possible or too expensive or there is a shortage of time. Then the model has to be adjusted and redesigned. Even in relatively simple models this may be an unfavourable complication. Therefore, it is necessary to compare the requirements on input data of the water resource system with the possibility of obtaining them in the first step of the system analysis, i.e. in the definition of the abstract system. In this way, the additional adjustments of the simulation model are reduced to a minimum.

The *outputs* of the simulation model are either *technical* or economic. The technical variables comprise the minimum values of reservoir storage, the deficits in the required reservoir releases at demand centres and diversion points, especially deficits in firm water supply. Some deficits cannot be allocated to a particular reservoir as they are covered by several reservoirs in the water resource system. Reliability indices are evaluated on the basis of volume deficits, e.g. their duration in time or their relative frequency (number of years without deficits divided by the number of years in simulation run).

For water power production the following inputs and outputs are used: the maximum and minimum operational storage of the reservoirs, the rated generating capacity, turbine water capacity, the number of hours at full generating capacity, the monthly firm and dump energy, the buffer storage, and in pumped-storage plants, energy storage capability, cycle efficiency, etc.

The recreational benefits are evaluated from the visitors-per-day attendance at recreational facilities and periods with reservoir pools at favourable levels (suitable for swimming, boating, fishing, camping, wind-surfing, etc.). For some purposes dynamic characteristics are important (e.g. the rate of the water pool fluctuation, the average duration of minimum water pool, etc).

The *economic* input parameters include costs of storage, irrigation diversion, power and recreation facilities, etc. The economic output values include the costs of operation, maintenance and replacement of facilities, benefits associated with various levels of firm water supply (for municipal, industrial and irrigational use), hydropower, recreation, reduction of flood flows and low-flow augmentation.

Water resource systems at certain periods fail to meet the targer outputs. Therefore, the simulation model outputs include the water supply deficits for various targets, energy deficits, reservoir level fluctuations, the duration, relative magnitude and total volume of the deficits, and the number of users concerned. The economic consequences of these deficits are evaluated by an economic loss function. With deficits these loss functions are the most important components as they determine the geatest differences between the alternatives. Construction of the loss functions, however, is complicated and practical application is hindered by lack of economic data. In view of this fact, the economic optimization in simulation models in Czechoslovakia has often been reduced to the method of relative efficiency. The costs have been minimized with respect to the claim to meet the demands on the system with a reliability given by certain standards. Now this method is being replaced by a differenciated approach; in the design of some water resource systems (e.g. with the primary goal of municipal water supply) this method is still used and will probably continue to be used in the near future. In other water resource systems (e.g. multipurpose systems with irrigation water demands and environmental objective) methods of decision analysis and total efficiency are implemented.

The greatest progress in economic data acquisition and processing was achieved in water power where costs and benefits from generating capacity and energy are available (Schmidt, 1976).

The methods of Cicchetti *et al.*, 1972, for recreational benefit determination were used in "Model of Comprehensive Water Resource Management, 1968 – 1976", and in the evaluation of the utilization of the South Moravian water resource system for recreation.

At best, the output of the simulation model is a value of a *scalar objective function*, where all the effects on the system have been expressed in financial terms. Such an objective function would be the best criterion for comparing the alternative simulation models. The existence of such a function may be supposed theoretically. In practical applications its construction is obstructed by a lack of data. The output of the simulation model should therefore include several parameters for a multicriterial evaluation by the method of decision analysis.

6.3.3 Description of Water Resource Systems, Simulation and Model Design

In the first place, the description of a water resource system includes a *specification of its elements*. In the design of the simulation model, the list of these elements is reduced to quantifiable elements. This step is not a closed one; during the process of model specification the list of elements is often extended by additional elements

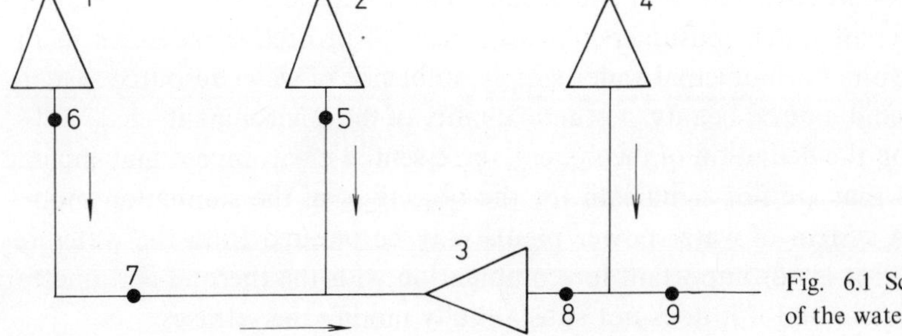

Fig. 6.1 Schematic diagram of the water resource system

(e.g. transfer of water from a neighbouring basin when the hydrological capacity of a particular basin has been exhausted etc.) or is reduced (e.g. some reservoirs are omitted as superfluous for the system).

The *hydrological relationships* between the elements of the system are the other components of the description of the water resource system (Kos, 1967). The schematic diagram of the system (Fig. 6.1) serves as a decription of the basic hydrological relationships especially the movement of water in the system. It contains the basic elements of the system and the directions of the inflow and outflow water. The analysis of water flow helps to identify the connections and cause-and-effect relationships that are used for the description and design of the model. The starting point is usually the demand for water and the water resources are allocated accordingly. In fairly simple systems this process can be carried out by engineering intuition, and, if more resources are available, they may be included in alternative simulation models. In complex systems like the water resource system of the River Delaware (Jacoby and Loucks, 1972) and in the basin of the River Sava (Study of the River Sava, Yugoslavia, 1972), etc., a screening model preceded the simulation modelling. For this purpose, *optimization* by linear programming can be used for water resource allocation to reduce the variety of possible alternatives. However, the linear allocation model is often oversimplified and it has to be complemented and verified by the simulation model.

The description of water resource systems contains the specifications of what is called its *essential environment*. This includes the catchment area or the region where the water resource system is defined, the nature and intensity of agricultural production, soil types (including the texture and structure of soils if water supply for irrigation is a goal of the water resource system) and a brief climatological description of the area.

Towns and cities, especially industrial centres with a *high population density and concentration of activities*, are also demand centres of water resource systems. Therefore, they are included in the essential environment of the systems. Topographical and geological conditions are important for the choice of reservoir sites. They influence the design of a water resource system, so they also form components of its essential environment. The description of the environment of the system includes technical, economic, environmental and social relationships (e.g. economic relationships with different water consumers, consequences of protective measures in the basins of reservoirs for municipal water supply, influence of water resource systems on ecosystems and natural beauty, aesthetic quality of the environment, etc.).

Depending on the definition of the system, the essential environment may include other elements that are not significant for the objectives of the simulation model. For instance, a system of water power plants may be omitted from the water resource system, if it is not important for combination with the thermal and nuclear power plant system and if it does not substantially modify the releases.

Water quality may be included in the environment of some systems. The question is not the importance of water quality in streams but the difficulties of combining quantitative and qualitative aspects in a single model. In such a case, the qualitative model serves for the determination of constraints that have to be considered in the releases in the quantitative simulation model.

The effects of the water resource systems on the environment, e.g. wildlife conservation, fishing, boating and other water-based recreational activities, navigation, river training, etc. are often expressed (for the purpose of simulation modelling) in the environment of the system.

The relationship of the system with its environment is translated into constraints of the model. This process is interactive. If the model runs show that these constraints cannot be compiled with or that they induce high costs, such constraints are often reduced.

In simulation models in Czechoslovakia, the *water supply* for different purposes and low-flow augmentation under certain constraints were the main objectives of the water resource systems. In some systems, water power generation was the objective, in others it formed the constraints. Flood control was investigated in separate simulation models with shorter time steps.

In the design of simulation models it is often advantageous, to subdivide large problems into *subsystems*. Sometimes, however, processing and output and results specification in a single system is easier. Modern computers have a sufficient inner and outer memory for the simulation models of relatively extensive water resource systems. However, systems analysis is very often concentrated on part of the system, and it is not efficient to use the simulation model of the whole system for this purpose. Ways of dividing the system into subsystems are therefore being investigated. Subsystems of large problems can be defined by the following methods:

(1) The first method uses an approach based on the *flow of water particles*. The subsystems are defined on the basis of flow modification. If some groups of elements show few and slight relationships to other elements (they are relatively independent) but the interactions inside these groups are relatively strong and numerous, then a subsystem and a corresponding simulation model can be defined for this group of elements.

(2) The second method of identifying the subsystems is the *functional approach*. This method is recommended if the flow approach fails due to strong interactions among all the system elements. The sequence of functions to be performed for different goals of water resource systems is investigated. The design of a simulation model with a functional approach is scarcely advantageous, since the interactions among the functional groups expressed in flow control complicate the problem, and a simulation model for the whole system seems to be more convenient. An exception is the simulation model for flood control in water resource systems that can be treated as subsystem isolated by the functional approach. A short-term operation has so

many characteristics that the functional approach dominates other views. Although the flood control objective is simulated separately in many water resource systems, methods are being sought to integrate this objective into a single simulation model with variable time advances (steps). In such a model, flood control would not be analysed in detail since the results of a separate functional system would be used, and more attention is paid to the interactions between the flood control and water supply objectives.

(3) The third method, based on the *rhythm of the changes* in the state of the system (state-change approach), is connected with the segmentation into subsystems. In water supply subsystems long-term characteristics are important, whereas in flood control subsystems, short-term characteristics predominate. The rhythm and time advances in the simulation model correspond to this fact. In some water resource systems, some elements (e.g. reservoirs with a long-term operating cycle) show such remarkable long-term effects that the annual flows are more important than their distribution within a year.

In modelling subsystems that are less complex than the systems themselves, some relationships can be expressed mathematically, i.e., analytically. An example is the subsystem consisting of two reservoirs on a river arranged in a close cascade. These reservoirs may be considered as a single reservoir with a storage capacity equal to the sum of the two individual storage capacities. The statistical methods developed for one reservoir (e.g. storage–probability curves), or some special approach, like the chance-constrained model for one reservoir, can be used.

If the system cannot be split into subsystems (where some problems may be investigated in advance), the process of design and application of the simulation model of the whole system is more complex. Even more difficult is the testing of the assumptions and premises of the simulation model and of the effect of parameter changes on the behaviour of the system.

In complex water resources systems, whose effective design and/or operation would be endangered by neglecting important relationships in the subdivision of the problem into subsystems, a simulation model has to be designed for the whole system.

During the design of the simulation model the managers and decision-makers should be informed about the assumptions and premises of the model. They can express certain demands, and the model can be adjusted to meet them. The model provides a new insight into the interrelations between the components of the system. The results of the simulation model can be better interpreted and possible objections forestalled.

6.3.4 Input Parameters of Simulation Models

The *input parameters* are defined as the values that can be changed by the planner, e.g. the design parameters of the reservoirs, capacities of water transfer facilities, values of minimum pools, some water requirements.

The simulation model does not implicitly include the optimization of the system. The optimum is therefore obtained by repeated runs of the model with altered parameters. The parameters are modified to improve the values of the objective function.

The number of possible alternatives is enormous and has to be reduced. Different procedures may be applied. One possibility is to use the results of isolated solutions, i.e. each water management facility is modelled separately to get an idea of its effects on the system. *Operation research* methods, mainly dynamic and linear programming are another possibility. With some simplifying assumptions, the screening of alternatives is carried out to reduce the range of values of the system parameters and the number of their combinations.

During this phase of preliminary investigation the proper definition of the abstract system uses a *rough discriminate level*, so that only substantial features of the problem will be reflected in the model. At the same time, the influence of unquantifiable (environmental, social, aesthetic, etc.) aims on the system parameters is investigated. The purpose of this analysis is to determine input parameters and their range.

The better the combinations of input parameters are estimated, the lower the number of interactive runs of the simulation model. The same number of runs allows either a better approximation of the optimum or the investigation of more alternatives in the case of multi-purpose optimization (the objective function is a vector).

The choice of parameters is complicated by the *stochastic relationship* between the input parameters and the corresponding response of the system. The stochastic nature of the problem is taken into account in the simulation model by the use of the synthetic flows (generated by the methods of stochastic hydrology) and in some WRS by the use of stochastic irrigation water requirements (Kos, 1982) for the input of the model or by combining the simulation model with a method allowing probabilistic evaluation, e.g. with the chance-constrained model. From this aspect, it is useful to start the simulation with the maximum values of the reservoir parameters (or values within the upper third of their range), as the yield of the system is often reduced in the stochastic model (as compared with the deterministic simulation model). This phenomenon may be observed in the combination of relatively high demands on the water resource system with high indices of reliability (e.g. 97–99%).

Further allowance should be made for the *dynamic character* of water resource systems. The dynamics of the development is modelled in the dynamic simulation models. In these models, not only the input parameters are considered but also the process of stepwise storage and transfer capacity expansion in relation to growing demands (Kos, 1968).

If relatively low values of some parameters of water resource systems are determined by a preceding investigation, the next step is often a design involving zero values of these parameters, i.e. omitting of the elements concerned. The choice of the correct input configuration of elements is difficult in water resource systems where the capacity considerably exceeds the demand; a large number of elements can be eliminated from the system.

The proposed Danube–Oder–Labe water resource and transport system may influence many water resource systems in Czechoslovakia. The parameters of water resource systems should therefore be designed in alternatives which take this possibility into account. The development of these water resource systems should be reviewed from the standpoint of their future function as subsystems of the general water resource system in Czechoslovakia with the Danube–Oder–Labe canal as the central element.

In hydroelectric simulation models the parameters of reservoirs (dead storage, active storage, flood control storage, reliable minimum release) are followed by parameters characterizing the water power plant. These are: the rated generating capacity, the turbine water capacity, and the maximum and minimum operational pool of the reservoir. Other parameters of the power plant, such as average annual energy production (firm energy and dump energy), number of hours at full generating capacity, service and out-of-service states for all generating units, are output parameters, and they are determined by the simulation runs.

6.3.5 Operation of Water Resource Systems

The operation of water resource system (WRS) requires continual adaptation of its function to variable conditions determined by the environment of these systems. The course of action to meet the objectives of WRS is also influenced by these conditions. This process has to be anticipated in the design of WRS. It is a complex prognostic problem; the more detailed the design is to be, the farther from reality it may turn out. However, an overall estimate of the operation of WRS (e.g. constant minimum releases from reservoirs in drought periods) may be a very rough reflection of the *real operational policy* of a future WRS.

Experience gained from the operation of existing WRS (e.g. WRS in the basins of the Rivers Odra nad Ohře) show that, in simulation models used for design, the operation of WRS can be modelled by a *simplified scheme* using two kinds of time steps: a simulation model with a main water supply objective (long-term operation) in monthly time steps and a simulation model with a flood control objective (short-term operation) in hourly time steps.

The long-term operational policy should contain an allocation of reservoirs in accordance with the objectives of WRS, e.g. which reservoir should release water for river flow regulation in demand centres. Often, there are several reservoirs that can meet the same objective (e.g. river flow regulation). In that case, the *order* (or an algorithm determining this order) in which the reservoirs can be used for this objective must be specified. The operational rule determining priority can be a fixed one, or it can be a function of the storage of reservoirs (absolute or relative), of the month, or of the demand centre concerned.

Water supply, as well as other objectives influence the operational rules, especially the *minimum pool* constraint in relation to month or season. For instance, for the months of July and August (and sometimes June and September) some minimum pool must be maintained for recreational purposes in WRS reservoirs. In the winter months, in order to secure an effective head for hydroelectric power production, some minimum pool must be maintained in certain reservoirs.

In simulation models the operation of WRS with a *reduction of functions*, e.g. the critical storage in reservoirs, is not usually considered. An exception may be the combined utilization of reservoirs in meeting the demands of users with different requirements concerning reliability, e.g. water supply for industry and irrigation. In this case, two steps of operational policy are often used: the first step without any reduction of irrigation withdrawals, and the second with these reductions in case the storage in the reservoir drops below the critical level. This level is a design parametr that is corrected by trial and error in simulation.

Other courses of action for operation during critical situations are difficult to determine in the design stage of WRS. For instance, in some situations it is advantageous to prolong the duration of reduction in order to lessen the relative magnitude of deficits (e.g. instead of a total failure, i.e. 100% deficit, in one month, three months with 33% deficits). This operation policy rests on an assumption of *non-linear response* of the loss function to the deficits when the economic losses grow more rapidly than the deficits.

Often the optimization of operational policy is carried out by simulation of existing WRS when the design parameters are given and a better operation policy is sought. In view of the uncertainty in some parameters, especially with regard to future requirements on WRS, e.g. till the year 2000, such opearational policy optimization in the design stage could be misleading.

In WRS with water power plants as elements of the system, the operation of power plants influences the overall operational policy in simulation models. Big water power plants are important elements not only of WRS but also of *power systems*. It is, therefore, advisable to consider their function in the power load in combination with thermal and nuclear power plants, which influences the operation of WRS.

Operational policy includes not only the rules determining the quantity and timing of releases, the priority of reservoirs in meeting the demands, etc., but also the rules for *transbasin transport of water*, related to the conditions limiting this transfer.

Depending on the mutual interrelations among the WRS elements in time and space, feed-back is considered in the design of operational policy of WRS, which results in an iterative process.

6.3.6 Assembling a Computer Program

The amount of computation involved in simulation models makes it impossible to perform without a computer. Only a limited part of such computation can be done with a simple calculator.This serves for debugging programs for a few periods. (However, programmable calculators like TI-59 and SHARP PC-1500 are used for simple WRS). Simulation models are therefore discussed in connection with computers and their programming.

A comparison of programs for computers for simulation models with scientific and technology tasks, on the one hand, and data for business information processing, on the other, reveals certain common properties. Simulation models are of a complexity similar to that encountered in scientific and technological problems, and the number of operations performed is similar for the two cases. However, simulation models have more input and output data than these problems. This property of an abundance of input and output data, is a common characteristic of business data processing. It is apparent that simulation models combine the disadvantages and difficulties of both types of computer tasks. *Fast computers* with a sufficient capacity of internal and external memory are therefore necessary for running simulation models.

Programming computers for WRS is a problem that is highly dependent on the model chosen. Therefore, the development of modelling WRS is connected with advances in languages like FORTRAN, ALGOL, etc. Nevertheless, the programming of computers, especially debugging using these languages, is difficult and time-consuming. Therefore, ways were investigated of subdividing the whole program into relatively independent parts that can be debugged separately. The program can be created by linking up these parts, coded as subroutines in FORTRAN or procedures in ALGOL. However, *subroutines*, as stated by Mass *et al.*, 1962, are still dependent on the structure of the model system. For a new WRS, new subroutines must be prepared even if the methodology of their assembly is common to all these tasks.

In the meantime, some *general-purpose simulation languages* (see Chapter 4) have been developed. These languages are more appropriate for simulation in a broader sense, and they are suitable for the representation of some types of production, communication and distribution systems (e.g. queueing problems). They have not found much application in simulation models of WRS.

Special simulation languages have been developed for WRS, e.g. HYDRO, HYCO, etc. In Czechoslovakia, SIM-WRS (Zeman, 1974) is the simulation language of this type. The basis of this language is a simulator package of ALGOL-60 procedures. These procedures have such interface that the properties of WRS and their changes can be modelled by choosing a correct sequence of these procedures and by the determination of their parameters. Details of the application and properties of this language are given in the following section 6.3.7.

Another method for programming the computer for simulation models of WRS used in Czechoslovakia is the development of a *universal simulation model program* that can be used for a wide range of systems (Study of the Sava River, 1972).

For implementation in case studies, some modification of the simulation program is often necessary, as it is not possible to anticipate the whole range of requirements. The symbolic simulation language SIM-WRS, used in many case studies in the General Water Plan, does away with most of the difficult routine programming activities and facilitated a better analysis of input data, better results of simulation models, and a better reflection of operational policy in simulation models.

6.3.7 Principles of the Symbolic Language SIM-WRS

The simulation language SIM-WRS (Zeman, 1974) is a procedure-oriented symbolic language based on a version of ALGOL-60, viz., Elliott-Algol. The procedures constituting the structure of this language can be classified into three groups. (Table 6.1).

Procedures concerning the *function of the reservoirs* or water power, if included in the system, are in the first group. These procedures can be used separately or can be joined into derived subgroups, which simplifies assembling the program. These procedures have the following functions:

MIREL (*P1*, *P2*). This procedure performs the minimum release operation, i.e. operation of the reservoir in such a way as to maintain a guranteed minimum release from reservoirs at points *P1* to *P2*. The corresponding changes in reservoir storage are computed within the limits 0 and WZA_i where WZA_i is the active storage; $i = P1$ to $P2$. (In computation, WZA_i is often the reduced value of active storage. This reduction takes into account losses due to seepage, evaporation and operation).

FLAD (*P3*, *P4*). This procedure performs the adjustment of flows on the basis of the difference DELTA and SPB at points $i = P3$ to *P4*. The DELTA value is equal to the change of storage in one time step. SPB equals zero if the preceding procedure was MIREL; it is equal to the sum of demands covered at points $i = P3$ to *P4* (demands satisfied by the function of WRS) if the preceding procedure was DESU.

Procedure FLAD often follows procedure MIREL. These two procedures can therefore be replaced by the joint procedure MF (*P1*, *P2*, *P3*, *P4*) that performs the function of MIREL (*P1*, *P2*) and FLAD (*P3*, *P4*).

Table 6.1 Basic procedures of the language SIM-WRS

No. of group	Goal of group	Procedure	Main goal of procedure
1	Modelling of functions of reservoirs	MIREL (*P1*, *P2*)	Minimum release operation
		FLAD (*P3*, *P4*)	Flow adjustment
		RES (*N*, *P1*, *P2*)	River flow regulation by water reservoir and minimum release operation
		DESU (*P1*, *P2*)	Demand-supply relationship
		ADDIT1 (*P1*, *P2*, *N1*, *N2*)	Additional release from reservoirs, option 1
		ADDIT2 (*P1*, *P2*, *N1*, *N2*)	Additional release from reservoirs, option 2
2	Input and output of data	DATA	Reading of basic data of WRS – File 1
		MGT	Reading of flows from the magnetic tape (disc) – File 2
		PT	Reading of flows from punched tape or punched cards – File 2
		PRINLP	Print of output on the line-printer
3	Service procedures	ZERO	Initialization of data
		YEAR	Computations at the end of the hydrological year
4	Derived procedures	RDF	Joined procedures RES, DESU, FLAD
		ADF	Joined procedures ADDIT1, DESU, FLAD
		MF	Joined procedures MIREL, FLAD

Procedure RES (N, $P1$, $P2$) performs the river flow regulation at points $i = P1$ to $P2$ by the release from reservoir N. The reservoir active storage is within the limits 0 and WZA_N and the change of storage DELTA is computed. If reservoir N is not able to cover the demand, the deficit CRP is computed.

DESU ($P1$, $P2$) – Demand-supply relationship, i.e. the comparison between the demands on the WRS and the possibilities of WRS to supply water for them is carried out in this procedure at points $i = P1$ to $P2$. In addition, the variables are computed for the determination of various forms of reliability of the WRS at these points, viz., occurrence-based reliability R_1, time-based reliability R_2 and quantity-based reliability R_3. They are defined as follows:

$$R_1 = \frac{m}{n + 1} \quad \text{or} \quad R_1 = \frac{m - 0.3}{n + 0.4}$$

where m is the number of years without any deficit,
n – the number of years under study,

$$R_2 = \frac{d}{t}$$

where d is duration of all periods with no deficit,
t – duration of all periods under study,

$$R_3 = \frac{w - v}{w} = 1 - \frac{v}{w}$$

where v is the total volume of all deficits under study,
w – the total volume of water that needs be supplied.

In addition, the sum of demands covered (SPB) and the above-mentioned value DELTA at points $i = P1$ to $P2$ are computed. The output of this procedure contains the values of discharge FP influenced by the releases, withdrawals and consumption of water.

Procedures RES, DESU and FLAD often follow in this sequence and the joint procedure RDF (N, $P1$, $P2$, $P3$, $P4$) can be used instead of these three procedures.

ADDIT1 ($P1$, $P2$, $N1$, $N2$) computes and allocates the additional releases from reservoirs $j = N1$ to $N2$ for demands at points $i = P1$ to $P2$. If procedure RES was used, then the total amount of release is computed. Otherwise it is known, and this value is only allocated among the reservoirs. In ADDIT1 option 1 is used, i.e. water is released from reservoirs in a predetermined order (if there is not enough water in the first reservoir, the second is used, etc.). Apart from the changes of reservoir storage the values DELTA and CRP are computed.

ADDIT2 ($P1$, $P2$, $N1$, $N2$). This procedure is similar to ADDIT1. Option 2, used in this procedure, involves a difference in the operational policy. The releases are determined so that the ratio of the actual reservoir storage to its maximum value (active storage) should be constant for all reservoirs $j = N1$ to $N2$.

ADDIT3 (*P1*, *P2*, *N1*, *N2*) uses option 3 for the same purpose. This operational policy is known as the space rule (Maass *et al*., 1962).

The sequence of procedures ADDIT, DESU, FLAD occurs very often, and it can therefore be replaced by a joint procedure ADF (*P1*, *P2*, *N1*, *N2*, *P3*, *P4*).

If water power generation is the objective of WRS, then procedures SVEN and PVEN are used. SVEN is used for peaking operation of power plant and PVEN for pumping-storage power plant.

The second group of procedures is used for input and output of data. In ALGOL-60 there is no definition of input and output procedures. Therefore, the procedures in the second group are dependent on the applied modification of ALGOL for the computer, its installation and input/output devices. In the SIM-WRS language the modification was Elliott-Algol with the following procedures: DATA – reads the basic data of WRS. Procedure DATA calls further procedures for reading of data, i.e. PFR, PFNR, PFPNR, CRS. If hydroelectric power is included, then procedures SVE and PVE are called. These procedures have the following function:

PRF, PFNR, PFPNR, CRS are used for the reading of data determined by various forms of demands and minimum flows in demand centres of WRS.

SVE, PVE – reading of input parameters of power plants for peaking operation and pumping-storage power plants, respectively.

MGT – reading of flows from magnetic tape or disc,

PT – reading of flows from punched tape or punched cards (especially in debugging),

PRINLP – printing of output on lineprinter. This procedure is called after the last step of stimulation by the procedure YEAR (see below).

The main service procedures consist of ZERO and YEAR:

ZERO – initialization of variables and indices at the start of computation,

YEAR – computations at the end of the hydrological year, such as the determination of initial storage for the next year, or computation of variables for the evaluation of various forms of reliability.

The application and possibilities of SIM-WRS are apparent from their description. The program for simulation of WRS operation is formed by the appropriate sequence of these procedures with adequate parameters. In this process some rules must be be followed, depending on the organization of input data and the functions of procedures. The main parameters describing the organization of data for simulation are the following:

The dimensions of the task are characterized by parameters:

NA – number of reservoirs in WRS,

CPPF – the total number of demand centres and reservoirs (points),

T – the number of hydrological years under study (the hydrological data at all points of WRS should be available for these years),

PPC – the number of rank indices.

Firstly, the reservoirs are given serial numbers from the natural series, i.e. numbers

1' 2, 3, ... NA, and the demand centres of WRS the serial numbers $NA + 1, NA + 2, \ldots$
.. $CPPF$. Numbering is performed downstream. In addition to the serial numbers, every point is given one or more rank indices.

In these procedures the reservoirs and demand centres must be assigned natural serial numbers (e.g. 15, 16, 17, 18 and not sequence 8, 12, 3, 7). Therefore, the rank indices in the array *P0* are used for this purpose. The parameters of reservoirs *N1* to *N2* and demand centres *P1* to *P2* and *P3* to *P4* are the elements of array *P0* (rank indices), not the serial number of points (with the one exception of parameter *N*). This explains the function of the array of rank indices *P0*. This array is read in input parameters. Further parameters are:

PHPF – the number of gauging stations in which data for computation of flows at points of the WRS are available,

PRIO – flag for function of procedure DESU.

Further input parameters are the pairs AN_i, KO_i that are read for all points of WRS. AN_i is the serial number of the gauging station and KO_i is a reduction coefficient for transformation of monthly flows from the gauging station AN_i to the point i of WRS.

Next, the parameters of reservoirs are read, i.e. for each reservoir, the active storage WZA_i (or reduced active storage taking losses into account) and the minimum guaranteed release MON_i. The demands in demand centres are read by procedures PFR, PFNR, PFPNR, CRS.

The demands are characterized in each demand centre by three parameters:

M_j – minimum flow that has to be maintained at this point j for environmental, water quality, aesthetic, ecological or other reasons apart from water supply,

$W_{m,j}$ – withdrawals of water in month m, at point j. The withdrawn discharge is returned into the river in the section between point j and the following point (often with altered quality or temperature),

$C_{m,j}$ – consumption in month m and at point j.

With the exception of the main parameters NA, $CPPF$, T, PPC, which are read at the beginning and facilitate the dynamic arrangement of arrays, all stated para-

Table 6.2 Scheme of operation policy

No. of reservoir	River flow regulation at points No.
1	6, 7
2	5
3	8
1, 2	8
4	9

meters are read by the procedure DATA. Having called this procedure, the program performs the initialization of variables by the procedure ZERO. Then, in a monthly cycle (return to the label REP), the flows in gauging stations are read by procedure MGT and transformed by the same procedure to the WRS points.

The output file consists of parameters of reservoirs in WRS, i.e. WZA_i, MON_i and the utilized active storage RWZ_i; in demand centres the parametres (M_j, W_j, C_j) are accompanied by the reliabilities in all three forms discussed above.

The sequence and parameters of procedures depend not only on the configuration of reservoirs and the demand centres but also on the operation of WRS. The system of points numbering and the record of operational policy are illustrated by an example of a program of a simulation model of a simple WRS comprising four reservoirs and nine points (five demand centres).

Table 6.3 Procedures and their parameters

Procedures	Reservoir (RES) No.	River flow regulation at point		Additional release from reservoir (ADDIT1)		Demand-supply integration (DESU) at points		Flow adjustment (FLAD) at points	
		No.	Rank ind.	No.	Rank ind.	No.	Rank ind.	No.	Rank ind.
RDF	1	6, 7	3, 4	–	–	6, 7	3, 4	3, 8, 9	5, 7
RDF	2	5	8, 8	–	–	5	8, 8	3, 8, 9	5, 7
RES, ADF	3	8	6, 6	1, 2	1, 2	8	6, 6	9	7, 7
RDF	4	9	7, 7	–	–	9	7, 7	9	7, 7

The scheme of the WRS is given in Fig. 6.1, and the scheme of its operational policy in Table 6.2. The application of procedures is shown in Table 6.3. The correspondence of serial numbers and rank indices is given in Table 6.4. This relationship is used in Table 6.3 with the limits of rank indices (e.g. instead of 5, 6, 7 only 5,7; instead of 8 the limits 8,8 are used). The sequence of procedures simulating the operation, together with other input, initialization, service and output procedures that

Table 6.4 Rank indices

Serial number i	1	2	3	4	5	6	7	8
Rank indices PO_i	1	2	6	7	3	8	9	5

form the user's program, is given in Table 6.5. The whole program consists of a file declaring the variables, the procedures, and the user's program which is appended to this file. At the end of the program the main and inner blocks are closed by the statements "END"; "END"; (or in reference Algol, end; end;).

Table 6.5 User's program

```
DATA;
ZERO;
REP: MGT;
RDF (1, 3, 4, 5, 7);
RDF (2, 8, 8, 5, 7);
RES (3, 6, 6);
ADF (6, 6, 1, 2, 7, 7);
RDF (4, 7, 7, 7, 7);
YEAR;
"END"; "END";
```

The simplicity of programming by the symbolic SIM-WRS – language is obvious. No time-consuming debugging of the program for new configurations of different WRS is necessary. A further advantage of the symbolic SIM-WRS language is the dynamic allocation of the inner memory for arrays according to the main parameters. That facilitates economical utilization of the capacity of the inner computer memory.

The file of procedures is an open system that can be supplemented by further procedures depending on the specific demands of simulation. The basic operational policy, however, is given by the stated procedures. The simplicity of the programming results in the possibility of multi-modelling. For the same problem, different models can be designed which simulate different WRS operational policy.

Depending on the arrangement of input data, the simulation model is not limited by the length of the hydrological time series. Therefore a program written in SIM-WRS can be used for synthetic sequences of stochastic hydrology in stochastic simulation models.

6.3.8 Verification and Validation of Models

The results of simulation of WRS require verification by various tests (Kos *et al.*, 1971). *Correctness tests* are the first. When the procedures have been debugged, these tests can be reduced to one alternative only. However, these tests must not be omitted, as a combination of procedures can activate certain parts that are not normally used and under some specific conditions the program does not necessarily give correct results.

Further tests analyse the *sensitivity* of the model to input data. These tests are carried out by varying the input data slightly; and observing the response of the output variables.

If the system meets all the demands with the required reliability, these tests can result in a reduction of surplus capacities in excess of the required goals. If, however, input parameters acquire their maximum values and undesirable deficits still occur, a correction in the WRS is necessary and new water resources must be incorporated to meet the demands. If there are no additional water resources, or if acquiring them would be too expensive or even impossible due to other reasons (e.g. legal), the requirements on the WRS must be reduced and the simulation repeated.

Verification tests include comparison with the results obtained by different approaches and methods, e.g. balancing, on the basis of the sum of the effects of individual reservoirs (a WRS should give better results), critical period approach (simulation performed for a drought year only), etc. The stochastic simulation model can be compared with deterministic simulation.

These *logical tests* include the service output, e.g. the print of the reservoir storage at the end of each month in the period analysed, the print of flows influenced by the functioning of the WRS at all its important points.

Testing the *validity* of simulation models is the most difficult methodological problem of each simulation technique. The critical issue is confirmation that all the important properties of the system investigated have been reflected in the model and that all the criteria have been properly defined.

The main aim of simulation models is the prediction of some characteristics of a WRS development plan. Each plan rests on a series of assumptions and a change in one of them can substantially influence the validity of the simulation model. The validity of the simulation model is therefore a relative concept and testing it is a continuous process in which convergent and sometimes divergent stages of the relationship between reality and the model may occur. The goal of the simulation model is to ensure that the results of the model can be used at all stages of this development and that they can be used to promote decisions that are correct in principle and can be adapted in future development.

6.4 ANALYSIS OF RESULTS AND DESIGN OF OPTIMIZATION METHODS

In the analysis of the results of simulation models criteria must be formulated for evaluation of simulation outputs. Sometimes these criteria are corrected and reformulated in the process of evaluation of simulation runs. In WRS there are elements whose function cannot be measured (e.g. in economic terms). In that case, an economic comparison of WRS alternatives for the determination of the optimal alternative cannot be performed. Sometimes these functions can be measured, but in units which are not comparable. Therefore, the results of a simulation model may be presented as a *set of alternatives*, optimal according to different criteria. Comparison of the alternatives in this set can be made by the methods of *decision analysis*, with due consideration for and weighing up of different criteria, e.g. costs, power output, economic efficiency, environmental impact, development adaptability, construction and operational stability and reliability, political considerations, etc.

Analysis of the results of simulation models in complicated by the stochastic nature of the problem and evaluation of the results includes the methods of decision-making in circumstances of *risk* and *uncertainty*. The stochastic nature is caused mainly by hydrological data and demands on WRS. The probability distribution of hydrological data is often known. However, the statistical properties of demands and their extrapolation into the future is a more complicated futurological problem that is difficult to solve (IIASA, 1979).

In some cases, probability characteristics of past events of this stochastic process are not a reliable basis for the prediction of demands (non-stationary stochastic process). Therefore hydrological data and demands lead to decision-making in circumstances of risk and uncertainty, respectively.

The risk of failure caused by hydrological conditions in connection with operational policy is the main component of the output of simulation models. Often its counterpart, viz. *reliability*, is examined instead of the risk. The limit of this risk, the acceptable risk, should be determined by an economic and environmental analysis of the WRS. However, the problems of multiobjective optimization and ill-defined WRS structures cannot be solved by these methods. The acceptable risk is then often given by standards or as a result of application of decision theory.

In the General Water Plan (GWP, 1976) in Czechoslovakia the problem of acceptable risk and reliability was solved in the following way: Based on the observed time sequence of flows in the system of gauging stations, operation without dangerous failures is required. Some small deficits were permitted, i.e. deficits such as do not cause an interruption in the functions of the WRS but only some reduction (e.g. a reduction of water supply for irrigation demands to 80% in a critical period, Kos, 1968). This aim was achieved by variation of the WRS parameters, especially of designed active storage of reservoirs. The minimum flow requirements were used as constraints. If this target was not reached with maximum values of active storage, the simulation model representing water quality was re-run with altered parameters reflecting a reduced requirement on water quality and reduced minimum flow requirements.

As stated earlier, the operational policy for flood control is modelled in two kinds of simulation models. The first of these uses the flood control storage in a monthly simulation for water supply and other goals, the second determines the flood control storage of reservoirs and the requirement for river training in an hourly simulation.

In some cases the requirement for flood control storage may conflict with the active storage requirements when the sum of active, flood control and dead storage is greater than the total storage given by the morphological and hydrological conditions of the reservoir site.

In simulation models for water supply, the flood control storage is often taken as a constant. If competition for use of reservoir storage is strong and the total storage is limited, an optimization model is necessary and a new simulation with a water supply operational policy involving reduced flood control storage is performed, or a reduction of water supply demand takes place (or some reduction in both objectives).

If the differences are relatively small, a better multiobjective operational policy may be found to enable the WRS to serve these objectives, i.e. water supply, low-flow augmentation and flood control.

One kind of operational policy that can be used is the dispatching operation. In simulation models this method is reflected by a rule curve giving the necessary storage (dispatching storage) for each month of the year (an annual cycle is considered). For the planning stage of WRS analysis this approach is too detailed and a fixed or flexible operational policy (Maass, 1962) is often tested. In most cases this operation gives a good approximation of the operational adaptability of the system.

If the level of water requirements on WRS is relatively high (e.g. more than 70% of the annual mean flow) and the required reliability is also high (e.g. 97%), the assumption that the observed flows provide a good basis for reliability and risk estimation, is questionable. For example, with a required 97% reliability the result of the simulation model corresponds to the most serious critical period that occurred once in 40 years. This single realization of a stochastic process may be a very inaccurate approximation of possible sequences of flows that can occur under stationary conditions of the stochastic process.

In such cases, a deterministic simulation model should be supplemented by a stochastic simulation model or by another model that includes a stochastic evaluation of the functions of WRS.

Analysis of the result of *stochastic simulation models* is more difficult than that of a deterministic model. If the sections (e.g. 50 years) of the synthetic flow series are investigated in separate simulation models, the output consists of a series of values that can be treated by the methods of mathematical statistics (i.e. parameters and functions such as mean values, standard deviations, cumulative distribution functions, probability density functions, etc. are investigated).

The second approach uses the complete hydrological series generated e.g. 500 years or more and the long-term characteristics of operational policy and design of WRS are investigated.

6.5 DESIGN OF SAMPLING STRATEGY

The basic sampling strategies in simulation models used for optimization are: systematic sampling, random sampling and the gradient methods (e.g. the steepest ascent method).

In *systematic sampling*, a survey of the response surface is carried out at points systematically distributed according to the input parameters. This response surface is given by the values of the objective function (e.g. the costs of the alternative). For example, an alternative in systematic sampling uses all combinations of parameters with their maximum values, and with values corresponding to one third and two thirds of their range, respectively. As the number of combinations in systematic sampling is very large, a combination of this sampling with *heuristic sampling* is often used. Obviously inefficient combinations of parameters are excluded in advance. This method can be used for relatively simple relationships in WRS, where the behaviour of the system can be estimated beforehand.

In *random sampling*, some points are chosen at random from a systematic grid and the objective function is computed. This method is suitable at the beginning of computation when the behaviour of a system is completely unknown and preliminary information is being acquired. Computation of an obviously bad combination is prevented by a heuristic approach or by some screening method of operation research for complex systems.

In a detailed investigation of the behaviour of a system and its response to input parameters, the *method of steepest ascent* is often used. By small variations of parameters the partial differentiations in the direction of the varied parameters are estimated and on the basis of these differences the steepest ascent on the response surface (i.e. the surface formed by the values of the objective function in n-dimensional space) is evaluated. The parameters are changed in the direction of the steepest ascent and a new value of the objective function is computed. This algorithm is repeated until the region of the optimum is reacted. In this area, systematic sampling is often preferred.

The method of steepest ascent requires computation of a large number of alternatives. Therefore, it is used in detailed WRS studies where the results obtained by preliminary investigation in planning models are corrected.

The above-mentioned sampling strategies are used for internal optimization of simulation models. Another type of sampling strategy concerns the relationships between the various WRS models, e.g. models of flood control, water quality, water power, water transport, etc.

In view of the difficulties involved in the explicit formulation of these relationships, the *method of trial and error* is often used in a computation process in which the constraints and prerequisites of individual models and the requirements of higher systems converge.

6.6 IMPLEMENTATION OF SIMULATION MODELS

The main problem in the implementation of models in the process of decision-making is to establish if the prediction obtained on the basis of simulation models is correct, or at least better than predictions made by the implementation of other models or methods of investigation. It is, in other words, a question of the *credibility of the model* and its results. Therefore, it is necessary to analyse how to persuade the decision-makers that the simulation model offers correct results that can be used in decision-making.

Simulation models of WRS have some advantages over mathematical models, as they use the same principles as the traditional methods of water resource research, especially in yield-storage relationships. This applies particularly to the deterministic simulation models where the results can be confirmed by partial calculations using some part of the historical records. For instance, it is well-known that the years 1933, 1934, 1947 and 1954 brought periods of drought in many regions in Czechoslovakia and they may, therefore, be a good basis for the verification of results and operational policies of water supply simulation models.

An actual operational policy of a WRS based on the same principles as the operational policy of the simulation model would be the best proof and simplest verification of the model's correctness. This can be done to some extent in simulation models of existing WRS when optimal operational policy is investigated. The assumption is a sufficiently frequent occurrence of drought periods. In WRS planning and in the design of structural changes in WRS where simulation models are used most, the modelled system does not exist and such a *direct test* cannot be carried out.

In such cases different types of verification tests should be performed. The *adaptability of the model* can be used as one such test, i.e. the ability of the model to react to various situations caused by hydrological input values and also by variations and *changes in demands.* The ability of a multipurpose WRS to adapt and react to such changes is much better than that of a single-purpose system.

In cases where the planned development of water supply demand is hindered by obstacles, the free capacity can be used for some other goal (e.g. flood control, recreation, etc.). In this way there is no unused investment. These properties can be easily reflected in a simulation model as alternatives of demands on WRS.

"Common sense" is often used for the verification of the results of simulation models. The results of simulation models rarely fall outside the limits assumed by engineering experience. If an exception occurs, the print of reservoir contents at the end of each month, releases from reservoirs, changes of flow at all points of the WRS, etc. help to detect possible errors and reveal the cause of the exceptional behaviour of the WRS. If no error is detected, several operational rules can be used and the reaction of the output can prove the correctness of the original operational policy.

The test of the sensitivity of output to input values helps to prove that a simulation model reacts logically, e.g. an increase in demand is related to an increase in storage etc.

An adequate *degree of simplification* is given by the assumptions of model implementation. If a simulation model reflects the main features of WRS only, no description of detailed behaviour can be obtained, e.g. a detailed comparison of different operational rules with and without some small reservoirs in the system, or the problems of weekly operation in a simulation model with a monthly time step. On the other hand, a high degree of detail can be misleading, if the main variables are not thoroughly analysed and all the necessary sensitivity tests are not performed.

At the beginning of the development of WRS simulation models, the validity tests were mainly carried out on the basis of their inner structure, modelled by a sequence of algorithms and coded in a computer program. An example of a simple situation, given in Chapter 4 (Fig. 4.5), illustrates this structure and the simulation method. Examples of simulation technique for more complicated WRS in river basins will be given in Chapter 12.

In simulation models of WRS not all the factors affecting WRS and all the relationships can be reflected; therefore, a simulation model cannot be considered the only tool in decision-making. It is obvious that in the design of WRS, where capital costs are very high, all the known models and their results are used. The results of simulation models are used as a very important element in the process of decision-making. The implementation of simulation models of WRS can therefore be considered a success.

7 INVENTORY MODELS

7.1 TYPES OF INVENTORY MODELS

Inventory theory deals with the problems of mathematical economic models that describe and solve problems connected with the storage of raw materials, intermediate goods, finished products, spare parts, etc. The aim of an inventory system is to balance the differences between production and demand that occur in almost all cases. Inventory theory concentrates on the problems of optimization of inventory systems. Investigation of unused redundant resources, e.g. storage, budget, labour, power production capacities, etc. (Walter and Lauber, 1975) is sometimes included in this theory.

Inventory theory does not apply a unique methodology, and it uses methods from various disciplines. The unifying element is the application of these methods to inventory problems and their optimization.

The basic feature of an inventory system is some space for storage or stock (usually of a finite capacity) that is required to meet the demand (supply of items – system output), and inventory replenishment (system input).

The construction and maintenance of stock requires funds – i.e. costs for setting up and holding the inventory. The objective of an optimal inventory policy is the minimization of overall costs.

The aim of the inventory model is an optimal inventory policy with various constraints. Various tasks may be investigated; e.g. the capacity of stock is given and the main goal is to determine the optimal system output (supply of items, dependent production or consumption process). Highly complex inventory models can occur if the items are stocked at several locations with various interrelationships; an inventory system is formed in this way. Difficulties occur if the stock can serve more than one purpose (this might apply to several kinds of items). The investigation of these systems requires great simplification to solve the problem analytically.

Inventory models are often classified according to the nature of the system variables. If no random effects occur in the inventory process, a single mathematical description is possible and the models are called *deterministic*. In stochastic models, on the other hand, random effects occur, and the models can be further classified by their stochastic characteristics. If they are time independent, the models are *stationary*; if time dependent, they are *non-stationary*, the characteristics of which are functions of time.

An important aspect of inventory models is the time factor. The model without an explicit time effect is called *static*. Usually it is an approximation of reality. A better solution of problems is given by a dynamic model which includes time-dependent relationships. Complex inventory models often require *dynamic* stochastic models.

Inventory models can be classified according to possibilities in the regulation of replenishment and supply, depending on the stochastic nature of one or both these features. All models can be classified according to the nature of parameters into discrete and continuous models.

Like models of operation research, inventory models can be classified by the methods used to find a solution. Most important are two groups of models: analytical and simulation models. The advantage of the latter group is the possibility of modelling complex situations, especially with random variables. Simulation modelling of inventory systems is performed by high-speed electronic computers using special simulation languages.

Different mathematical models can be used in inventory theory. Their choice depends on the result of systems analysis of the problem and its character.

7.2 DETERMINISTIC INVENTORY MODELS

A characteristic feature of deterministic models is a unique description of the inventory process where no random input occurs. Solution of a deterministic problem is often easy but differences occur, depending on the type of model, in relation to maximum storage, e.g. its time dependence and variability of the replenishment, interval, etc. (Dráb, 1973; Wagner 1973).

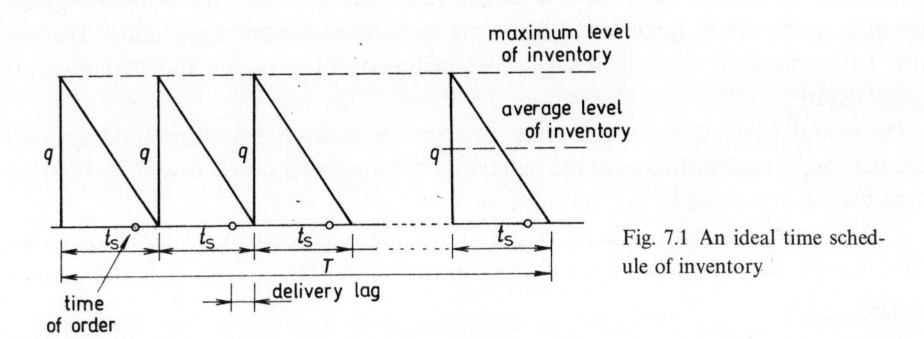

Fig. 7.1 An ideal time schedule of inventory

The simplest model, with an ideal replenishment and state of inventory, is given in Fig. 7.1. The maximum inventory is constant; demands of customers do not change and the replenishment frequency and delivery time-lag remain constant. In reality these characteristics can take different values.

Figure 7.2 illustrates examples of inventory situations when the supply from the inventory depends on the demand of customers and is approximated by different functions: an outline of a nonexhausted inventory is given in Fig. 7.2a. If this situation is repeated regularly in the period investigated, the capacity of stock is greater than necessary and not effective. Figure 7.2b shows that an inventory level can vary as a result of uneven replenishment or withdrawal. In Fig. 7.2c an important situation is described, i.e. when the inventory was exhausted before replenishment; in period t_2 losses occur (e.g. losses in production). This replenishment policy may be deliberate in a model with a non-linear relationship, if a small loss at the end of the replenishment cycle is compared with the greater cost of increasing stock capacity.

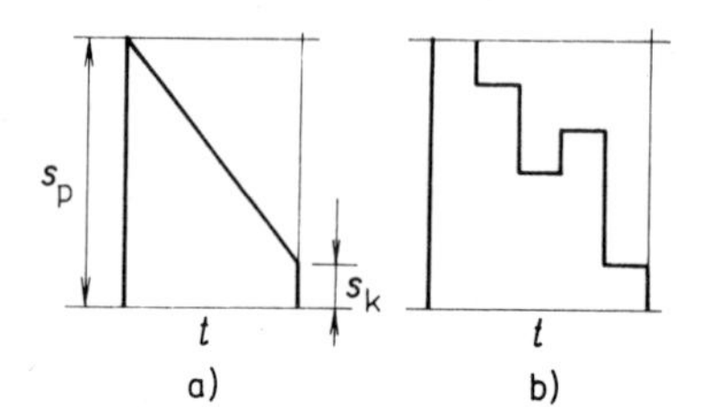

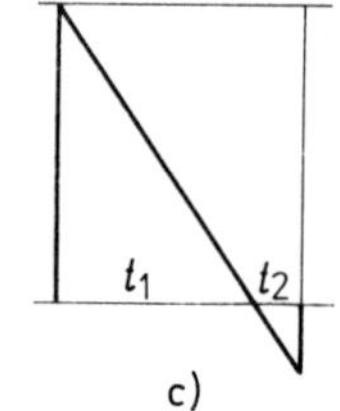

Fig. 7.2 Characteristic examples of inventory situations

This problem is solved by economic analysis which determines which losses or penalties due to exhausted inventory are acceptable as compared with the increased cost of a higher maximum inventory level.

The solution of problems connected with the inventory process is illustrated in an example of a simple deterministic model. In the plan in Fig. 7.1, it is assumed that the stock is regularly replenished by items in batches containing q items. During time T the amount Q is supplied in Q/q replenishments of items, scheduled at intervals $t_s = T/(Q/q)$.

The overall costs of the inventory process are composed of inventory holding costs and the cost of replenishment of the batch of q items (reorder costs). Inventory holding costs can be expressed in the following way:

The average inventory level is $q/2$, assuming a regular, continuous demand. The costs of storing one item are C_1 per unit of time. The inventory holding costs for time t_s are

$$\frac{q}{2} C_1 t_s \tag{7.1}$$

and during the whole production period T

$$\frac{q}{2} C_1 T \tag{7.2}$$

The costs for delivery of one batch of q items are C_s. The costs for delivery of Q/q batches are

$$C_s \frac{Q}{q} \tag{7.3}$$

The total costs are the sum of expressions (7.2) and (7.3)

$$N = \frac{q}{2} C_1 T + C_s \frac{Q}{q} \tag{7.4}$$

Figure 7.3 illustrates the relationship of functions (7.2), (7.3) and (7.4) to batch q.

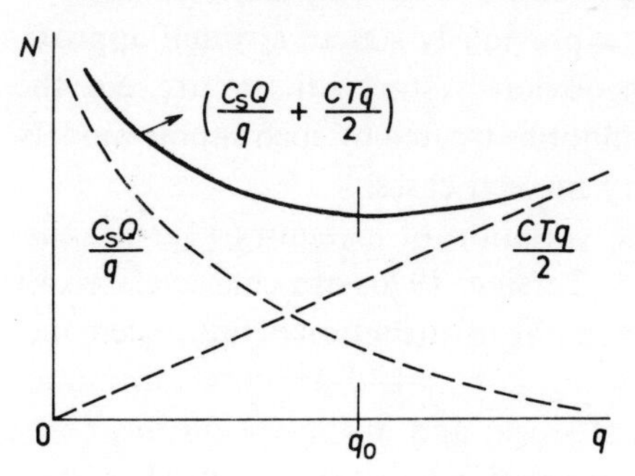

Fig. 7.3 Relation between the total costs and the batch size

The batch size q_0 that minimizes costs, is given by the zero value of the first derivative of function $N(q)$ with respect to q

$$\frac{dN}{dq} = \frac{1}{2} C_1 T - \frac{1}{q^2} C_s Q = 0 \tag{7.5}$$

It is solved for q_0

$$q_0 = \sqrt{\frac{2C_s Q}{C_1 T}} \tag{7.6}$$

Minimum costs N_0 are obtained by the substitution of (7.6) into (7.4)

$$N_0 = \frac{q_0}{2} C_1 T + \frac{C_s Q}{q_0} = \sqrt{2C_1 C_s Q T} \tag{7.7}$$

From the basic relationship for the replenishment interval t_s

$$t_s = \frac{T}{Q} q$$

the optimal interval t_0 that corresponds to minimum costs is obtained by substitution of q_0 for q

$$t_0 = \frac{T}{Q} q_0 = \sqrt{\frac{2C_s T}{C_1 Q}} \tag{7.8}$$

Equations (7.6) and (7.8) show that variables q_0 and t_0 depend on the ratio of costs C_s/C_1 (dimensionless value) and on ratio Q/T or its reciprocal value.

The type of model presented above explains the technique of model building and the method of derivation of optimal parameters that can be applied to more complex models (Ter-Manueliane, 1968). These complex models can describe actual inventory systems in big organisations with automatic inventory control using modern electronic computers.

7.3 STOCHASTIC INVENTORY MODELS

Simple static or deterministic models are, as previously stated, a rough approximation of reality. Inventory problems of an obviously stochastic nature, e.g. the problems of the operational policy of WRS, cannot be treated by such simple models. Dynamic and stochastic models are necessary in such cases.

Dynamic inventory models with stochastic variation of inventory (Ter-Manueliane, 1968; Hanssmann, 1962; Prabhu, 1965; Tersine, 1976) are characterized by resupplies (constant or variable) and by the probability distribution of future demand. Models of such problems include the following cost: build-up costs, inventory holding costs, penalty costs for an item out of stock, and inventory surplus costs.

In dynamic and stochastic models (as compared with static models), certain new features appear. For instance, the surplus inventory costs do not have only a negative impact as in static models, since the period of inventory holding is not assumed to be finite. Therefore, inventory surplus in one period can be compensated in the following period by a reduction in re-order. Therefore, the surplus inventory costs are equal to the increment of inventory holding costs.

The stochastic nature of demand results in variation in inventory consumption. If a stortage of inventory stock results in high penalty costs, the inventory level that corresponds to the average consumption is increased by safety (buffer) stock that should guarantee the inventory system against stochastic deviation of demand to higher values. The safety stock diminishes the risk of running out of stock, but it ties up the capital that otherwise could profitably be used, it requires inventory space, inventory maintenance, etc., and in brief, it represents additional inventory holding costs. It is clear that in dynamic stochastic models the penalty costs induced by a shortage of inventory stock should be balanced by the costs of safety stock.

In dynamic stochastic models the most important parameters are the intervals of replenishment. In a system with deterministic demand, the replenishment interval was computed without difficulty. On the other hand, in systems with stochastic demand the time required for replenishment is the main reason for the safety stock. The replenishment intervals can vary and during these intervals no items can be delivered. Therefore, increments in consumption during these intervals can be met only by safety stock.

The inventory policy in dynamic models consists of two types of decisions: the timing and magnitude of the resupply decision, i.e. indications when to replenish, and the amount to replenish.

In WRS, often a reversed alternative of this problem is considered. This alternative is also treated by inventory theory, and the solution of problems is analogous to that described below: usually the stochastic replenishment of items is assumed it corresponds to stochastic inflow into a reservoir), and the demand is governed by some rules (withdrawal of water from a reservoir according to some operational policy).

A further important difference between stochastic and deterministic models is the relationship between frequency and magnitude of orders that in deterministic models was given by the equation (7.9)

$$q = \frac{Q}{n} \tag{7.9}$$

where n is the frequency of orders Q/q

q – the batch size ordered,

Q – amount of demand during the period investigated.

In stochastic models where demand is a random variable, equation (7.9) holds true for mean values only. In individual periods there are random deviations in the actual demand from this mean value, and therefore a random deviation in the actual inventory level from the mean value occurs. The effect of demands on the actual inventory level must be balanced. This can be achieved by a change in the ordering frequency with constant batch size or by a change in batch size with a constant interval between replenishments. These procedures are the basis of the two main inventory policies in models with stochastic demand, viz., the Q-system and the P-system inventory policies, respectively (Ter-Manuelianc, 1968; Whithin, 1957; Wagner, 1962; Hadley-Within, 1963).

7.3.1 Q-System Inventory Policy

The Q-system (Fig. 7.4) uses a constant batch size, and variation in demand is balanced by changes in the frequency of orders. The safety stock is defined; it is used as a buffer for meeting demand during the delivery time-lag. As soon as the bulk of the stock has been used up, i.e. inventory reaches the signal level, a new batch is ordered and the safety stock is used till its arrival. Determination of the signal level takes into account the random variation of demand and penalties for items out of stock. As all batches are of a constant quantity, these systems are called Q-systems, where Q refers to the constant quantity. The safety stock is first refilled from the delivered batch and the remainder forms the bulk of the stock.

Variation in demand is met by the changes in the ordering frequency and therefore no safety stock is necessary for random increases of demand during the re-order cycle. It is sufficient to order the quantity that corresponds to average demand in the period investigated. If the actual demand is higher, the inventory level will drop sooner to the level of the safety stock and a new batch is ordered sooner. This automatic regulation of demand variation cannot be performed during the delivery time-lag. In Q-systems the safety stock safeguards the system from penalties due to items being cut of stock because of higher demand during the delivery time-lag.

The relationship between the level of the safety stock and the delivery time-lag is apparent from the equation for total inventory costs.

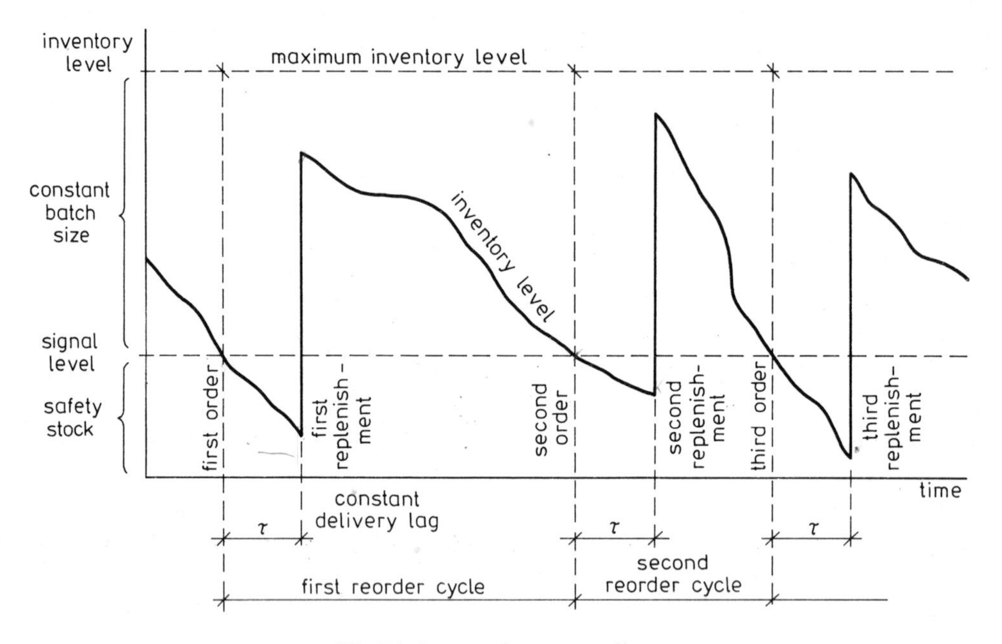

Fig. 7.4 Q-system inventory policy

The methods for the Q-system:

Notation

Q – the total demand of items per year,

q – the stochastic demand during the delivery time-lag with probability density $f(q)$,

z – safety stock in items,

C_s – costs of one batch,

c_1 – price of one unit in batch with costs C_s,

$c^{(1)}$ – annual inventory holding costs as a percentage of price of mean inventory level,

$c_1c^{(1)}$ – expected annual holding costs of one item,
C – penalty costs during the delivery time-lag (due to shortage of items); these costs do not depend on the level of item shortage,
t – mean duration of re-order cycle in years (expected interval between successive orders),
τ – delivery time-lag,
q_τ – average demand during delivery time-lag,
$1/t$ – number of re-order cycles per year,
C_s/t – costs of all batches per year.

Inventory holding costs for annual demand Q are

$$\frac{1}{2}Qc_1c^{(1)}t \tag{7.10}$$

Inventory holding costs of safety stock are

$$c_1c^{(1)}z \tag{7.11}$$

The computed costs of risk of inventory shortage for one re-order cycle are

$$C\int_{q_\tau+z}^{\infty} f(q)\,\mathrm{d}q \tag{7.12}$$

The costs of risk per year are

$$\frac{C}{t}\int_{q_\tau+z}^{\infty} f(q)\,\mathrm{d}q \tag{7.13}$$

The total annual inventory costs are given by the expression

$$N_{(t,z)} = \frac{C_s}{t} + \frac{1}{2}Qc_1c^{(1)}t + c_1c^{(1)}z + \frac{C}{t}\int_{q_\tau+z}^{\infty} f(q)\,\mathrm{d}q \tag{7.14}$$

The values of t and z that minimize the total costs were derived from equation (7.14) using the partial derivative with respect to z and taking its zero value; then t is expressed by:

$$t = \frac{C\,f(q_\tau + z)}{c_1c^{(1)}} \tag{7.15}$$

The value (7.15) is substituted into (7.14) and some rearrangement

$$[f(q_\tau + z)]^2 = \frac{2c_1c^{(1)}[C_s + C(1 - F(q_\tau + z))]}{C^2Q} \tag{7.16}$$

With the given probability density $f(q)$ the above equation can be used for computation of the optimum value of the safety stock $z = z^*$ with given values q_τ, c_1, $c^{(1)}$, C_s, C and Q, and from the equation (7.15) the optimum average length of the re-order cycle $t = t^*$ is determined. The optimum batch size is then determined by $o^* = Qt^*$.

7.3.2 *P*-System Inventory Policy

The second type of inventory policy in a dynamic and stochastic system is based on firm re-order time intervals with variable batch size. The batch size is determined on the following assumption: The sum of the batch size and the actual inventory level at the re-order point should equal the given quantity, i.e. the re-order inventory level, that is determined according to the variation in demand. The change in the batch size balances the variation of actual demand. The symbol *P* was derived from the word period as the model uses firm periods of re-order cycles (Fig. 7.5).

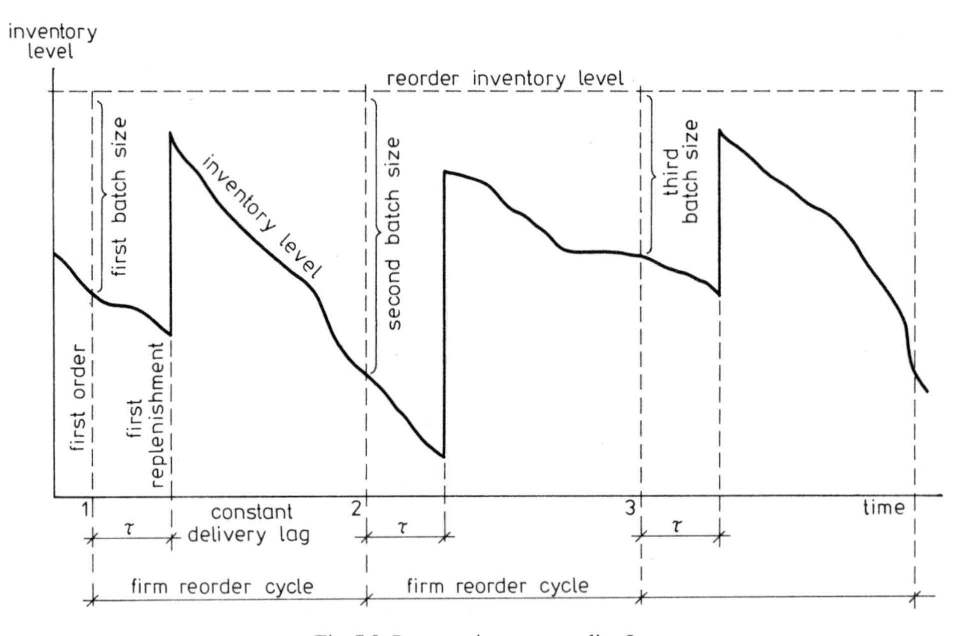

Fig. 7.5 *P*-system inventory policy I

In practice, the actual inventory level is monitored at regular intervals and the items are re-ordered accordingly. The model is used for the determination of the optimum re-order cycle duration and the optimum re-order inventory level, i.e. the level that should be reached at re-order point. The batch size is given as the difference between the re-order inventory level and the actual inventory level at re-order point. This quantity may be increased by the batch size of orders that have not yet arrived in stock. Such a situation (Fig. 7.6) is possible if the delivery time-lag is longer than the re-order cycle. The two main parameters, defining the *P*-system, are the duration of the re-order cycle and the re-order inventory level. The main difference between the *Q*-system and the *P*-system is in the method of dealing with a stochastic variation of demand. In *Q*-systems, a higher demand requires shortening of the re-order cycle, and the safety stock is used to meet the higher demand during the delivery time-lag. In P-systems, the safety stock has to cover variations in demand during the whole re-order period. There is some relationship between the batch size in one period and the risk of items being out of in all subsequent periods. For simplification of the P-system model, the safety stock is often determined in relation to one re-order cycle plus the following delivery lag.

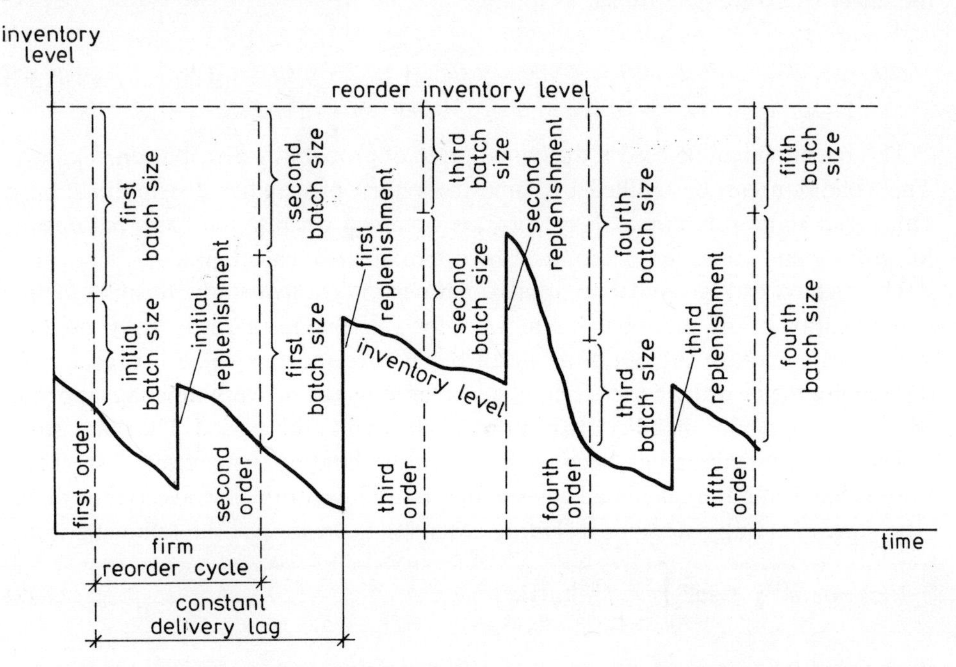

Fig. 7.6 P-system inventory policy II (if the delivery lag is longer than the reorder cycle)

The method of P-system optimization (notation is similar to that in the Q-system):

Q – the total demand of items per year,

q – demand per week (stochastic variable); one week was chosen as the basic time interval,

$f(q)$ – probability density of weekly demand q (more precisely v-th convolution of probability density),

C_s – costs of one batch,

c_1 – price of unit in batch with costs C_s,

$c^{(1)}$ – annual inventory holding costs as a percentage of price of the average inventory level,

$c_1 c^{(1)}$ – annual holding costs of one item,

C – penalty costs during delivery time-lag (due to shortage of items); these costs do not depend on the level of item shortage,

t – duration of re-order cycle in weeks,
τ – delivery time-lag in weeks,
q_{week} – average demand per week,
z – safety stock.

The total annual inventory costs related to the duration of the re-order cycle and the safety stock are determined as follows:

$$N(t, z) = \frac{52C_s}{t} + \frac{1}{2} c_1 c^{(1)} t + c_1 c^{(1)} z + \frac{52C}{t} \int_{(t+\tau)q_{\text{week}}+z}^{\infty} f^{t+\tau}(q) \, dq \tag{7.17}$$

This equation can be used if the convolution of probability distribution is known. This condition can be fulfilled by some theoretical probability distributions only. Otherwise, some suboptimal P-systems are obtained because the basic parameters for given input values are determined by approximative calculations.

The method can be similar to approximations in Q-systems. The formula for the determination of optimum batch size, valid for a deterministic case, can be used for the determination of the optimum mean batch size, and this value is used for the duration of the re-order cycle. Then it is increased by the delivery time-lag, applying the convolution of probability distribution of demand to this period. The safety stock is determined by taking into account the inventory holding costs of safety stock, on the one hand, and the calculated costs of the risk of inventory shortage, on the other. The corresponding equation, including costs, for this case has the following form:

$$N(z) = c_1 c^{(1)} z + nC \int_{q_0+z}^{\infty} f^{t+\tau}(q) \, dq \tag{7.18}$$

where n – number orders per year,
q_0 – demand during the re-order cycle,
other symbols as previously defined.

The optimum level of the safety stock is computed by a similar method to equation (7.15), using the partial derivative with respect to z and taking its zero value:

$$f^{t+\tau}(q_0 + z) = \frac{c_1 c^{(1)}}{nC} \tag{7.19}$$

Up to this point, it has been assumed that the safety stock is always available. Actually, this inventory may be partly or completely exhausted. In that case, it is not correct to compute the inventory holding costs of the whole safety stock. If this fact is considered in the cost function, the expected average cost of the safety stock should be applied. In this way a further alternative of the P-system is formed. Similarly, other assumptions for the optimization of parameters of P-system inventory policies could be used.

7.3.3 Application of Markovian Stochastic Processes

The application of inventory theory is easier if a certain type of stochastic process is assumed, called the Markovian process – see Chapter 3 (Walter, 1970, 1973; Votruba and Nacházel, 1971).

A simple stochastic model of an inventory policy for a Markovian process will be demonstrated. The ordering points are fixed, the batch sizes differ; the *P*-system is therefore assumed. Two economically different inventory policies will be compared (see Table 7.1).

Table 7.1 Two inventory policies

Inventory level at the beginning of the week	Batch size	
	1st policy	2nd policy
0	1	2
1	1	0
2	0	0

The weekly demand is given by the following probability distribution:

demand q (in items)	0	1	2
probability $p(q)$ of occurrence of demand q	0.2	0.5	0.3

A weekly replenishment interval is assumed. The replenishment costs of one item are 30, of two items 60. Inventory holding costs of one item per week are 20. Sale of one item yields 100.

Evaluation of the first policy

The state of the system is defined by the inventory level at the beginning of the week. The re-order is placed at the beginning of the week, and it arrives at the beginning of the following week. First, the matrix of transition probabilities is set up, i.e. matrix

$$\mathbf{P} = \begin{Vmatrix} p_{00} & p_{01} & p_{02} \\ p_{10} & p_{11} & p_{12} \\ p_{20} & p_{21} & p_{22} \end{Vmatrix}$$

where e.g. p_{12} is the probability that the system will move from state 1 to state 2. For our case the matrix of transition probabilities is

$$\mathbf{P} = \begin{Vmatrix} 0 & 1 & 0 \\ 0 & 0.8 & 0.2 \\ 0.3 & 0.5 & 0.2 \end{Vmatrix}$$

This matrix was constructed in the following way:

The system can move from state 0 to state 1 only if one single item is ordered in this state (see Table 7.1) and no demand can be met. State 1 gives more possibilities: $p_{10} = 0$ as the inventory may be exhausted, but one item was ordered, and it is not possible that the state will be 0 at the beginning of the next week. The system can move in two ways from state 1 to state 1 if demand is 1 or 2, then $p_{11} = 0.8$. The system can move from state 1 to state 2 on condition that demand is 0, then $p_{12} = 0.2$. Similarly, the last row was filled in.

As this transition matrix $\mathbf{P}$ is neither divisible nor periodic, a limiting state vector V (steady-state probability vector) exists. It is defined by the conditions:

$$V = V\mathbf{P}$$

and

$$\sum_{i=0}^{2} p_i = 1$$

where p_i are components of this vector (steady state or stationary probabilities).

Solution of the system

$$\begin{aligned} p_0 &= \phantom{p_0 + 0.8p_1 +{}} 0.3p_2 \\ p_1 &= p_0 + 0.8p_1 + 0.5p_2 \\ p_2 &= \phantom{p_0 +{}} 0.2p_1 + 0.2p_2 \\ p_0 &+ p_1 + p_2 = 1 \end{aligned}$$

produces the result

$$\begin{aligned} p_0 &= 0.06 \\ p_1 &= 0.75 \\ p_2 &= 0.19 \end{aligned}$$

The probabilities can be interpreted in the following way: During long-term activity of the system with the first inventory policy, the stock could be found to be zero at the beginning of the week in 6% of the weeks, and the inventory states 1 and 2 could be achieved in 75% and 19% of the weeks, respectively.

Next, the costs that result from the first inventory policy are computed. One item is ordered for states 0 or 1 that occur in 81% of the weeks. Then the expected weekly value of replenishment costs is $0.81 \cdot 30 = 24.3$.

The computation of the expected gross benefits for each possible state is as follows: if the system is in state 0, there are no inventory holding costs and no benefits from sold items; if the system is in state 1, then, with probability $p_{11} = 0.8$, one item will be sold that week, and with probability $p_{12} = 0.2$ one item will stay in stock all week. State 1 results in gross benefits

$$0.8 \cdot 100 - 0.2 \cdot 20 = 76$$

Similarly, for state 2 the gross benefits are

$$0.3 \cdot 200 + 0.5 \cdot 100 - 0.5 \cdot 20 - 0.2 \cdot 40 = 92$$

In each week the following gross benefits are expected for the individual states:

Initial state in the week	Expected benefits in the week
0	0
1	76
2	92

The computed steady-state vector determines how often these initial states occur in a long period. Therefore, the net weekly benefits for the first inventory policy are

$$0.06 \cdot 0 + 0.75 \cdot 76 + 0.19 \cdot 92 - 24.3 = 50.2$$

Evaluation of the second inventory policy (the method is similar)

The transition probability matrix is

$$\mathbf{P} = \begin{Vmatrix} 0 & 0 & 1 \\ 0.8 & 0.2 & 0 \\ 0.3 & 0.5 & 0.2 \end{Vmatrix}$$

The steady-state vector is

$$p_0 = 0.33$$
$$p_1 = 0.26$$
$$p_2 = 0.41$$

Using the second inventory policy, two items are ordered for the initial weekly state 0, which occurs with probability $p_0 = 0.33$. Replenishment costs of two items are 60. Therefore, the expected weekly replenishment costs are

$$0.33 \cdot 60 = 19.8$$

The expected gross weekly benefits are the same as with the first inventory policy. The expected net benefits of the second inventory policy are

$$0.33 \cdot 0 + 0.26 \cdot 76 + 0.41 \cdot 92 - 19{,}8 = 37.7$$

Since the first inventory policy yields higher weekly net benefits (50.2), it is more advantageous.

7.4 APPLICATION OF INVENTORY THEORY IN WRS

In principle, many problems of inventory control are similar to problems of WRS. The mathematical description of inventory issues of a stock with stochastic replenishment is analogous to that of active storage of a reservoir with stochastic inflow. Also, the inventory operational policy and operational rules for the release of water from reservoirs are similar (Fig. 7.7).

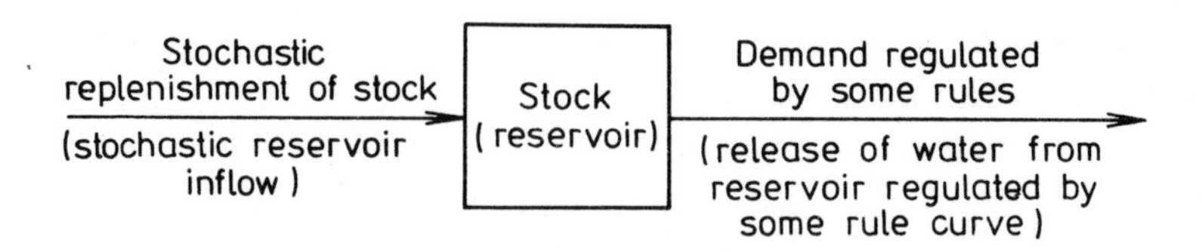

Fig. 7.7 Scheme of similarities of problems of the inventory theory and of WRS

The objective may be determination of an optimum capacity (e.g. storage of reservoir) with given demand (water requirement), investigation of optimum operating policy of inventory (reservoir) or determination of optimum reliability of an inventory system (WRS).

The mathematical similarity extends to other aspects of these problems, e.g. the inventory operating policy with stochastic demand for some items and water resource operational policy with stochastic demand for water.

These, and other possible analogies, are the basis of a series of approaches related to discrete or continuous time and items with finite or infinite inventory capacity. The common property of these problems is their random nature allowing application of the theory of probability.

Inventory theory and its applications have not been developed to such a degree as to solve the problems of complex WRS. It is confined to individual reservoirs or systems with a few reservoirs. The model presented in the following section is, therefore, a very rough approximation of a WRS, and the system is reduced to a system of a few reservoirs.

The mathematical background of inventory theory is similar to that of queue theory. Therefore, the application of both theories in WRS have similar characteristics and sometimes it is difficult to decide whether the WRS model is an application of inventory theory or queue theory. Some models will be described that use the principles of both theories, but they can scarcely be called applications of these theories. These models were developed by Kritsky, Menkel, Kartvelishvili, Cvetkov and others. In view of this fact, the model classification as inventory models or queue models is approximate but useful.

Methods of reservoir storage analysis based on inventory theory and queue theory and some other specialized methods can collectively be termed "the reservoir operational policy theory". The characteristic feature of this theory is its reflection of the stochastic nature of phenomena. It has produced many positive results, although it does not constitute a complete and finished theory. It is represented by a series of models, of various degrees of simplification, of WRS problems. The numerical solutions involved in the application of these methods are difficult, especially in models with many reservoirs, even if big computers are used. The Monte Carlo methods, i.e. numerical solution by repetitive computation using many realizations of stochastic processes for the solution of mathematical problems, can often help in such cases. Instead of an analytical solution, the Monte Carlo experiments are used for the determination of probability values.

The aim of the following examples of investigation of reservoir function by methods of the inventory, queue or similar theories is to determine the probability distribution of water storage at some time point, e.g. at the end of some chosen time interval (at the end of the year). The simplest models assume one reservoir, discrete time and discrete probability distribution; the more complicated models assume continuous time and probability distribution and a system of reservoirs.

Examples that greatly simplify the assumptions of WRS, have little significance for practical application. They are used as explanations of more complicated models. Complicated mathematical considerations leading to results that cannot be used in practice were omitted, and only methods of practical applicability have been concentrated on.

7.4.1 Operational Policy of a Single Reservoir at a Discrete Time

Firstly, the operational policy of a single reservoir is investigated (Moran, 1959). Inflow into the reservoir is a random variable. Release from the reservoir depends on the volume of water in the reservoir and the operational policy that determines the quantity and timing of the release. A reservoir with limited active storage is assumed. Time t is assumed in discrete time steps $t = 0, 1, 2 \dots$ (Fig. 7.8).

Annual steps are assumed, and water is released once a year at a given fixed time. These simplifying assumptions are far from reality; they produce a rough approximation of the operation of a reservoir that is filled regularly in one season of the year (melting snow in the upper catchment area) and is released regularly in another season.

The total volume of water inflow in year t is denoted as X_t. Let us assume that X_t is a random variable at the interval $(0, \infty)$, it has a discrete probability distribution, and the values X_t for different t are mutually independent and have the same distribution[1]). The storage of water in the reservoir before the inflow X_t is denoted by Z_t. If $X_t + Z_t$ is greater than the active storage of the reservoir V, the volume $Z_t + X_t - V$ will be spilt uselessly. The distribution of this spill can be determined directly from the distribution of variables X_t and Z_t. At the end of the year water is released according to a fixed rule. The simplifying assumption is used that the same amount of water $M < V$, is released every year if this amount remained in the reservoir, or the amount $X_t + Z_t$, if $X_t + Z_t < M$. The storage Z_{t+1} remains in the reservoir.

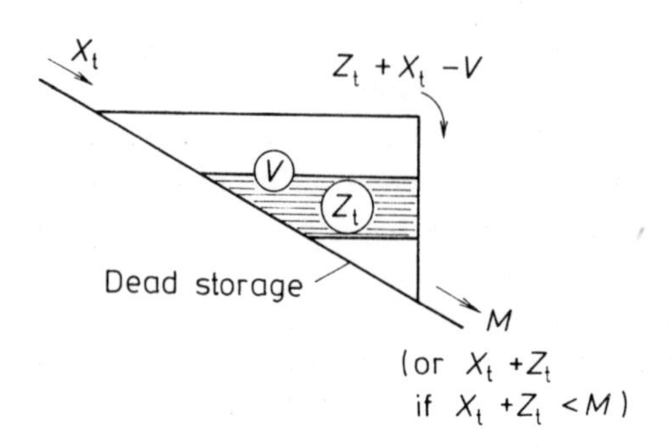

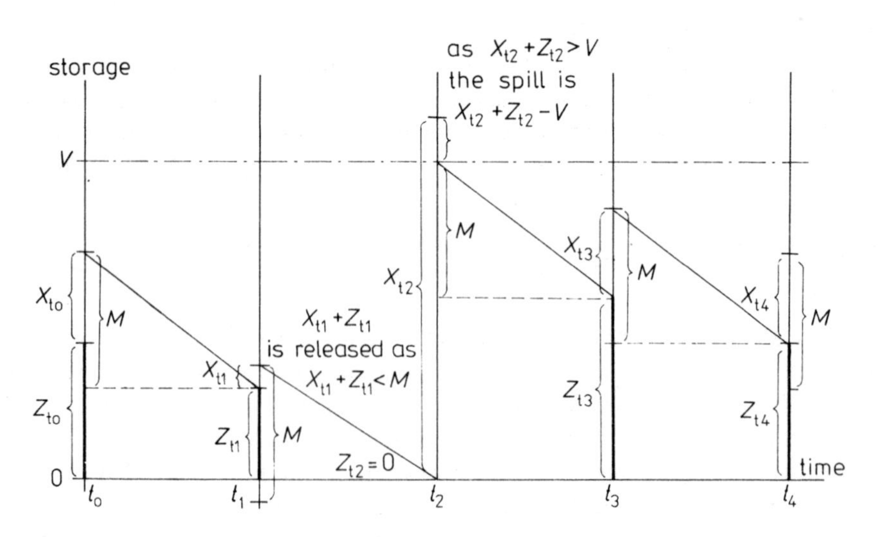

Fig. 7.8 Schematic representation of computation for the discrete time and the discrete probability distribution

V – reservoir active storage volume, X_t – reservoir inflow volume in the year t, Z_t – reservoir storage before the inflow X_t, M – release volume

---

[1]) The assumption of independence and constant distribution is not fulfilled in short time steps – e.g. in daily reservoir operation according to seasonal variability and serial correlation among daily flows.

The main task is the determination of the probability distribution of the variable Z_t (i.e. the storage in the reservoir at the end of interval t) for the stationary state of the stochastic process, defined by this system of rules, i.e. a state which is not dependent on the initial conditions. It can be proved that such a state of the process can be reached. This is an ordinary Markovian process, i.e. the conditional distribution of variable Z_{t+1}, given Z_t, does not depend on the past, i.e. the period before time t.

For numerical solution of this example a finite system of approximating linear equations can be used to determine the probability distribution of variable Z_t.

Let us assume that in time t X_t acquires the values 0, 1, 2, ..., with probabilities $p_0, p_1, \ldots$. At the same time, variable Z_t acquires the values 0, 1, ... with probabilities $P_0, P_1, \ldots, P_k$ and in time $t + 1$ it acquires the same values with probabilities $P'_0, P'_1, \ldots P'_k$. It is also assumed that $V \geqq M$. Then the following system of equations is valid

$$
\begin{aligned}
P'_0 &= P_0(p_0 + p_1 + \ldots + p_M) + P_1(p_0 + p_1 + \ldots + p_{M-1}) + \ldots + P_M p_0 \\
P'_1 &= P_0 p_{M+1} + P_1 p_M + \ldots + P_{M+1} p_0 \qquad (7.20) \\
&\ldots\ldots\ldots\ldots\ldots\ldots\ldots\ldots\ldots\ldots\ldots \\
P'_{V-M} &= P_0(p_V + \ldots) + P_1(p_{V-1} + \ldots) + \ldots + P_{V-M}(p_M + \ldots)
\end{aligned}
$$

Solution of this system gives the probability distribution of variable Z_{t+1}, given Z_t. We can then proceed to $Z_{t+2}, Z_{t+3}, \ldots, Z_{t+n}$. These values form a Markov chain. For $n \to \infty$, a stationary solution is obtained if $p_i > 0$ $(i = 0, \ldots, V)$.

Stationary probabilities are determined from (7.20) assuming $P'_i = P_i$ $(i = 0, \ldots, \ldots, V - M)$ and taking into account the relationship $\sum_{i=1}^{V-M} P_i = 1$. This task can be solved by matrix methods or (for big problems) by the Monte Carlo method. The computation of the simplified case of a single reservoir is presented. The same procedure can be used for models with several reservoirs and a more complex inflow pattern and operational policy. These more complicated cases require the use of computers.

The computation by the matrix method is performed for the following input values: active storage of the reservoir $V = 4$, release from reservoir $M = 2$. The discrete probability distribution of the reservoir inflow X_i $(X_i = 0, 1, \ldots)$

$$
\begin{aligned}
p_0 &= 0.2 \\
p_1 &= 0.4 \\
p_2 &= 0.2 \\
p_3 &= 0.1 \\
p_4 + p_5 + p_6 + \ldots &= 0.1 \\
\hline
\textstyle\sum p_i &= 1.0
\end{aligned}
$$

The aim of computation is the determination of the stationary probabilities P_i, $i = 0, 1, \ldots, V - M = 2$ so that the storage at the end of the year will be $Z = 0, 1, 2$.

The input values are substituted in equations (7.20) and a set of $V - M + 1 = 3$ equations with three unknowns is obtained. The matrix of transition probabilities is

$$\mathbf{P} = \left\| \begin{matrix} 0.8 & 0.6 & 0.2 \\ 0.1 & 0.2 & 0.4 \\ 0.1 & 0.2 & 0.4 \end{matrix} \right\|$$

This matrix fulfils the conditions for the existence of a stationary state. To find the solution for the steady state, the following equations are used: $P_i' = P_i$ $(i = 0, 1, \ldots, \ldots, V - M)$ and $\sum P_i = 1$. Then the following system will be solved

$$\begin{aligned} &P_0 + P_1 + P_2 = 1 \\ &P_0 = 0.8P_0 + 0.6P_1 + 0.2P_2 \\ &P_1 = 0.1P_0 + 0.2P_1 + 0.4P_2 \\ &P_2 = 0.1P_0 + 0.2P_1 + 0.4P_2 \end{aligned}$$

The resulting probabilities are $P_0 = 0.666$, $P_1 = 0.167$, $P_2 = 0.167$ so that storage at the end of the year will be $Z = 0, 1, 2$.

The problem can be solved, for example, by the Monte Carlo method in this way (Moran, 1959): Random two-digit numbers are used for the determination of transition probabilities from one state to another (i.e. for storage at the beginning and at the end of the year). Starting from zero, i.e. assuming an empty reservoir at the

Table 7.2 Monte Carlo method in tabular form

Hundreds of realizations of the random process	Observed frequencies of states		
	State (content of reservoir)		
	0	1	2
1	65	18	17
2	65	16	19
3	66	18	17
4	68	17	15
5	69	13	18
6	67	18	15
7	67	14	19
8	72	15	13
9	64	19	17
10	63	18	19
Sum of frequencies	666	166	168

beginning, the transition probabilities into states 0, 1, 2 are given by the first column of the given matrix. For example, if the first random number is A_1, the next state will be 0 if $A_1 \leqq 80$, it will be 1 if $80 \leqq A_1 \leqq 80 + 10$ or it will be 2 if $80 + 10 \leqq$ $\leqq A_1 \leqq 80 + 10 + 10$.

In the next step, the column of the matrix of transition probabilities related to the state reached is used. The resulting values are presented in tabular form (Table 7.2). For example, ten sets with 100 realizations are used in every row and the resulting frequencies are listed. These frequencies approximate the required probabilities P_0, P_1, P_2. The sum of the frequencies in the last row gives a better approximation based on 1000 realizations. The number of realizations necessary for the required degree of accuracy is computed from the estimate of the standard error. In the simple case presented, a very good fit of accurate results and the frequencies for 1000 realizations are achieved.

7.4.2 Operational Policy in Continuous Time with Continuous Probability Distribution

The model in the previous section can be transformed by certain mathematical operations into a more general model that can better reflect reality by assuming continuous time and continuous probability distribution (Moran, 1959).

In this transformation the following procedure was used: First, a modification in the discrete model is carried out. We assume that in discrete probability distribution the sum of remaining storage in reservoir Z_t and the inflow X_t is Y_t, i.e. $Z_t + X_t = Y_t$. Let R_i denote the probability that $Y_t = i$. Moreover, we assume that the process is stationary, and that it does not depend on the initial conditions; with this assumption, the required probability distribution of Z_t can be found for a given distribution of $Y_t = Z_t + X_t$ and X_t. By substitution of R in (7.20) and rearrangement an infinite system of linear equations is obtained describing the distributions of random variables R_i:

$$
\begin{aligned}
R_0 &= p_0R_0 + p_0R_1 + \ldots + p_0R_M \\
R_1 &= p_1R_0 + p_1R_1 + \ldots + p_1R_M + p_0R_{M+1} \qquad (7.21) \\
&\ldots\ldots\ldots\ldots\ldots\ldots \\
R_{V-M-1} &= p_{V-M-1}R_0 + \ldots + p_{V-M-1}R_M + p_{V-M-2}R_{M+1} + \ldots + p_0R_{V-1} \\
R_{V-M} &= p_{V-M}R_0 + \ldots + p_{V-M}R_M + p_{V-M-1}R_{M+1} \ldots + p_0(R_V + R_{V+1} + \ldots) \\
&\ldots\ldots\ldots\ldots\ldots\ldots \\
R_{V-M+S} &= p_{V-M+S}R_0 + \ldots + p_{V-M+S}R_M + p_{V-M+S-1}R_{M+1} + \ldots \\
&\quad + \ldots + p_s(R_V + R_{V+1} + \ldots) \\
&\ldots\ldots\ldots\ldots\ldots\ldots
\end{aligned}
$$

where, as in equations (7.20), M represents the amount of water released from the reservoir.

Since

$$P_0 = R_0 + \dots + R_M$$
$$P_{V-M} = R_V + R_{V+1} + \dots$$
$$R_{V-M+S} = 0 \quad \text{for } S > 0$$

the equations for the stationary state can be obtained from equations (7.21).

The case of continuous probability distributions is analogous. If the probability density function of a random variable is

$$X_t = f(x) \quad \text{for } x \geqq 0$$
$$X_t = 0 \qquad \text{for } x < 0$$

and the density function of $X_t + Z_t$ is $g(x)$, the following equation is valid:

$$g(x) = f(x) \int_0^M g(y)\,\mathrm{d}y + \int_M^V g(y)\, f(x + M - y)\,\mathrm{d}y + f(x - V + M) \int_V^\infty g(y)\,\mathrm{d}y \tag{7.22}$$

The three terms of the right-hand side of the equation correspond to the following conditions:

$0 \leqq X_t + Z_t \leqq M$ ... the reservoir is empty
$M < X_t + Z_t \leqq V$ ... the reservoir is not empty and no spillage occurs
$V < X_t + Z_t$ ... spillage is $(X_t + Z_t) - V$

The probability distribution of variable Z_t consists of three parts, a discrete component p_0 if $Z_t = 0$, the continuous probability density function $g\,(x + M)$ if $Z_t = x$ where $0 < x < V - M$ a discrete component p_1 if $Z_t = V - M$, and

$$p_0 = \int_0^M g(y)\,\mathrm{d}y \quad p_1 = \int_V^\infty g(y)\,\mathrm{d}y \tag{7.23}$$

Although the function $f(x)$ has a simple form, solution of equation (7.22) is complicated, and it has an analytical solution in strips of width M. Therefore, Moran and Prabhu (Moran, 1959) tried to find an explicit solution of equation (7.22) for a class of distributions that can describe the empirical distributions of flows. They concentrated their attention on Pearson's distribution type III and geometrical distribution. Although they used many sophisticated mathematical operations, they obtained solution for integer p only. Moran, therefore, suggested the application of the known results for Pearson's distribution type III with integer p with interpolation for other p. Kartvelishvili, 1963, argues that these types of distribution do no represent all actual cases.

Most difficulties in an analytical solution are caused by the boundary conditions $Z = 0$ and $Z = V - M$. Therefore, a possible simplification of computation is to assume a reservoir with infinite storage. Two cases are possible: a situation involving an almost full reservoir can be investigated if the probability distribution of variable Z gives very small probabilities that Z_t will approach zero. In this case a reservoir of infinite depth can be assumed. The opposite case assumes the very small probability of completely filling the reservoir, and the probability distribution of storage near the bottom of the reservoir is investigated with no limitation in the upper direction.

These problems, assuming an infinite reservoir storage, are beyond the scope of the application of inventory theory and are dealt with by queue theory (see section 8.6).

7.4.3 Operational Policy of a Single Reservoir with Carry-Over Storage in Discrete Time

The previous models applied inventory theory. Other methods (Kartvelishvili, 1967) are similar to these models; however, their classification as applications of inventory theory will be a simplification.

The origin of these models is older than inventory theory; they were developed independently of this theory, and as late as the last decade, some common features can be observed. These models are applied practically and compared with the previous simple models, they reflect more accurately the complexity of reservoir operational policy, taking into consideration its stochastic character.

The simplest model of a single reservoir can be described as follows: the annual flows and annual releases are assumed to be integer variables. This approximation is acceptable if the seasonal component is negligible in comparison with the carry-over storage of the reservoir. The losses are considered as a reduction in inflow or an increase in withdrawal. River flows are assumed to form a stochastic process with discrete time and an annual time step; the stochastic serial correlation of flows is neglected.

The model uses very simple assumptions, similar to those used in the first model. However, it has some advantages over Moran's model. It does not require the release to take place at a predetermined moment, and the release need not be the same every year. Relatively simple graphical methods of solution have been developed. Kartvelishvili emphasizes the methodological importance of this model, since the more general models of the theory of storage, with the release deterministic function of time or stochastic release function of time or in a form of operational rule (e.g. during reservoir filling), use this simple model of a single reservoir with carry-over storage as a starting point.

The method is explained graphically. Let α denote the annual release from the reservoir, and β the reservoir storage.

Both these values are related to the mathematical expectation of the annual flow. Further, $X(z) = 1 - F(z)$ is the reliability of the annual flow (in volume units), i.e. the probability that the ratio of its actual and expected values will not be less than z.

Let us start with a constant release α, i.e. $\alpha = \text{const} > \beta$. Assume (Fig. 7.9) $OD = 1$, the $ABCD$ is the curve of reliability of filling up the reservoir at the end of the k-th year of operation. The ordinates of this curve are $\Theta_k(y)$.

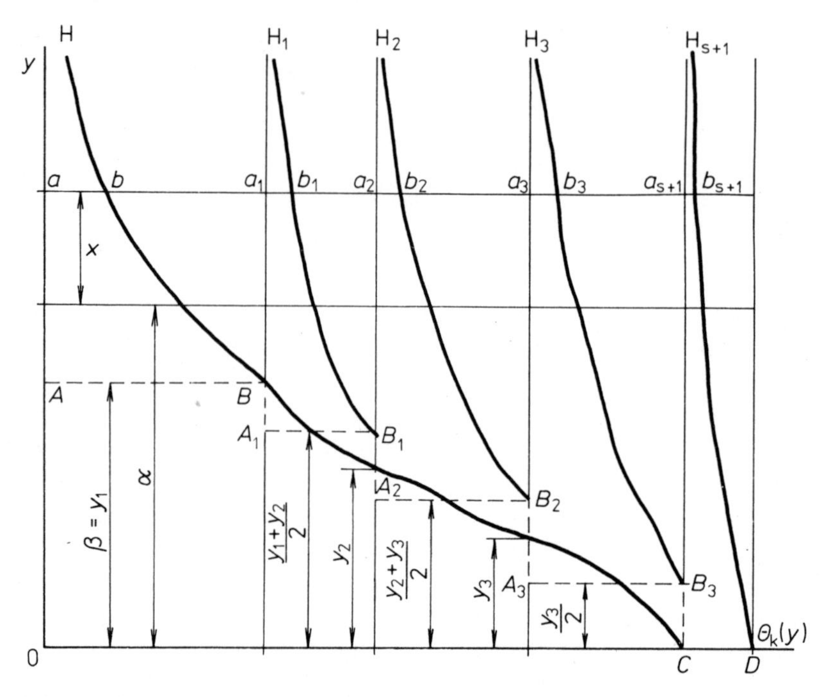

Fig. 7.9 The graphic method for the determination of the operation policy of a single reservoir with carry-over and discrete time

No serial stochastic relationship between flows is assumed; the filling up of the reservoir at the end of the year and the release in the following year are, therefore, stochastically independent. Using these assumptions, Kritsky and Menkel designed a method for the determination of function Θ_{k+1} of reliability of filling up of the reservoir at the end of the year $k + 1$. The curve BC is replaced by a graph with a finite number of steps, $BA_1B_1A_2 \ldots A_sB_sC$. Curves HB, H_1B_1, ... H_sB_s, $H_{s+1}\,D$ are also constructed. Their functional values relative to points A, A_1, ..., A_sC are equal to z and their abscissas relative to the same point are $AB_X(z)$, $A_1B_{1X}(z) \ldots A_sB_{sX}(z)$, $CD_X(z)$. Then with annual release $\alpha = \text{const}$, the reliability of the reservoir filling up at the end of the year $k + 1$ by volume x $(0 \leqq x \leqq \beta)$ will be approximated by

$$\Theta_{k+1}(x) = ab + a_1b_1 + \ldots + a_sb_s + a_{s+1}b_{s+1} \tag{7.24}$$

It is apparent that abscissa ab is the probability that the reservoir content at the end of the year $(k + 1)$ will exceed the value x on condition that it was full at the end of the year k (the probability of the reservoir being full at the end of the year k is AB). The abscissa $a_1 b_1$ is the probability that at the end of the year k the reservoir content was $(y_1 + y_2)/2$ etc. The abscissa a_{s+1} is the probability that at the end of the year $k + 1$ the reservoir content will exceed x on condition that it was empty at the end of the year k (the probability of the reservoir being empty at the end of the year is CD). Since the phenomena whose probabilities are given by abscissas ab, $a_1 b_1, \ldots,$ $\ldots, a_{s+1} b_{s+1}$ are mutually exclusive, probability $\Theta_{k+1}(x)$ will be given by the sum of the corresponding probabilities, as expressed in equation (7.24). A precise analytical solution can be obtained by division of the curve BC into an infinite number of steps (Kartvelishvili, 1958). The reliability of a full reservoir was determined by a simple integral Fredholm equation, type II, the theory and solution of which are known.

This model of operational policy of a single reservoir was generalized by Kartvelishvili, 1967, for two or more reservoirs. It takes into account the stochastic relationship between reservoir inflows, but it does not assume a serial correlation in a one-reservoir site. This relatively simple model requires a system of complicated integral equations of probability distribution of reservoir filling.

Similar models of the same and related problems were investigated by Kritsky and Menkel, 1952, 1959; Kartvelishvili, 1967 and Reznikovsky, 1964. Some models are oriented towards hydropower; the objective of some models is the maximization of reliability, and, therefore, these models are, in fact, optimization models. One group of models includes seasonal operational policy.

All these models of operational policies with given release requirements (e.g. expressed by a probability function) involve the solution of integral equations or a set of these equations. In the simplest case (e.g. at the beginning of section 7.4.3) the solution of mathematical integral equations is the required probability characteristic of the operational policy. In other models the solution of these equations is the basis for the determination of these characteristics.

In principle, three different procedures are possible in practical computation:

– the direct solution of integral equations or a set of equations,

– the approximation of integral equations by a system of linear algebraic equations (e.g. by the methods of moments or by numerical integration),

– the Monte Carlo method.

All these procedures were described by Kartvelishvili, 1967. The application of the Monte Carlo method to these models was described by Svanidze, 1964, Reznikovsky, 1964, and Vicens, 1963.

8 QUEUING THEORY MODELS

8.1 KINDS OF QUEUES

Queuing theory (the theory of waiting lines) is a discipline of operational research, the subjects of which are mathematical models and quantitative analysis of processes involving waiting for the service of some technical equipment. Common to all these processes are the arrivals of people or objects requiring service and the attendant delays when the service mechanism is busy. The aim of the theory is the identification of these characteristics of the queue processes that facilitate technical and economic analysis of the given system and the attainment of optimal parameters of these systems.

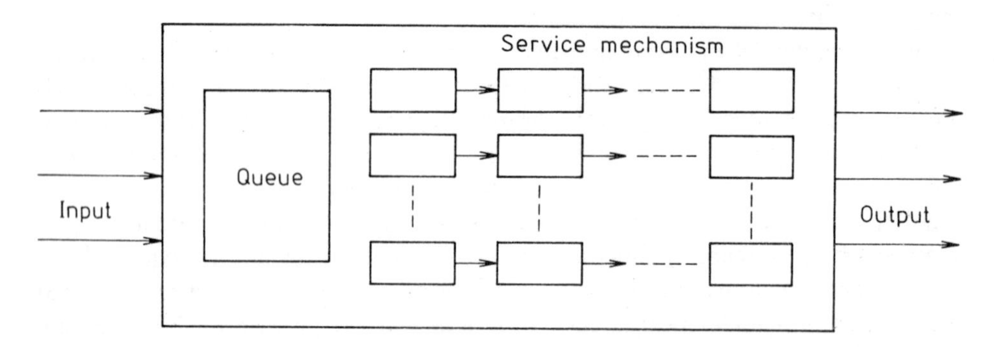

Fig. 8.1 Schematic representation of the queuing process

The first application of queue theory were to problems in the operation of telephone and telegraph lines. Now queue theory is applied in various fields, e.g. in the organisation of production, design operations, transport, services, maintenance, etc.

Problems using queue theory have the structure represented in Fig. 8.1. The mathematical description in queue models requires specification in this form: a mathematical description of the input process, its time dependence, queue discipline, service mechanism, and the output process resulting from the previous stages.

According to the given characteristics queue processes and their mathematical models can be classified in different ways. Stochastic models with random variables are frequently applied in queue theory.

The queue models are investigated with different input processes, queue discipline and service mechanism. Input is characterized by the pattern of arrivals into the system (regular or random patterns), individual or bulk arrivals, etc. A random pattern is described by its probability characteristics: a mean number of items arriving in a given unit of time, the probability that no unit arrives during a certain time interval, the mean number of items in the bulk arrivals, etc. It is important to make an analysis of the stationarity or non-stationarity of probability characteristics of the input process.

Queue discipline describes the order in which customers entering the system are eventually served. It can have different forms given by the input process and the ways of transition from queue to service. In the simplest case, the discipline is first come, first served (first in, first out – FIFO). However, some customers may become impatient and decide to leave the system before being served (reneging), or a cus-

Table 8.1 Kendall's notation of queuing systems

Letter	is substituted for	
	X	Y
M	Exponentially distributed (independent) interarrival time, (M = Markovian), Poisson input	Exponentially distributed service time
E_n	Erlangian distribution of order n for interarrival time (with parameters λ and n)	Erlangian distribution of order n for service time (with parameters μ and n)
K_n	Distribution χ^2 for interarrival time (with n degrees of freedom)	Distribution χ^2 for service time
D	Regular, deterministic interarrival time	Constant, deterministic service time
G	General distribution of interarrival time (no assumption for distribution)	General distribution of service time
GI	General independent distribution of interarrival time	

tomer seeing a long line may balk, i.e. not join the queue if it is too long. In more complicated cases a different queue discipline may be applied, e.g. last come, first served (last in, first out – LIFO) or a priority discipline if certain customers are served preferentially.

Similarly, there can be different types of service systems. Service can be offered along one or several parallel lines (servers in parallel), service time can be constant (for all customers) or it can be random. The probability characteristics of service time are investigated and its probability distribution can be stationary or time dependent.

A concise representation of queue systems was given by Kendall. In his notation, the systems are classified according to three main aspects: (1) the type of input stochastic process describing the arrival of customers; (2) the distribution of service time and (3) the number of servers. Information concerning these characteristics is signified by three symbols: $X/Y/c$, where X and Y are specified by capital letters and c is a natural number (or ∞) denoting the number of servers. For X and Y the letters M, D, G, E_n, K_n are substituted here (for X the couple GI may be substituted); an explanation of these symbols is given in Table 8.1.

Kendall's classification is not exhaustive, as some important characteristics are not included in the symbolic notation (e.g. the existence and length of the queue, queue discipline, etc.). Therefore, these data must be added in each case.

8.2 MARKOVIAN AND OTHER PROCESSES IN QUEUING MODELS

Markovian processes form a separate class of stochastic processes. A number of publications deal with this type of process (Walter, 1970; Zítek, 1969; Wagner, 1975) and its application to WRS problems (Votruba and Nacházel, 1971). In queue models the Markovian properties of processes are very important. The solution is simpler if the process is Markovian, or if it can be approximated by a Markovian process. In other cases, formidable computational difficulties occur and some sophisticated methods are used, e.g. the method of imbedded Markov chains.

In section 8.2 the relationship between Markovian properties of processes and the type of queue model is explained; in section 8.3 the models applicable to WRS are analyzed.

According to Kendall's notation of queue models (section 8.1) e.g. the system designated $M/M/1$ has the following features:

- Poisson input, i.e. exponentially distributed (independent) inter-arrival time
- exponentially distributed service time,
- single server.

Similarly, the system designated $M/E_n/1$ is characterized in the following way:

- Poisson input (see previous type),

– Erlangian distribution of order n for service time (with parameters μ and n),
– single server.

The theory of Markovian processes can be applied to queue systems with Poisson input, i.e. exponentially distributed inter-arrival time and exponentially distributed service time. In Kendall's notation these systems fall into the first row i.e. systems designated $M/M/n$. These systems have been intensively investigated and have been applied successfully in practice. Consequently, queue systems are often divided into two basic groups:

– Markovian systems (type $M/M/n$),
– other systems (non-Markovian, e.g. types $M/D/1$, $M/G/1$, $M/E_n/1$, $GI/M/1$, etc.).

8.2.1 Markovian Queuing Systems

System $M/M/1$

The model of this system is the simplest one. It can be described as follows: customers arrive in the system individually, and their arrivals are mutually independent and independent of the service mechanism. The inter-arrival time is an independent random variable with an identical exponential distribution. The customers who cannot be served because the single server is occupied, wait in a single, unlimited queue. The queue discipline is first come, first served, without priorities.

The solution gives the following parameters: the mean number of customers in the queue, the mean number of customers in the system, the probability distribution of the waiting time of any customer, the mean time spent by a customer in the system, the system utilization factor (fraction of the time the server is busy).

System $M/M/n$

When the requirements of customers exceed the capacity of a single server, a system with more servers, $n > 1$, can be used. (However, some other systems are possible, e.g. with a higher intensity of the service mechanism or with a reduction in the number of customers).

In the simplest form of this system, a single queue is assumed, common to all servers. The customers wait in the queue when all the servers are occupied. Whenever a server becomes available, the customer at the head of the queue is served. The solution of problems of this system is similar to that in the $M/M/1$ system. Special problems occur if the solution is to determine the number of servers n. In some alternatives a variable number of servers is possible providing flexibility in handling the requirements of customers.

Systems $M/M/1$ and $M/M/n$ can be relatively easily handled by the theory of Markovian processes, and various alternatives of these systems can be considered.

For example, different queue disciplines can be assumed, i.e. the rules determining the order in which customers entering the system are eventually served. Frequent types of queue discipline are FIFO (first in, first out), LIFO (Last in, first out; last come, first served) when the last customer is served when the server has just finished processing the previous one, random serving of customers in the queue, or *priority queue discipline* when the customers are classified into several types and priorities are assigned to these types in decreasing order of importance.

The theory of Markovian processes can be applied to systems $M/M/n$ with a finite queue. A loss of customers is assumed, i.e. the supposition that all customers will be served is not valid.

In solving problems of Markovian systems no major computational difficulties occur. The tasks can be reduced to the solution of a set of linear equations. This computational simplicity rests on the following assumptions:

– steady-state systems (i.e. we are not interested in the initial stages of the system before its stabilization),

– homogeneous Poisson input (exponential arrivals of customers),

– exponentially distributed service time.

These assumptions can be accepted in many practical cases and they are a good approximation of the real situation; however, there are many cases when they are not. Then a non-Markovian queue system with more general properties must be used.

8.2.2 Other Queuing Systems

In Kendall's notation in section 8.1 only the processes in the first row are Markovian. The applicability of other processes is more general, but their solution is more complicated. Often, a reduction to the Markovian case is attempted to allow solution by simpler methods (Zítek, 1969). A well-known procedure is called the *method of the imbedded Markov chain*[1]).

---

[1]) The principle of the imbedded Markov chain is as follows: Assume that the process being investigated is not Markovian, i.e. the condition

$$P\left\{X(t) = \frac{k}{X(s_1)} = j_1, X(s_2) = j_2 \ldots, X(s_n) = j_n\right\} = P\left\{X(t) = \frac{k}{X(s_1)} = j_1\right\} \tag{I}$$

is not met for some choice of numbers $n, t > s_1 > s_2 \ldots > s_n \geqq 0$ and $k, j_1, j_2, \ldots, j_n$. Then a certain sequence of times

$$0 \leqq t_1^* < t_2^* \ldots < t_m^* \ldots, \lim_{m \to \infty} t_m^* = \infty \tag{II}$$

can be formed in such a way that condition(I) is met if the numbers $t, s_1, \ldots, s_n$ are chosen from this time sequence (II). Therefore, a time sequence is sought where, for any $t = t_M^*$ and for natural n for any n-tuple (vector with n components) $s_i = t^*{}_{m_i}$ $(i = 1, 2, \ldots, n)$, $0 \leqq s_n < s_{n-1} \ldots < s_1 < t$, this condition is met.

Another method for transformation to Markovian-type processes is based on an approximate, *broader definition of the process.* Apart from reduction to the Markovian process, certain processes can be investigated directly, e.g. by the theory of *semi-Markovian stochastic processes.* Often the *Monte Carlo methods* are used for queue processes that are not Markovian. They are based on a simulation of the investigated systems, and they exemplify some of the experimental methods of probability theory. They can be used for any type of process but they require a good deal of computer time and capacity.

In the literature (Zítek, 1969) the following types of non-Markovian queue systems are listed.

System $M/D/1$

In system $M/M/n$ customer arrivals and service time are random. These assumptions may be an acceptable simplification of reality. In some queue systems model $M/D/1$ with constant service time may be more appropriate: each customer spends a fixed time interval in the serving stage.

System $M/E_n/1$

The advantage of this system as compared with the $M/M/1$ system is the Erlangian distribution of service time. This distribution is determined by two parameters, and therefore it can fit the observed frequency better than the exponential distribution determined by one parameter only. Some other modifications of this system are possible, e.g. the service can be composed of several independent, successive stages; it can include the assumption that more than one customer enters the system at the same time in the form of bulk arrivals.

System $M/G/1$

This system has been developed in a number of alternative forms that try to make the rules for customer arrivals and service time more flexible in order to produce a more general model. In computation the imbedded Markov chain is often used.

$$P\left\{X(t_M^*) = \frac{k}{X(t_{m_i}^*)} = j_i,\, i = 1, 2, \ldots, n\right\} = P\left\{X(t_M^*) = \frac{k}{X(t_{m_1}^*)} = j_1\right\}$$

for all integer non-negative numbers $j_1, j_2, \ldots, j_n, k$. The time sequence (II) is determined in such a way that the sequence of random numbers $X(t_m^*)$ for $m = 1, 2, \ldots$ forms a Markov chain. If such a sequence (II) has been found, the investigation is restricted to this chain $\{X(t_m^*)\}$ and used only the theory of Markov chains. The resulting properties of this process form the basis for the estimation of properties of the original process. This imbedded chain cannot reflect all the properties of the original process. It gives information concerning the momentary states of the process at times t_m^*. This is often sufficient, e.g. in investigation of the limiting behaviour of the process for $t \to \infty$.

System GI/M/1

This class of systems with one server is characterized by a generally independent distribution of inter-arrival time and exponential service time.

System GI/GI/1

This notation concerns general systems with one server and no assumptions about the inter-arrival time and service time distribution (with the exception of the assumption of the homogeneity of the process). In computation the method of the imbedded Markov chain is used and the aim is the distribution of steady-state waiting time.

8.3 QUEUING SYSTEMS MODELS

The basic queue model is the simple steady-state exponential channel. It is a single-server model with exponential inter-arrival and service times, and the queue discipline is first come, first served. The system can remain in some situation or it can move on to the adjoining situation. If n is the state of the system, i.e. n items are in the system (one customer is being served and $n-1$ are in the queue) only transitions from state n to states $n-1$, n, and $n+1$ are possible.

On this assumption, the probability density function of the time that a customer spends in the queue $p(w)$ is given by (Walter *et al.*, 1973)

$$p(w) = \frac{\lambda}{\mu} e^{-w(\mu-\lambda)} \tag{8.1}$$

where λ is arrival rate per unit of time, i.e. probability rate, of transition from state n to state $n+1$ ($1/\lambda$ is the mean time between arrivals);

μ – service rate per unit of time that the server is busy ($1/\mu$ is the mean service time)

The ratio $\varrho = \lambda/\mu$ of arrival rate to service rate is frequently called the traffic intensity.

The basic operating characteristics of the system are: the mean number of customers in the system and its variation, mean line length, mean number of customers served per busy period and mean time in the system.

The mean number of customers in the system with an infinite length of queue is given by

$$\bar{n} = \sum_{n=0}^{\infty} np_n = \sum_{n=0}^{\infty} n(1-\varrho)\varrho^n = (1-\varrho)\,\varrho \sum_{n=1}^{\infty} n\varrho^{n-1} \tag{8.2}$$

The expression $n\varrho^{n-1}$ is the first derivative of ϱ^n; the sum of derivatives is the derivative of the sum (the progression is convergent) and the following transcription is possible:

$$\bar{n} = (1 - \varrho)\,\varrho\,\frac{\mathrm{d}}{\mathrm{d}\varrho}\sum_{n=1}^{\infty}\varrho^n = (1 - \varrho)\,\varrho\,\frac{\mathrm{d}}{\mathrm{d}\varrho}\,\frac{\varrho}{1 - \varrho} = \frac{\varrho}{1 - \varrho} \tag{8.3}$$

In original values λ and μ, the expression (8.3) will be

$$\bar{n} = \frac{\lambda}{\mu - \lambda} \tag{8.4}$$

The mean number of items in the queue (mean line length) is

$$\bar{n}_f = \sum_{n=1}^{\infty}(n - 1)\,p_n = \sum_{n=1}^{\infty}np_n - \sum_{n=1}^{\infty}p_n = \bar{n} = (1 - p_0) = \frac{\varrho}{1 - \varrho} - (1 - 1 + \varrho) =$$

$$= \frac{\varrho^2}{1 - \varrho} \tag{8.5}$$

The mean number of items in the queue is equal to the mean number of items in the system reduced by the quantity $\varrho = 1 - p_0$, i.e. the fraction of the time the server is busy. For a single server the quantity $\bar{n}_f$ is not very important but it is in the case of several servers.

For the variance of the number of items in the system the following expression is valid:

$$\sigma_{\bar{n}}^2 = \bar{n}^2 + \bar{n} = \frac{\varrho}{(1 - \varrho)^2} \tag{8.6}$$

or, in original quantities,

$$\sigma_{\bar{n}}^2 = \frac{\lambda\mu}{(\mu - \lambda)^2} \tag{8.6'}$$

The mean time in the system can be computed using the following consideration. In the system, $\bar{n}$ items, on average, are waiting and an average of λ, items per time unit arrive in the system. The mean time in the system can be derived by the division of $\bar{n}$ by λ, i.e.

$$\bar{T}_s = \frac{\bar{n}}{\lambda} = \frac{1}{\lambda}\,\frac{\lambda}{\mu - \lambda} = \frac{1}{\mu - \lambda} \tag{8.7}$$

Similarly for the mean time in line

$$\bar{T}_f = \frac{\bar{n}_f}{\lambda} = \frac{1}{\mu - \lambda}\,\frac{\lambda}{\mu} \tag{8.8}$$

By these methods a simple queue model can be handled that uses a Poisson exponential distribution or Erlangian distribution. Erlangian distribution[1]) of k-th order is defined by the probability function (with parameters $\lambda > 0$ and $k \geqq 1$ integer):

$$f(x) = \lambda\, e^{-\lambda x}(\lambda x)^{k-1}/(k-1)! \quad \text{for } x \geqq 0$$
$$f(x) = 0 \quad \text{for } x < 0$$

It can be shown that exponential distribution is a special case of the Erlangian distribution for $k = 1$.

In the queuing theory, more complicated models occur and special methods have to be used (Dráb, 1973; Ventcel, 1966; Walter-Lauber, 1975).

The models of queuing systems can be applied for the solution of problems of various technological processes, the rational organization of the health service, traffic problems and of problems of water resources systems (see section 8.6).

8.4 UNRELIABLE SYSTEMS

So far, the queuing systems have been assumed to be reliable. This assumption is frequently unrealistic in practice. Therefore, much work has been devoted to the investigation of unreliable systems (Zítek, 1969; Gnedenko-Kovalenko, 1966; Klimov, 1966).

In models of queuing systems, unreliability is caused by the intermittent failure of servers that cannot, at some stage, serve the customers.

Often, the unreliable systems are modelled as *systems with a priority queue discipline*, as in the following example: from time to time a customer with absolute priority arrives (= failure) and the service of other customers continues only after this customer has left the system (i.e., after restarting the service).

In principle, the following systems with priority queue discipline are possible (characterized by the treatment of the customer that is being served at the moment when a customer with higher priority arrives, i.e., at the moment when the failure occurs):

a) the service of the customer is normally finished (a system with a weak priority),

b) the service is immediately interrupted (a system with strong or absolute priority),

ba) the customer whose service was interrupted immediately leaves the system (without being served)

---

[1]) The Erlangian cumulative distribution function is defined by

$$F(x) = 1 - e^{-\lambda x} \sum_{j=0}^{k-1} (\lambda x)^j/j! \quad \text{for } x \geqq 0$$
$$F(x) = 0 \quad \text{for } x < 0$$

bb) the customer with interrupted service returns to the queue and is then served
bba) the service continues at the point of interruption,
bbb) the service is restarted from the beginning.
Unreliable systems can similarly be classified by these types of failures.

Certain differences exist between unreliable systems and systems with priority discipline:

– in unreliable systems the server stops in the middle of carrying out the repair and no further failure is possible; it corresponds to the assumption that there is only one customer with absolute priority;

– the failure can occur at any time, even if the server is not busy and when no customer is being served. However, there are systems where failure is possible only when the server is busy or when the server is idle, or the service is in any case finished (this last case corresponds to the system with weak priority, the failure is modelled for a customer with weak priority).

In all unreliable systems, new data are added to the basic data describing the organization and operation of the system. These new data characterize the probability distribution of failure occurrence and the probability distribution of repair time. The probability of failure can depend on different factors, e.g., the time from starting the system after the last failure, the time spent by the customer in being served, the number of customers that have been served after the last failure, the total service time, the time that was necessary for the last repair, etc. Using these data, the probability of service without failure, the probability distribution of the possible failure, etc., can be determined. The problems are far more complicated in cases of unreliable systems with a larger number of servers.

8.5 QUEUING SYSTEMS SIMULATION

The principles of the simulation of queuing systems have been explained in the references (Buslenko and Shreyder, 1961; Zítek, 1969; Saaty, 1961; Votruba *et al.*, 1974; Kaufman and Cruon, 1961, etc.). Only the basic ideas of this approach are mentioned in this section.

Although some outstanding results have been achieved in the queuing theory, this theory is not (and probably never will be) able to solve the problems that arise in practical life. The models oversimplify the complex problems of reality, or if the models are a good approximation of reality they are mathematically intractable as there are no effective analytical methods for solution. If a rough approximation is not acceptable, simulation methods and mainly the Monte Carlo simulation models have to be used for this more complicated case. The main advantage of these methods is their universality; they can be used for any process and any model. However, they need a lot of computation and computer time and great memory capacity.

The application of simulation methods to the tasks of the queuing theory is based on a simple principle (Zítek, 1969): that the unknown probability of a random event is estimated by repeating many independent experiments, and the relative frequency of this event is the estimate of this probability. A similar method is used in the estimation of characteristics of distribution of the random variable, e.g., the mean value is approximated by the average of the observed values. Both of these tasks occur in the queuing theory; the probability of some event is looked for (e.g., service without queuing), i.e., the probability that x will get a value from some interval, or its mean value.

The practical possibility of the application of simulation models is determined at sufficient speed in repeating these experiments. Repetition of experiments is often necessary to obtain the probability (or mean values) with sufficient accuracy. These estimations are based on the law of large numbers.

The determination of the probability of one phenomenon is often not sufficient; a set of phenomena is required. For example, in searching for the probability distribution function of a random variable x, we want to know the values $F(x) = P(X \leq x)$ for all real x values; in practice the values $F(x_j)$ are determined only for some samples x_j (in finite number) and the concept of the diagram of the whole function $F(x)$ is based on these samples. The more values $F(x_j)$ that are found, the more accurate is the approximation of function $F(x)$, but more experiment are required. In addition, we are interested in changes of function $F(x)$ related to changes in basic characteristics and parameters of the system. It is apparent that without computers the practical application of simulation procedures would not be possible.

8.6 APPLICATION OF QUEUING THEORY IN WRS

In section 7.4 the mathematical similarity of the inventory theory and the queuing theory was mentioned, resulting in the application of these theories in WRS. Now this similarity is clarified. In section 7.4.2 the general analytical solution of sets of equations (7.21) was not reached, i.e., the set

$$
\begin{aligned}
R_0 &= p_0 R_0 + p_0 R_1 + \dots + p_0 R_M \qquad (8.9)\\
R_1 &= p_1 R_0 + p_1 R_1 + \dots + p_1 R_M + p_0 R_{M+1}\\
&\dots\dots\dots\dots\dots\dots\dots\dots\dots\dots\\
R_{V-M-1} &= p_{V-M-1} R_0 + \dots + p_{V-M-1} R_M + p_{V-M-2} R_{M+1} + \dots + p_0 R_{V-1}\\
R_{V-M} &= p_{V-M} R_0 + \dots + p_{V-M} R_M + p_{V-M-1} R_{M+1} + \dots + p_0 (R_V + R_{V+1} + \dots)\\
&\dots\dots\dots\dots\dots\dots\dots\dots\dots\dots\\
R_{V-M+S} &= p_{V-M+S} R_0 + \dots + p_{V-M+S} R_M + p_{V-M+S-1} R_{M+1} + \dots +\\
&\quad + \dots + p_S (R_V + R_{V+1} + \dots)\\
&\dots\dots\dots\dots\dots\dots\dots\dots\dots\dots
\end{aligned}
$$

where (Fig. 7.8) V is the active storage of the reservoir, and M is the release from the reservoir.

The definitions of values p_i and R_i is as follows: if X_t is the inflow in the reservoir in the year t and Z_t is the storage before the inflow X_t then in time t the variable X_t is equal to i with probability p_i and the sum of the volume of water remaining in reservoir Z_t and the inflow X_t will equal i $(Z_t + X_t = i)$, with probability R_i. By this infinite system of linear equations the probability distribution of Z_t is determined, given the distribution of $Z_t + X_t$ and X_t.

The main difficulties in the analytical solution of the system of these equations was caused by the boundary conditions $Z = 0$ and $Z = V - M$. Moran *et al.*, 1959, investigated the assumption of infinite storage in two cases:

(1) the phenomena near the top of the active storage in cases where the distribution of variable Z_t was such that it was very improbable that Z_t would be equal to values near zero, and they assumed an infinitely deep reservoir;

(2) in the opposite case where the probability of a full reservoir was very small, they assumed the reservoir to be infinite in the upward direction and a distribution of Z_t near zero values was considered.

Assuming the infinite active storage $V = \infty$ and steady state and discrete variables, the system of equations (8.9) has the following form:

$$\begin{aligned} R_0 &= p_0R_0 + p_0R_1 + \ldots + p_0R_M \\ R_1 &= p_1R_0 + p_1R_1 + \ldots + p_1R_M + p_0R_{M+1} \\ &\ldots\ldots\ldots\ldots\ldots\ldots\ldots\ldots\ldots\ldots \\ R_S &= p_SR_0 + p_SR_1 + \ldots + p_SR_M + p_{S-1}R_{M+1} + \ldots + p_0R_{M+S} \end{aligned} \tag{8.10}$$

for $S = 2, 3 \ldots$.

Using Foster's method (Foster, 1953) it can be proved that this system has a solution different from zero, i.e., there exists a probability distribution (different from zero) near the level of the infinitely deep reservoir. The solution, however, is not easy, and some methods of the queuing theory can be used for it. The queue with bulk service is used (Bailey, 1954): the active storage is represented by the length of the queue and the service is assumed in time $\ldots t_{i-1}, t_i, t_{i+1}, \ldots$ so that intervals $v = (t_{i+1} - t_i)$ are independently distributed with probability distribution $\mathrm{d}B(v)$.

The customers' arrivals are random, forming a Poisson process with the mean value λ; in a time t_i, a bulk of customers, M (or the whole queue if its length is shorter than M) is served. Knowing the distribution $\mathrm{d}B(v)$ and the parameter λ, the probabilities p_i can be computed, i.e., the probabilities that at one interval, i, customers will arrive in the queue. The length of the queue just before the beginning of the service is a random variable of an infinite Markov chain, the transition and stationary probabilities of which are given by the set of equations (8.10).

Using this analogy with the queuing theory, the set of equations (8.10) can be solved. The solution by generating functions was first published by Bailey (1954) and was described in detail by Moran (1959).

The queuing theory can be used for the similar problem of the operating policy for release, where continuous variables are used and the input is a negative exponential distribution.

Let us investigate the possibility of the direct solution of a continuous alternative of the set of equations (8.10), i.e., the equation (7.22) of section 7.4.2. For an infinite V the set of equations (8.9) resp. the equations (7.22) become

$$g(x) = f(x) \int_0^M g(x)\,dt + \int_M^{M+x} f(M + x - t)\,g(t)\,dt \tag{8.11}$$

where $g(x)$ is the probability density function of variable Y_t, $f(x)$ is the probability density function of X_t. For this equation the Laplace transform can be used for the cumulative probability function of variable X_t and $V_t = X_t + Z_t$

$$F(y) = \int_0^y f(t)\,dt, \qquad G(y) = \int_0^y g(t)\,dt \tag{8.12}$$

and in simple form

$$G(y) = \int_0^y G(M + y - t)\,dF(t) \tag{8.13}$$

This type of integral equation is well-known from the queuing theory. Lindley, 1952, derived the solution and proved that the steady-state solution exists if the mean input value is less than M, and if the initial state is zero. The distribution of variable Y_t converges on the stationary state of the process.

Smith, 1953, has shown that another useful analogy between infinite reservoirs and the queuing theory is possible, and it can be seen in the following case of the queuing theory. A queue is assumed with one server only and with regular inter-arrival time M. In Lindley's general theory these intervals are random, in this application it is possible to use fixed intervals. If no queue has been formed, the customer is served immediately, otherwise he must wait till all the customers in front of him are served.

The service time, denoted X_i, is a random variable with the cumulative distribution function $F(x)$, X_i being mutually independent and independent of queue characteristics. The time interval necessary for serving the whole queue that had formed immediately before the arrival time t_i, is denoted Z_i. This is the waiting time of a new customer in the queue between the time of his arrival and the beginning of his service. $Z_i + X_i$ is then the total time that a new customer spends in the system, i.e. the time interval between the arrival of a customer and the end of his service. The whole time interval necessary for serving the queue immediately before the time t_{i+1} is Z_{i+1}. It is equal to

$$\begin{aligned} Z_{i+1} &= Z_i + X_i - M \quad \text{if } Z_i + X_i - M > 0 \\ Z_{i+1} &= 0 \qquad\qquad\quad\;\; \text{if } Z_i + X_i - M \leqq 0 \end{aligned}$$

This queuing problem can be applied for an infinite reservoir, assuming that X_i represents the random inflow X_t into reservoir and reservoir storage just before the withdrawal is represented by the waiting time of the last customer. The withdrawal M from reservoir is represented by the reduction in waiting time of the new customer during the time interval (t_i, t_{i+1}). In this model (Smith, 1953) the release is not realized at once, but in regular steps. The assumption of regular withdrawals does not change the underlying theory, as the equation (8.13) is valid for this example.

It is apparent that the way the problems of reservoir operation and queuing correspond to each other is quite different from the previous case, where the length of the queue taken as a discrete variable correspond to storage of the reservoir, and bulk service was assumed. In this case, the service of a single customer is assumed and storage of reservoir is modelled by the waiting time of a new customer in the queue.

The numerical solution of this case is that of equation (8.13). It was published by Lindley, 1952, for cases where the input variable has a Pearson type III probability distribution. The resulting probability distribution (Moran, 1959) is the distribution in this case. From a mathematical point of view, this latter case of the application of the queuing theory for WRS is a continuous analogue of the previous case, often called Bailey's queuing theory with bulk service. However, the correspondence between the WRS model and queuing theory is different in each case.

Gani and Prabhu, 1957, derived the solution of equations for the infinite reservoir for the case when storage in the reservoir has a negative exponential distribution.

Moran, 1959, Kendal, 1957, and others described a number of possible approaches, the latter used the operation of reservoirs with continuous time. Input is a continuous inflow, withdrawal is also continuous, and takes place until the reservoir is empty. Both finite and infinite reservoirs were investigated. In addition, Kendall, 1957, investigated a non-stationary case of distribution of storage: the initial state in time $t = 0$ is given by storage y and the probability distribution of time interval T in which the reservoir is emptied for the first time is looked for. This method approaches the theory of common risk. All these studies are stimulative and use sophisticated mathematical theory, but the stage of development is not sufficient for its practical application. In addition, most models use an unrealistic assumption that the inflow is an additive process with a Pearson type III distribution.

The numerical calculation in these models is similar to that in the inventory theory, illustrated by a simple example in section 7.4.1.

Another approach to the application of the queuing theory for reservoir operation was described by Chorafas, 1965. Binomial probability distribution is assumed for random inflow to the reservoir in time-sequenced stages. The release from the reservoir, in each stage, is predetermined, where there is an empty reservoir, or where spillage has occurred. These models yield relationships among the following quantities: mean values and variances of reservoir inflows, reservoir storage, the chosen values of reservoir release, probability of deficits and spillage.

These models can be described by

$$Z_i = Z_{i-1} + (\bar{x}_i + es_{x_i}) - M_i \tag{8.14}$$

where Z_i is content of reservoir at the end of stage i, Z_{i-1} is the remaining content of reservoir from the previous stage, $\bar{x}_i$ is the mean value of inflow at stage i, s_{x_i} is its standard deviation, e is a random number with a normal distribution N (0, 1) (zero mean and unit variance), and M_i is the reservoir release at stage i.

In solution the initial state is determined by the given or chosen storage of reservoir and release. In equation (8.14) other values are computed. The calculation is carried out several times on a computer with different input values. The results of such calculations are taken to be a sample of a set of physical reactions of the system that forms a response surface for various possible solutions. If the economic parameters are considered in the model, namely initial and operational costs, the benefits from different values of withdrawals, eventual losses due to deficits in water supply, the physical response surface can be transformed into an economical response surface or a net benefit one. These methods can be used to define operation of reservoirs with carry-over; some results, however, may be sub-optimal.

To summarize the application of the queuing theory in WRS: the idea of using the queuing theory for the operation of reservoirs is promising; it is, however, obvious that the people behind this idea are mathematicians and not water resource engineers. The studies use complicated and sophisticated mathematical operations, but the applicability of the results in the practice of WRS is limited for two main reasons: (i) the assumptions oversimplify the reality, and (ii) the difficulties in applying the calculations to real situations are neglected. Moran admitted this situation and stated that all the methods investigated can, in principle, be used for a system of reservoirs; however, the calculations involved for application to a single reservoir are so complicated that the recommended numerical method is the Monte Carlo procedure.

Kartvelishvili, 1963, criticises these applications because they use the simplified assumption that the river flow has a Pearson type III distribution, and he underlines the assumption that infinite active reservoir storage can be accepted in a limited number of cases. Buras, 1972, in his WRS monograph devotes little space to the application of the queuing theory in WRS.

9 GRAPH AND NETWORK THEORY

9.1 APPLICABILITY OF NETWORK ANALYSIS METHODS

The mathematical theory of graphs is the theoretical basis of network analysis methods used in problems of sequential processes (Alexy, 1967). It is an extensive and highly developed theory of finite mathematics, including linear and non-linear programming as a group of problems.

Historically, the oldest task is the problem of the shortest path, which was the origin of some modifications of the transportation problem (transportation network). The applicability of these methods and their implementation is now much broader.

Let us assume a process that consists of partial processes or, stages with their qualitative and quantitative evaluation. Assume that these partial processes have a defined linkage in real time, which may, but need not be serial and direct for each partial process. Therefore indirect linkage in a combination of serial and parallel relationships exists in a group of partial processes (stages). The objective of solving problems of such complex linkage of processes in time and space is frequently a maximum reduction in total processing time.

Such a formulation is often encountered in problems of organization and scheduling of the construction of complex objects and systems, in the management of investment activities, in planning delivery and the set-up of intricate facilities, in planning maintenance, replacement, research and technology developments.

Very often problems of parametric linear programming with methods of flow in networks, mainly the maximum flow in networks with deterministic or stochastic inputs are applied.

In technical and economic activity some parts of the theory and the application of these methods and their modifications are called CPA: *Critical Path Analysis*; what they have in common is the set-up of networks and their analysis.

The *Critical Path Method* (CPM) can be used if these complex activities can be split up into partial activities. It is applied in planning the replacement times of large-scale and complex production units, and for the organization of building large-scale systems.

Program Evaluation and Review Technique (PERT) is also an alternative to the method of critical path analysis; it was successfully used for planning large-scale development and construction programmes. Its first modification was developed for time scheduling of these programmes and was called PERT/TIME, a second modification called PERT/COST is used for planning the costs of the realization of these programmes and projects.

Together with these two basic methods, the method of *Resource Allocation and Multi-Project Scheduling* (RAMPS) is also used. The allocation of limited resources in space and time in different branches (WRS included) is, however, designed by different methods, e.g., by dynamic programming with relatively more possibilities than by methods derived from linear programming.

9.2 BASIC TERMS OF GRAPH THEORY

The *node* (vertex) is an isolated point in a plane or in space, or it can be the starting, or ending, point of an arc in a graph. The *arc* (edge) is any curve connecting various pairs of nodes. It does not include any node. The *graph* is a geometric form consisting of a set of points called nodes (vertices) and set of line segments joining any two nodes (vertices) and called arcs (edges). These arcs can be the segments of a line or of a curve.

In the theory of graphs, the definition of a graph G is possible without its geometrical representation as a diagram (as used in this chapter) by the set of nodes N and a set of arcs $G \equiv (N, \mathrm{A})$.

In the graph theory different kinds and types of graphs are investigated and only a part of them, under certain conditions, are useful for the schematic representation of WRS or critical path analysis of complex processes.

The *null graph* consists of a finite number of nodes that are isolated, i.e., they are not connected by arcs.

The *simple graph* is a graph that has neither self-loops (i.e., arcs having the same node as both their end nodes) nor parallel arcs (i.e., more than one arc associated with a given pair of nodes). The graph that is not simple is called a general graph (multi-graph).

The *complete graph* is a simple graph in which every pair of nodes is connected by an arc.

The *directed graph* (*digraph*, orientated graph) is a graph where every arc is represented by a line segment between the node n_i and n_j with an arrow directed from n_i to n_j.

Two graphs G_1 and G_2 are called *isomorphic* if there is a one-to-one correspondence between their nodes and arcs, such that a pair of nodes in the graph G_1 are joined by arcs, if and only if, the corresponding pair of nodes in the graph G_2 are also joined. In directed graphs the orientation is also maintained.

Planar graphs have some geometric representation which can be drawn on a plane in such a manner that no two of its edges intersect (they only meet in the node). When a node n_i is an end (or starting) node of an arc a_j, n_i and a_j are said to be *incident* to each other. The number of arcs incident to a node is called the degree of this node.

The number of arcs is then equal to half the sum of the degrees of all the nodes.

The graph is termed *connected* if there is at least one chain between every pair of nodes in this graph. The *chain* (walk) is a broken line in the graph, such that no arc is traversed more than once.

A closed chain is called a *cycle*. A cycle in which no node (except the initial and final node) appears more than once is called an *elementary cycle* (circuit).

If a closed chain in a graph traverses every arc of the graph exactly once, the chain is called a *Euler line* (named after Euler, who in 1736 developed the basis of the graph theory). Each graph that contains this line is called a *Euler graph*, and it has some special properties (e.g., it is connected).

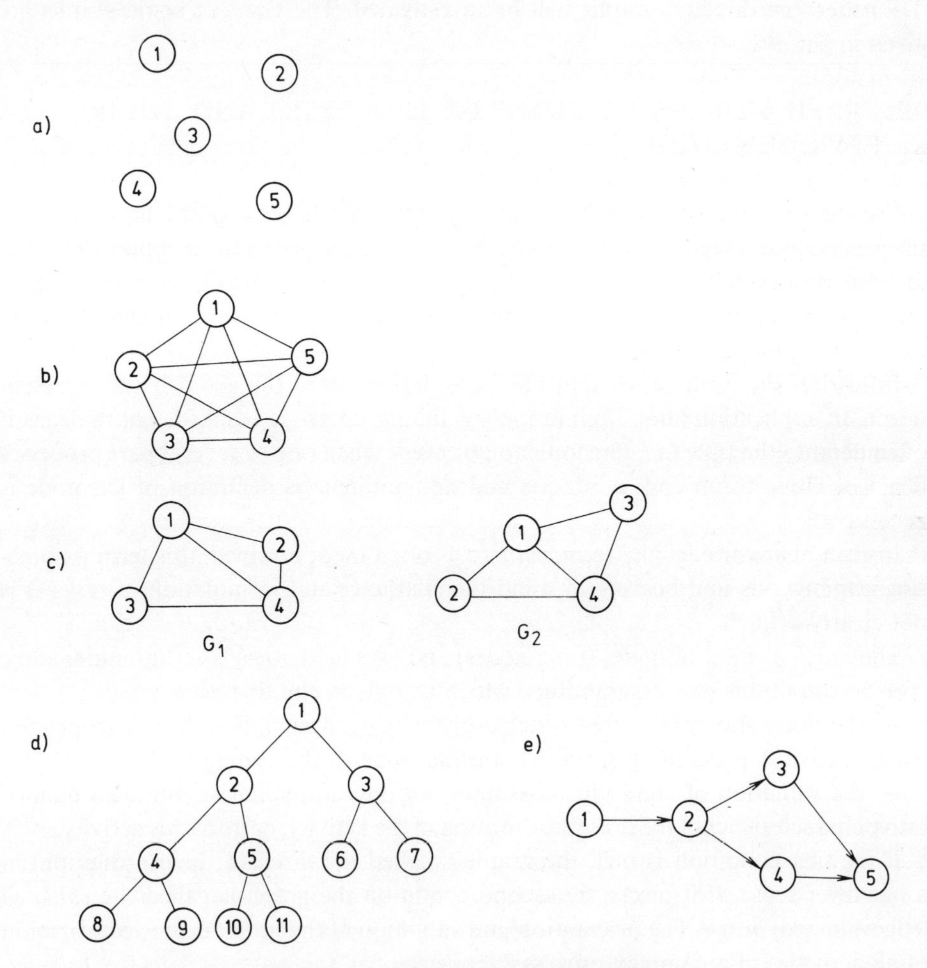

Fig. 9.1 Examples of some types of graphs
a – a null graph, *b* – a complete graph, *c* – isomorphic graphs G_1 and G_2, *d* – a tree, *e* – a directed graph (digraph)

If the chain traverses every node of the graph, and if this chain is an elementary cycle, it is called the *Hamiltonian circuit*.

A connected graph without any cycle is called a *tree*. A tree with n nodes has $(n - 1)$ arcs. If the tree is not directed, it can start in any node – each node of the graph can be the *root* of the tree. The remaining components of a tree graph form the *arborescence*.

Sometimes a combination of cycles and trees is encountered. Then the following task can be carried out: the determination of the number of arcs that should be dropped, in order to exclude all cycles. This number is called the *cyclomatic number* or *nullity* of the graph.

Further, the directed graphs will be investigated. The types of some graphs are given in Fig. 9.1.

9.3 GRAPH MODELS OF COMPLEX PROCESSES AND THEIR REPRESENTATION

For the CPA methods that are often applied to WRS, some of the basic terms of the general graph theory are important. An important aspect is the complex character of these processes, which require not only qualitative but also quantitative representation of the problem dealt with. Therefore directed graphs with some valuation of arcs are often applied.

Instead of the term vertex that has been derived from the geometrical representation, in application (and often in theory) the term *node* is used, or *time node*, as it often denotes the time of a phenomenon or event when one or several part processes begin or come to an end. A unique and non-ambiguous definition of the node is necessary.

Instead of arc or edge the term '*activity*' is often used; however, this term is sometimes ambiguous and bearing in mind the character and formulation of tasks it is not clearly defined.

The *graph* as a set of nodes (time nodes) and arcs (activities) is set up under some specific conditions and assumptions, which include in the first place:

– the orientation of the arcs, which is given by the direction of the partial process (stage, activity), proceeding from the starting node to the ending node;

– the valuation of arcs, which assumes not only a qualitative, but also quantitative characteristics of the arcs (the duration of the activity, costs of this activity, etc.).

If the first condition is met, the graph is called the *directed graph* (digraph) in a narrower sense. If it meets the second condition the graph is called the *valuated* (edgevaluated) *graph*. The orientation and valuation of the graph is the vectorization of all activities of a complex process.

Besides edge-valuation node-valuation is sometimes performed, and the graph is called a *node-valuated graph*. In principle it is an exchange of the function of nodes and edges. Such an approach may sometimes be more advantageous. The node can then represent an activity, it is valuated by the duration of the activity, its costs, etc.

In the generation of a graph representing a complex process the following rules have to be obeyed:

– each activity needs to have a uniquely defined starting and terminating point,

– no activity starting in one node can start before all the activities ending in this node have been completed,

– the graph can have only one initial and one terminal node; the initial node is a node in which no activity ends, the terminal node is a node in which no activity starts,

– two nodes can be connected by one activity only (this is not valid in a multigraph),

– there is no sequence of partial processes (activities) following each other, which starts and ends in the same node; therefore no cycle (or loop, feedback) can appear in the graph.

These conditions are used for an edge-valued digraph (directed graph); some corrections would be necessary for a node-valued graph.

A network prepared for the solution of problems can be represented by data, presented in a tabular form, on the nodes and arcs, their connection, orientation and valuation in a matrix of the network called the *incidence matrix*. Computer processing of the problem cannot be done without this matrix. Some very complicated processes use this convenient and useful matrix representation, as the graphical representation is cumbersome and can be a source of errors.

In Fig. 9.2 there is a simple general scheme of this matrix for the network in Fig. 9.3.

$t_i^{(0)}$	i \ j	0	1	2
$t_0^{(0)}$	0		τ_{01}	τ_{02}
$t_1^{(0)}$	1			τ_{12}
$t_2^{(0)}$	2			
$t_j^{(1)}$		$t_0^{(1)}$	$t_1^{(1)}$	$t_2^{(1)}$
$t_j^{(1)} - t_i^{(0)}$				

Fig. 9.2 The scheme of the incidence matrix of a simple network

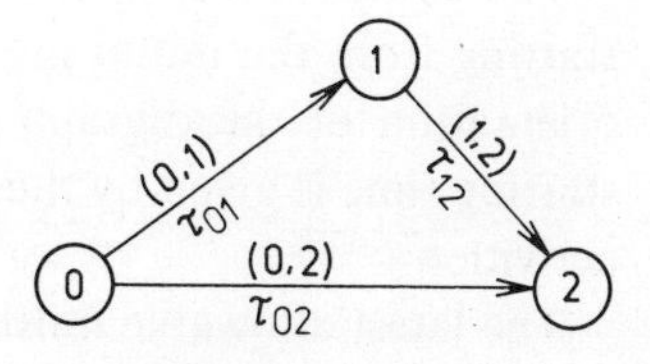

Fig. 9.3 The scheme of a simple network

The graph has three nodes and three arcs. In the first column the earliest possible starting points of activities $t_i^{(0)}$ (earliest event time) are listed.

Similarly in the penultimate row of the matrix the latest event times of activities $t_j^{(1)}$ and in the last row the differences $t_j^{(1)} - t_i^{(0)}$ are listed, which are important for the computation of the total slack (see 9.4.1).

In the body of the matrix (in Fig. 9.2 inside the heavy lines) there are the non-zero elements in the cross-section of rows denoting the starting nodes and columns denoting the terminating nodes of arcs of this network. Each row of the matrix represents the valuation of this arc and in the critical path analysis it is the duration of the activities.

The calculation of the values $t_i^{(0)}$ and $t_j^{(1)}$ and of the total slack is done in section 9.4.1. In this matrix the earliest event times (in the upper right hand corner) and the latest event times (in the lower left hand corner) are often listed.

9.4 TIME ANALYSIS OF NETWORKS

9.4.1 Critical Path Analysis

The first stage of the time analysis by the method Critical Path Analysis (CPA) is the development of the network represented by a graph, its orientation, and valuation. It requires proper definition and notation of activities, their specification and description. For each activity the following items are listed (usually in tabular form):

– all the preceding activities,

– the estimation of the normal duration of the activities, (possibly their standard values).

The shortest time to complete all the activities consecutively is given by the longest directed path (walk) in the network called the *critical path.* The determination of the critical path consists of the determination of the earliest (expected) starting time and the earliest finishing time of each activity and the latest allowable starting time and the latest allowable finishing time, which are necessary for completing the process plan (complex activity). Their values are often listed directly in the network.

The computation of the earliest starting times is the aim of the forward pass starting from the initial project event (node) of the network in the direction of the orientation of the digraph (logical sequential activity dependencies). The earliest starting time is given by the maximum of the earliest finishing times of all preceding activities.

The latest allowable finishing time is the minimum of all latest allowable starting times of the successors of this activity, and it is determined by the backward pass calculation, beginning with the project terminal event (node).

The critical path is then the sequence of the dependent activities where the difference between the earliest starting time and the latest allowable starting time (or finishing time) is zero.

Each increase of the duration of the critical activities (activities on the critical path) will delay the entire project. If a reduction of project time is required, a reduction of the duration of critical activities is necessary. If the critical activities follow immediately upon each other, then the non-critical activities have some *time reserve*.

The positive difference between the earliest expected and the latest allowable starting time form a time reserve called the *total activity slack*. Its use can reduce the time reserve of the succeeding activities. Therefore the total activity slack should be used by the planner who is responsible for the whole project.

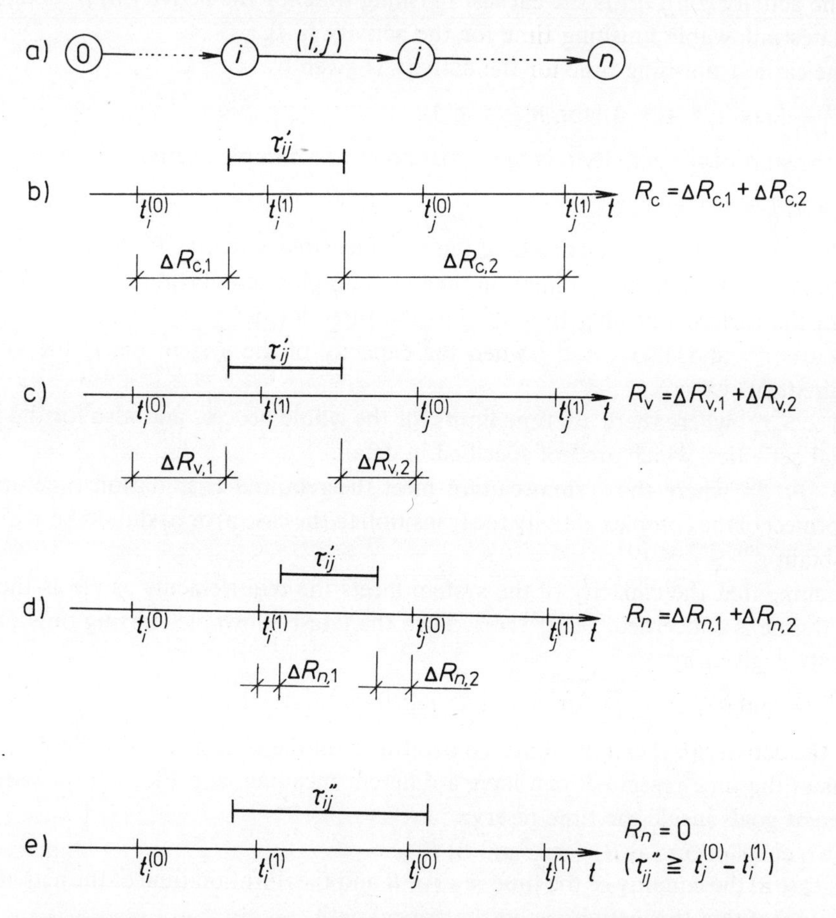

Fig. 9.4 The representation of activity slacks in analysis of the network by CPM
a – scheme of the activity (i, j), *b* – total activity slack, *c* – free activity slack, *d* – activity independent slack for $\tau'_{i,j} < t_i^{(0)} - t_i^{(1)}$, *e* – activity independent slack for $\tau''_{i,j} \geqq t_j^{(0)} - t_i^{(1)}$

There is a part of the total activity slack that does not influence the earliest starting time of the succeeding activities. This part is called *activity-free slack*. It can be used by the planner responsible for the individual activities as the use of this activity cannot influence the succeeding activities.

The activity-independent slack does not depend on the use of the proceeding activities, and therefore it can be used independently.

The method outlined for the determination of the critical path and the activity slack and the rule of their use can be described mathematically (Fig. 9.4a), where i is the starting node of the activity (i, j); j is the ending node of the activity (i, j); $\tau_{i,j}$ is the estimate of the mean (standard) duration time for the activity (i, j); $t_i^{(0)}$ is the earliest starting time for the activity (i, j); $t_i^{(1)}$ is the latest allowable starting time for the activity (i, j); $t_j^{(0)}$ is the earliest finishing time for the activity (i, j); and $t_j^{(1)}$ is the latest allowable finishing time for the activity (i, j).

The earliest finishing time for the activity is given by

$$t_j^{(0)} = \max \left(t_i^{(0)} + \tau_{i,j}\right) \quad \text{for } n \geqq j \geqq 1 \tag{9.1}$$

If the start of the activity t_i is the initial point of a complex activity (process), then

$$t_i = t_i^{(0)} = t_0^{(0)} = 0$$

The activities (i, j) are elements of the set of feasible activities P, i.e. $(i, j) \in P$. The relationship between the termination time of a complex activity (of the whole process) λ and the earliest finishing time $t_n^{(0)}$ has the three forms:

a) $\lambda = t_n^{(0)}$ and also $\lambda = t_n^{(1)}$ when the capacity of the system meets the system requirements exactly;

b) $\lambda > t_n^{(0)}$ where there are time limits for the whole process and also for the individual activities, which are not specified in detail;

c) $\lambda < t_n^{(0)}$ where the system cannot meet the required termination time and in the project of the complex activity the transition to the case a) or b) should be achieved to obtain $t_n^{(0)} \leqq \lambda \leqq t_n^{(1)}$.

Assume that the capacity of the system meets the requirements as far as the terminal time is concerned, i.e., $t_n^{(1)} = \lambda$. Then the latest allowable starting time of any activity is given by:

$$t_i^{(1)} = \min \left(t_j^{(1)} - \tau_{i,j}\right) \quad \text{for } n - 1 \geqq i \geqq 0 \tag{9.2}$$

and the activity (i, j) can be allocated arbitrarily in the time interval $\langle t_i^{(0)}; t_j^{(1)} \rangle$. The parts of the time reserve R can have a different meaning, and they can be used for different goals inside the time reserve.

Two cases occur: a) $R = \tau_{i,j}$, and b) $R > \tau_{i,j}$.

In case a) the equality of the time reserve R and the duration time of the activity $\tau_{i,j}$ is the proof that this activity is on the critical path.

It can be proved that for $\tau_n^{(0)} = \lambda$ the network contains at least one critical path composed of activities where the condition $t_j^{(1)} - t_i^{(0)} = \tau_{i,j}$ is met, i.e., at least one critical path. For the activities on this path it is essential that $t_j^{(1)} = t_j^{(0)}$ and $t_i^{(0)} = t_i^{(1)}$. In case b) when $R > \tau_{i,j}$ these activities are not on the critical path. The time reserve

$$t_j^{(1)} - t_i^{(0)} - \tau_{i,j} = R_c \tag{9.3}$$

is called the *activity total slack.*

Activity (i, j) can be arbitrarily allocated inside the interval $\langle t_i^{(0)}; t_j^{(1)} \rangle$ (Fig. 9.4b). Using the total activity slack R_c, this activity becomes a critical one: it is on the critical path. This means the development of an additional critical path, which is not desirable.

If the total activity slack $(R_c = \Delta R_{c1} + \Delta R_{c2})$ is not used, but only a part of it given by time intervals $\Delta R_{v,1}$, $\Delta R_{v,2}$ (Fig. 9.4c), a more convenient state will be developed not affecting the earliest finish time for this activity (i, j). This part of the total activity slack

$$t_j^{(0)} - t_i^{(0)} - \tau_{i,j} = R_v \tag{9.4}$$

is called the *free activity slack*, and it does not influence the successors in the network.

If neither the activity's earliest finish time nor the latest allowable start time are not to be influenced, then the activity-independent slack is used,that for $t_j^{(0)} - t_i^{(1)} > \tau_{i,j}$ is equal (see Fig. 9.4d).

$$R_n = t_j^{(0)} - t_i^{(1)} - \tau_{i,j} \tag{9.5}$$

and for $t_j^{(0)} - t_i^{(1)} \leqq \tau_{i,j}$ is equal to zero (Fig. 9.4e).

Some simple examples are explained by Alexy, 1967.

9.4.2 The PERT Method

The PERT method is suitable for the time analysis of complex processes where the time dimension and its exploitation is the most important feature. In this method two basic interrelated components appear: the planning and the control components.

The basic methodological approach to the treatment of complex problems by the PERT method is similar to that by the CPM method: a network is developed, the duration of the activities is estimated and a critical path is sought. The difference between the two methods is in the estimation of the duration times of the activities. In CPM this estimation is deterministic, as it is assumed that the duration time can adequately be characterized by a single value. In PERT these inputs are stochastic: the duration time for the activities in the network is characterized by a probability distribution of the time values. Therefore CPM can be considered to be a special case of the PERT method if the estimates of the expected duration times have a probability equal to one, or, the PERT method can be considered as a generalization of the CPM method, if the probability of estimation of the mean value is less than one, i.e., there is some uncertainty in its occurrence. The PERT analysis can be considered as a *stochastic method.*

In practical applications of this method the whole probability distribution of the duration time is not used; only some values (usually three) are applied. These values are:

– optimistic (minimum) estimate of the duration time for the activity τ_a (in the

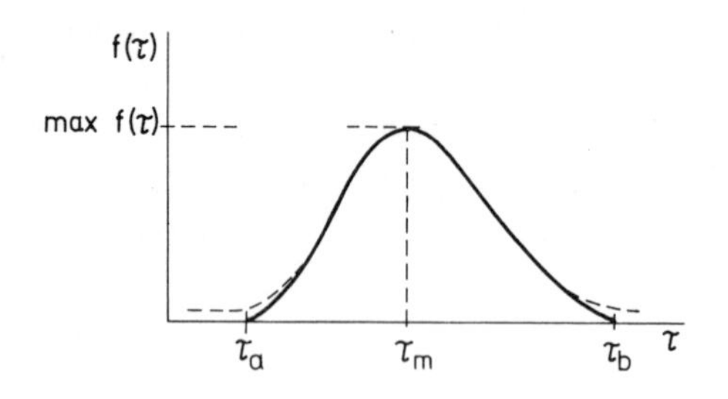

Fig. 9.5 Probability density function of the duration time for the activity in the time analysis of a network by the PERT method

graph in Fig. 9.5) is the lowest value in the case of the lower-bounded probability distribution or the value τ'_a given by a very small portion of the area under the probability density function (e.g., 1% or 5%),

– pessimistic (maximum) estimate of the duration time for activity τ_b (Fig. 9.5) or τ'_b similarly defined as an optimistic value (e.g. 99 or 95%),

– the most likely estimation of the duration time τ_m – the modal value of the distribution.

It is assumed that the duration time of activities in a complex process has a normal distribution or a beta distribution[1]). In the case of a beta distribution, the probability density function can be written in the form

$$f(\tau) = k(\tau - \tau_a)^{\alpha} (\tau_b - \tau)^{\gamma} \tag{9.6}$$

where k is a constant and α and γ are functions of τ_a, τ_b and τ_m.

The value used in computation is estimated by

$$\tau_0 = \frac{\tau_a + 4\tau_m + \tau_b}{6} \tag{9.7}$$

[1]) The beta distribution is defined by the cumulative distribution function, which uses the Euler integral of the first kind

$$B_x(\alpha, \beta) = \int_0^x x^{\alpha-1}(1 - x)^{\beta-1}\,\mathrm{d}x$$

with two parameters α, β. For some combinations of α, β it is similar to the Pearson type I distribution (see 3.1.2.7).

with variation

$$\sigma^2(\tau_0) = \left(\frac{\tau_b - \tau_a}{6}\right)^2 \qquad (9.8)$$

This method is used for processing all input data of the activities in time analysis. The further procedure is similar to that used in CPM.

The earliest expected and latest allowable activity starting times $t_i^{(0)}$ and $t_i^{(1)}$ and their variances $\sigma^2(t_i^{(0)})$ and $\sigma^2(t_i^{(1)})$ are computed. The finishing time of the whole process can be given by a deterministic value, otherwise it is unknown. In the latter case it is made equal to the latest allowable time of the latest activity of the complex process in the critical path $t_n^{(1)}$.

A specific feature of the PERT method is the determination of slack probability at the end of the complex process. It includes assessing the probability of completing the project on target with the probability estimation that does not appear in CPM.

Let us analyse the duration of the activities. The value x is given by

$$x = \frac{t_j^{(0)} - t_i^{(1)}}{\sqrt{\sigma^2(t_j^{(0)}) + \sigma^2(t_j^{(1)})}} \qquad (9.9)$$

where $t_j^{(0)}$ is the estimation of the earliest finishing time for the activity, $t_j^{(1)}$ is the estimation of the latest allowable finishing time for the activity, and $\sigma^2(t_j^{(0)})$; $\sigma^2(t_j^{(1)})$ are the respective variations.

The value x gives a relative deviation from the negative value of the total activity slack for this activity, i.e., from the difference $t_j^{(0)} - t_j^{(1)}$. To x, a probability of $F(x)$ can be assigned from the cumulative distribution function that is equal to the probability of the activity slack. Using these values, the slack for the whole process can be determined.

Three cases can be classified in the practical applications of the PERT method:

a) $F(x) \leqq 0.25$, then the time target (the duration of the whole project within a given time limit) is endangered,

b) $0.25 < F(x) \leqq 0.60$ with a realistic assumption of meeting the time target,

c) $F(x) > 0.60$ with a time reserve.

Comparison of the two basic methods for the critical path analysis of the network shows that the PERT method which makes the development of a stochastic model possible is a more detailed and sophisticated tool of the network analysis with a more general scope of application.

9.4.3 Other Methods of Critical Path Analysis

All other methods of critical path analysis (CPA) are modifications of the CPM and PERT methods.

If there are a great number of activities with a complicated dependence, the development of the diagram of the network may be tedious and cumbersome. Then the matrix representation is often used with better insight and prevention of possible errors. The calculation of the incidence matrix is usually done by computer. The critical path programs are often provided in standard computer software.

The CPA can also be carried out by the simplex method of linear programming. The CPM is related to the transportation problem, and it can be used for the search for the Hamiltonian path as well as the travelling salesman problem.

Linear programming with the CPM method is used for decisions about the reduction of activity times, which are not on the critical path where a suboptimal solution can cause a non-effective use of funds.

There are programs for the optimization of postponing the starts of some activities under budgetary constraints to achieve the minimum duration of the whole process.

A general advantage of the CPM is the possibility of adapting to new conditions: to re-run alternatives on computers using new parameters, to new formulations of the model, and changes in the network. This advantage is used in practice and, in addition, it has some theoretical significance: the model is more sophisticated, and it can react flexibly to dynamic changes in time. These methods facilitate effective and flexible decision-making.

These methods are not suitable, however, for long-term planning. The general problem common to all methods of CPA is collecting reliable input data.

The analysis of the possibilities of reducing the duration of the whole process is done by a method somewhere between the deterministic CPM and the stochastic PERT method. In practice it means a flexible increase of funds for particular activities on the critical path in order to achieve the necessary reduction in time.

Then the duration time $\tau_{i,j}$ of an activity (i, j) is not characterized by one constant value, but by two independent variables:

$\vartheta_{i,j}$ = minimum duration for the activity (i, j), and

$\Theta_{i,j}$ = maximum duration for the activity (i, j).

Therefore:

$$0 \leqq \vartheta_{i,j} \leqq \tau_{i,j} \leqq \Theta_{i,j} \tag{9.10}$$

The value $\Theta_{i,j}$ is often called the expected (standard) duration for the activity (under standard conditions) using minimum costs. It is possible to reduce this value to as low as $\vartheta_{i,j}$ by the development of better conditions and at greater costs.

The reduction of the duration of the whole process is accompanied by the requirement of minimum costs. For each activity (i, j), a cost curve $N_{i,j} = f(\tau_{i,j})$ is constructed, which is generally a non-linear, convex, and decreasing function. In practical applications it is linearly approximated in the interval $\tau_{i,j} \in \langle \vartheta_{i,j}, \Theta_{i,j} \rangle$ so that

$$N_{i,j} = a_{i,j}\tau_{i,j} + b_{i,j} \tag{9.11}$$

The cost function of the whole process can be written in the form

$$N_p = \sum_{(i,j)} N_{i,j} = \sum_{(i,j)} (a_{i,j}\tau_{i,j} + b_{i,j}) \tag{9.12}$$

The duration of each activity (i, j) $\tau_{i,j}$ is a function of the required duration of the whole process λ; therefore $N_p = N_p(\lambda)$. If the constraints $\tau_{i,j} < \Theta_{i,j}$ and $\lambda < t_n^{(0)}$ occur, then the duration of some activities has to be curtailed.

Often, there are several choices and an optimum is chosen which requires some increase of total costs for the realization of the complex process. Such an optimization problem is often solved by linear programming.

Minimize the objective function

$$\sum_{i,j} (a_{i,j}\tau_{i,j} + b_{i,j}) = \min \tag{9.13}$$

under constraints

$$\tau_{i,j} \leqq \Theta_{i,j}; \; \tau_{i,j} \geqq \vartheta_{i,j}; \; \tau_{i,j} = t_j - t_i;$$

$t_0 = 0$ and $t_n = \lambda$
and for

$$(i, j) \in P$$

where $i = 0, 1, \ldots, n - 1$; $j = 1, 2, \ldots, n$, and $\tau_{i,j} \geqq 0$.

The reduction of the duration of the whole process implies the reduction of those activities where it is most effective.

Assume that the complex process represents a project of some actual system. In real applications the optimization of the reduction of the total duration time is to include losses of production, operation, maintenance and replacement costs, etc., which would occur if the original project duration were assumed.

9.5 GRAPH THEORY APPLICATION IN WRS

The methods of network analysis by the graph theory have been extensively used in municipal water supply networks and irrigation network systems. The application of the graph theory with the representation of a complex problem by an incidence matrix is used for the simplification of the mathematical formulation of the task. The electrical network analysis was an original model for this application of this network and matrix analysis.

The representation of the topological structure[1]) of the river network by a graph was investigated by Schneidegger, 1967. The matrix analysis of municipal water supply networks was studied in Czechoslovakia by Šerek, 1968, 1972, and was used by Ošlejšek, 1975.

The topological structure of WRS is modelled by a digraph, the arcs of which represent the water transportation facilities and their valuation represents the capacity of these facilities. The nodes form the points of connection of these arcs, storage volumes, points of beginning and ending of water diversions, mains, etc. The orientation of the digraph is given by the character of the problem or some chosen convention. The incidence matrix represents the topological structure of WRS in an algebraic form that is convenient for algorithmization for computation of the problem.

9.5.1 Description and Analysis of River Networks by the Graph Theory and Matrix Analysis

A part of the river network is represented in Fig. 9.6 by a digraph (Šerek, 1972). The river network does not form cycles and its graph is a tree, whose arcs correspond to the branches. The graph is orientated in the direction of flow in the branches.

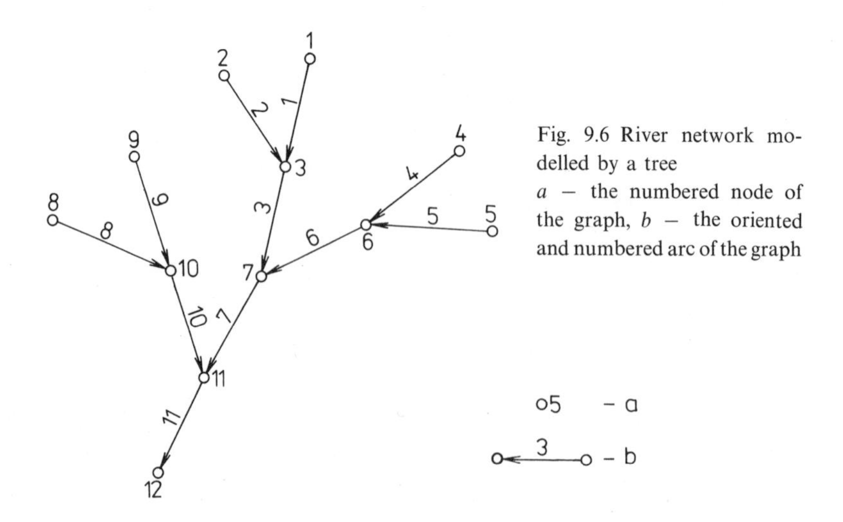

Fig. 9.6 River network modelled by a tree
a – the numbered node of the graph, *b* – the oriented and numbered arc of the graph

---

[1]) Topological structure is the system structure of an abstract system defined on the WRS in some topological space – in this case on some graph topological space – where the characteristics of the system are invariant to continuous, one-to-one correspondence transformations.

If the graph has u nodes and v arcs (branches), then $v = u - 1$. As a formality, the terminating node of the graph is not used in the incidence matrix **A** and only $(u - 1)$ nodes are considered. Each row represents one arc (branch) and each column one node. Then the incidence matrix will be a square matrix (v rows and $u - 1 = v$ columns) and a regular[1]) matrix.

For the graph in Fig. 9.6, the incidence matrix has the following form

$$\mathbf{A} = \begin{Vmatrix} 1 & 0 & -1 & 0 & 0 & 0 & 0 & 0 & 0 & 0 & 0 \\ 0 & 1 & -1 & 0 & 0 & 0 & 0 & 0 & 0 & 0 & 0 \\ 0 & 0 & 1 & 0 & 0 & 0 & -1 & 0 & 0 & 0 & 0 \\ 0 & 0 & 0 & 1 & 0 & -1 & 0 & 0 & 0 & 0 & 0 \\ 0 & 0 & 0 & 0 & 1 & -1 & 0 & 0 & 0 & 0 & 0 \\ 0 & 0 & 0 & 0 & 0 & 1 & -1 & 0 & 0 & 0 & 0 \\ 0 & 0 & 0 & 0 & 0 & 0 & 1 & 0 & 0 & 0 & -1 \\ 0 & 0 & 0 & 0 & 0 & 0 & 0 & 1 & 0 & -1 & 0 \\ 0 & 0 & 0 & 0 & 0 & 0 & 0 & 0 & 1 & -1 & 0 \\ 0 & 0 & 0 & 0 & 0 & 0 & 0 & 0 & 0 & 1 & -1 \\ 0 & 0 & 0 & 0 & 0 & 0 & 0 & 0 & 0 & 0 & 1 \end{Vmatrix} \tag{9.14}$$

The elements of the matrix have the following values:

$a_{i,j} = 0$ if the node j is not the ending point of the i-th arc; $a_{i,j} = 1$ if the node j is the starting node of the i-th arc[2]); $a_{i,j} = -1$ if the node j is the end node of the i-th arc.

Let us read the matrix (9.14) row by row: arc 1 starts in node 1 ($a_{11} = 1$) and ends in node 3 ($a_{13} = -1$), arc 2 starts in node 2 ($a_{22} = 1$) and ends in node 3 ($a_{23} = -1$). etc.

Assume the task of flow determination in arcs (branches) of this river network with an assumed stationary flow pattern without storage in the river bed and flood plains. The unknown flows in v arcs and withdrawals and inflows in $(u - 1)$ nodes are expressed as column vectors

$$\boldsymbol{q} = \begin{Vmatrix} Q_1 \\ Q_2 \\ \vdots \\ Q_i \\ \vdots \\ Q_v \end{Vmatrix} \tag{9.15}$$

$$\boldsymbol{r} = \begin{Vmatrix} R_1 \\ R_2 \\ \vdots \\ R_j \\ \vdots \\ R_{u-1} \end{Vmatrix} \tag{9.16}$$

---

[1]) In a regular (non-singular) matrix, the rank is equal to the order, i.e., the number of the rows or columns; in a square matrix, the rank is defined.

[2]) In the sense of graph orientation.

where Q_i is the flow in the i-th arc, $R_j > 0$ is the withdrawal from the j-th node, $R_j < 0$ is the inflow to the j-th node. Using the mass balance equation[1]) we get

$$^{T}\mathbf{A}\boldsymbol{q} + \boldsymbol{r} = \boldsymbol{o} \tag{9.17}$$

where $^{T}\mathbf{A}$ is transposed[2]) matrix of matrix $\mathbf{A}$, $\boldsymbol{q}$, $\boldsymbol{r}$ are the column vectors defined by (9.15), (9.16), and $\boldsymbol{o}$ is a zero vector. The flow vector $\boldsymbol{q}$ is given by the relationship

$$\boldsymbol{q} = -\boldsymbol{r}^{T}\mathbf{A}^{-1} \tag{9.18}$$

where $^{T}\mathbf{A}^{-1}$ is the inverse[3]) matrix of the matrix $^{T}\mathbf{A}$.

Reservoirs can be added to the river network described (Šerek, 1972) in Fig. 9.7.

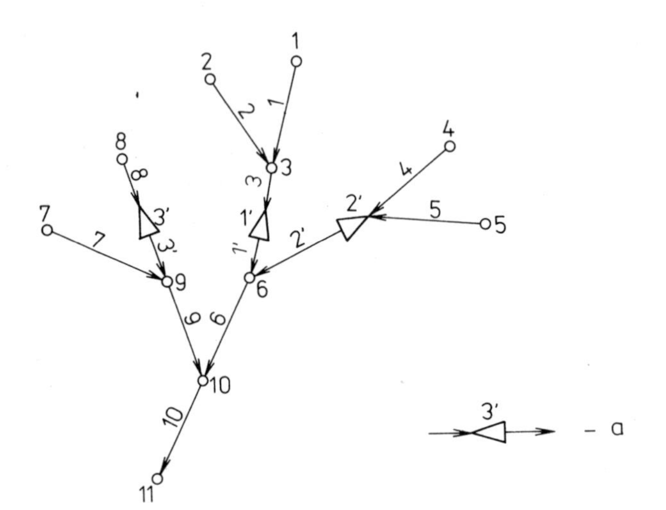

Fig. 9.7 A river network with reservoirs modelled by a tree a – a reservoir and its number

It is then necessary to use a new numbering system and the added reservoirs are given new numbers, e.g., nodes 1′, 2′, 3′ up to $n' = 3$. Thus new arcs of the graph will originate at 1′, 2′, 3′ (the total sum being $v' = n' = 3$).

---

[1]) The mass balance equation, or, in electrical networks, Kirchhoff's law, states that the sum of the flows entering the node is equal to the sum of the flows exiting from the node.

[2]) A transposed matrix is generated by the change of rows and columns and vice-versa. Therefore it satisfies $a_{i,j} = T_{a_{j,i}}$ and vice-versa for all i, j where $a_{i,j}$ is the element of the original matrix and $T_{a_{j,i}}$ is the element of the transposed matrix.

[3]) Inverse matrix $\mathbf{A}^{-1}$ (of matrix $\mathbf{A}$) is such a matrix that satisfies $\mathbf{A} . \mathbf{A}^{-1} = \mathbf{I}$ where $\mathbf{I}$ is the identity matrix (unit matrix with non-zero elements on the main diagonal equal to one).

The incidence matrix has the following form

$$
\mathbf{A} = \begin{array}{c|ccccccccccccc|}
 & 1 & 2 & 3 & 4 & 5 & 6 & 7 & 8 & 9 & 10 & 1' & 2' & 3' \\
1 & 1 & 0 & -1 & 0 & 0 & 0 & 0 & 0 & 0 & 0 & 0 & 0 & 0 \\
2 & 0 & 1 & -1 & 0 & 0 & 0 & 0 & 0 & 0 & 0 & 0 & 0 & 0 \\
3 & 0 & 0 & 1 & 0 & 0 & 0 & 0 & 0 & 0 & 0 & -1 & 0 & 0 \\
4 & 0 & 0 & 0 & 1 & 0 & 0 & 0 & 0 & 0 & 0 & 0 & -1 & 0 \\
5 & 0 & 0 & 0 & 0 & 1 & 0 & 0 & 0 & 0 & 0 & 0 & -1 & 0 \\
6 & 0 & 0 & 0 & 0 & 0 & 1 & 0 & 0 & 0 & -1 & 0 & 0 & 0 \\
7 & 0 & 0 & 0 & 0 & 0 & 0 & 1 & 0 & -1 & 0 & 0 & 0 & 0 \\
8 & 0 & 0 & 0 & 0 & 0 & 0 & 0 & 1 & 0 & 0 & 0 & 0 & -1 \\
9 & 0 & 0 & 0 & 0 & 0 & 0 & 0 & 0 & 1 & -1 & 0 & 0 & 0 \\
10 & 0 & 0 & 0 & 0 & 0 & 0 & 0 & 0 & 0 & 1 & 0 & 0 & 0 \\
\hline
1' & 0 & 0 & 0 & 0 & 0 & -1 & 0 & 0 & 0 & 0 & 1 & 0 & 0 \\
2' & 0 & 0 & 0 & 0 & 0 & -1 & 0 & 0 & 0 & 0 & 0 & 1 & 0 \\
3' & 0 & 0 & 0 & 0 & 0 & 0 & 0 & 0 & -1 & 0 & 0 & 0 & 1
\end{array} \tag{9.19}
$$

A non-stationary flow pattern is assumed, with a vector function for the storage of water V_j in reservoirs in time t calculated in the following manner:

$$
\mathbf{v}(t) = \begin{Vmatrix} V_{1'}(t) \\ V_{2'}(t) \\ \vdots \\ V_{j'}(t) \\ \vdots \\ V_{u'}(t) \end{Vmatrix} = \begin{Vmatrix} V_{1'}(t) \\ V_{2'}(t) \\ V_{3'}(t) \end{Vmatrix} \tag{9.20}
$$

The time dependent vector function of withdrawals from the j-th node or of inflow into this node for reservoirs and other nodes have the following form

$$
\mathbf{r}(t) = \begin{Vmatrix} R_{1'}(t) \\ R_{2'}(t) \\ \vdots \\ R_{j'}(t) \\ \vdots \\ R_{u'}(t) \end{Vmatrix} = \begin{Vmatrix} R_1(t) \\ R_2(t) \\ R_3(t) \\ \vdots \\ R_8(t) \end{Vmatrix} ; \; \mathbf{r}'(t) = \begin{Vmatrix} R_{1'}(t) \\ R_{2'}(t) \\ \vdots \\ R_{j'}(t) \\ \vdots \\ R_{u'}(t) \end{Vmatrix} = \begin{Vmatrix} R_{1'}(t) \\ R_{2'}(t) \\ R_{3'}(t) \end{Vmatrix} \qquad \begin{matrix}(9.21)\\(9.22)\end{matrix}
$$

The relationship of the flow in the i'-th arc under the j'-th reservoir in time t, representing the reservoir release is similarly expressed by a vector function

$$\boldsymbol{s}(t) = \left\| \begin{matrix} S_{1'}(t) \\ S_{2'}(t) \\ \vdots \\ S_{i'}(t) \\ \vdots \\ S_{v'}(t) \end{matrix} \right\| = \left\| \begin{matrix} S'_1(t) \\ S'_2(t) \\ S'_3(t) \end{matrix} \right\| \tag{9.23}$$

The vector function of flows in the arcs is

$$\boldsymbol{q}(t) = \left\| \begin{matrix} q_1(t) \\ q_2(t) \\ \vdots \\ q_i(t) \\ \vdots \\ q_v(t) \end{matrix} \right\| = \left\| \begin{matrix} q_1(t) \\ q_2(t) \\ \vdots \\ q_8(t) \end{matrix} \right\| \tag{9.24}$$

Matrix $\mathbf{A}$ can be partitioned as shown in Fig. 9.9. Submatrix $\mathbf{A}_0$ is similarly to matrix (9.14): a square and regular matrix. As in nodes $1, 2, \ldots, j, \ldots, n-1$ the mass balance is satisfied, it is

$${}^{\mathrm{T}}\mathbf{A}_0\, \boldsymbol{q}(t) + {}^{\mathrm{T}}\mathbf{A}_2\, \boldsymbol{s}(t) + \boldsymbol{r}(t) = \boldsymbol{o} \tag{9.25}$$

and the function sought is

$$\boldsymbol{q}(t) = -{}^{\mathrm{T}}\mathbf{A}_0^{-1}[{}^{\mathrm{T}}\mathbf{A}_2\, \boldsymbol{s}(t) + \boldsymbol{r}(t)] \tag{9.26}$$

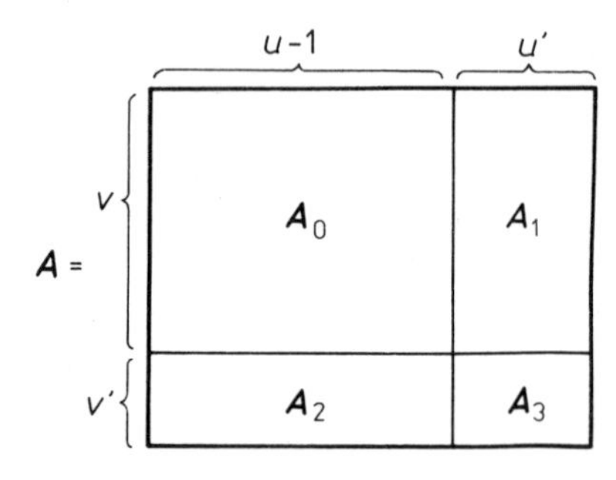

Fig. 9.8 Partition of the incidence matrix of the river network with reservoirs into submatrices

In the nodes representing the reservoirs $(1', 2', \ldots, j', \ldots, u')$ the mass balance is not satisfied (in time t) due to storage. The difference of inflow and the outflow in the node is not necessarily equal to zero, and it can be described by a vector function for the whole system

$$\mathbf{z}(t) = \begin{Vmatrix} z_{1'}(t) \\ z_{2'}(t) \\ \vdots \\ z_{j'}(t) \\ \vdots \\ z_{u'}(t) \end{Vmatrix} = \begin{Vmatrix} z_{1'}(t) \\ z_{2'}(t) \\ z_{3'}(t) \end{Vmatrix} \tag{9.27}$$

This function equals

$$\mathbf{z}(t) = -[{}^{\mathrm{T}}\mathbf{A}_1\, \mathbf{q}(t) + {}^{\mathrm{T}}\mathbf{A}_3\, \mathbf{s}(t) + \mathbf{r}'(t)] \tag{9.28}$$

Substituting $\mathbf{q}(t)$ from the equation (9.26) the following expression is obtained

$$\mathbf{z}(t) = {}^{\mathrm{T}}\mathbf{A}_1\{{}^{\mathrm{T}}\mathbf{A}_0^{-1}[{}^{\mathrm{T}}\mathbf{A}_2\, \mathbf{s}(t) + \mathbf{r}(t)]\} - {}^{\mathrm{T}}\mathbf{A}_3\, \mathbf{s}(t) - \mathbf{r}'(t) \tag{9.29}$$

The time-related change of reservoir storage can be expressed by

$$\mathrm{d}\mathbf{v}(t) = \mathbf{z}(t)\, \mathrm{d}t \tag{9.30}$$

and the differential equation for the determination of the vector function $\mathbf{v}(t)$ has the following form

$$\frac{\mathrm{d}\mathbf{v}(t)}{\mathrm{d}t} = {}^{\mathrm{T}}\mathbf{A}_1\{{}^{\mathrm{T}}\mathbf{A}_0^{-1}[{}^{\mathrm{T}}\mathbf{A}_2\, \mathbf{s}(t) + \mathbf{r}(t)]\} - {}^{\mathrm{T}}\mathbf{A}_3\, \mathbf{s}(t) - \mathbf{r}'(t) \tag{9.31}$$

For the given boundary condition in the point $t = t_0$: $\mathbf{v}(t_0) = \mathbf{v}(t_0)$ the equation (9.31) can be solved numerically.

9.5.2 Matrix Analysis of Cyclic Public Water Supply Networks

The methods of matrix analysis of water supply networks were developed by Šerek (1968, 1972, 1975). The algorithms for computers were set up for the solution of practical problems.

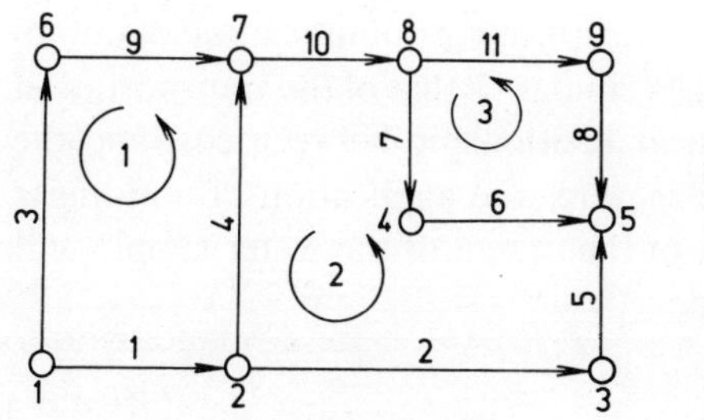

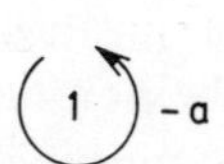

Fig. 9.9 Network model of the cyclic water supply network
a – orientation and notation of the cycle

The topological structure of the graph of a cyclic water supply network is characterized by cycles with a chosen orientation in the graph. The incidence matrix of the graph is not generally a square matrix. The rows represent the arcs of the graph and the columns represent the cycles. For the graph in Fig. 9.9 the incidence matrix has the following form

$$\mathbf{B} = \begin{array}{r} \\ 1 \\ 2 \\ 3 \\ 4 \\ 5 \\ 6 \\ 7 \\ 8 \\ 9 \\ 10 \\ 11 \end{array} \begin{Vmatrix} 1 & 2 & 3 \\ 1 & 0 & 0 \\ 0 & 1 & 0 \\ -1 & 0 & 0 \\ 1 & -1 & 0 \\ 0 & 1 & 0 \\ 0 & -1 & 1 \\ 0 & -1 & 1 \\ 0 & 0 & -1 \\ -1 & 0 & 0 \\ 0 & -1 & 0 \\ 0 & 0 & -1 \end{Vmatrix} \tag{9.32}$$

Then the vector of flows $\boldsymbol{q}$ and the vector of pressure head losses $\boldsymbol{h}$ are computed in each arc (sections of the water supply network) which are the vector function of the flows, i.e.

$$\boldsymbol{h} = \boldsymbol{f}(\boldsymbol{q}) \tag{9.33}$$

The vectors $\boldsymbol{h}$ and $\boldsymbol{q}$ must satisfy both Kirchhoff's laws:

$$^{\mathrm{T}}\mathbf{A}\boldsymbol{q} + \boldsymbol{r} = \boldsymbol{o} \tag{9.34}$$

$$^{\mathrm{T}}\mathbf{B}\boldsymbol{h} = \boldsymbol{o} \tag{9.35}$$

where the matrix **A** is set up as in section 9.5.1.

An irrigation network can be treated similarly. The solution of problems of water supply for municipalities, or irrigation, is the determination of optimal sites for water mains and other facilities, using an economic objective (minimization of initial) and operation costs). The task of optimization is similar to that of the transportation problem in linear programming. The non-linear relationship between cost and the amount of water transported in the pipeline requires the application of non-linear programming (Šerek, 1975). The automation of the design of the water supply and irrigation networks is a great help in this task.

9.5.3 Network Model of WRS

A WRS as a set of water management facilities and resources with mutual relationships, can be represented as a valuated digraph network. The algebraic description is done in the form of an incidence matrix that is set up using the method described in 9.5.1.

For the investigation of WRS by optimization models, the optimization of flow in networks can be used (Štichová, 1972). The valuation of the arcs of the graph is given by the capacity of the transportation or transfer facilities, possibly the valuation of the nodes is given by (a) the capacity of the facilities and water resources objects (waterworks reservoir, water and waste water treatment stations, etc.), or (b) by the required withdrawal in the node.

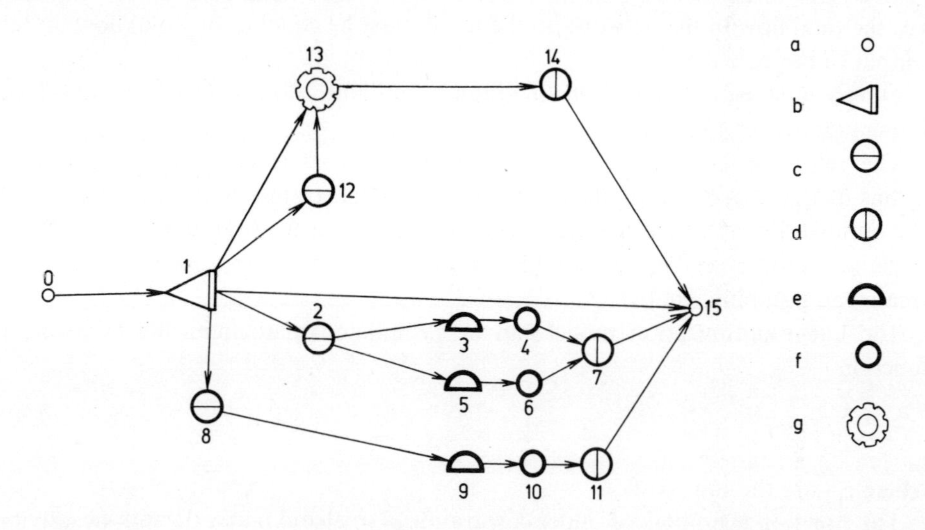

Fig. 9.10 The network model of the WRS
a – input and output, *b* – reservoir, *c* – water treatment station, *d* – waste water treatment station, *e* – waterworks reservoirs, *f* – withdrawal of drinking water, *g* – withdrawal of water for industry

An example of a model of WRS is given in Fig. 9.10. It is a static model with no time dimension and using one stage only. The incidence matrix of the graph can be set up as in section 9.5.1. The following boundary conditions must be satisfied:

a) $x_{i,j} \leqq k_j$, i.e., the flow in the arc j[1]) from node i to node j must be less than, or equal to, the capacity of the node (in the case of node valuation (a));

---

[1]) The arc connecting nodes i and j is denoted $x_{i,j}$.

b) $x_{i,j} \geqq k_j$, i.e., the flow in arc j from node i to node j, where this arc ends, must be greater than, or equal to, the withdrawal in this node (in case of node valuation (b));

c) $\sum_{i=1}^{m} x_{i,p} = \sum_{j=1}^{n} x_{p,j} + \xi$

i.e., the sum of all inflows to the node must be equal to the sum of the outflows, withdrawals and losses (Kirchhoff's law), where $x_{p,j}$ are outflows and withdrawals and ζ is the losses;

d) $\sum_{j=1}^{n'} x_{0,j} = \sum_{i=1}^{m'} x_{i,k}$,

i.e., the total flow of the network on the input must be equal to the total flow on the output of the network;

e) $x_{1,k} \geqq d$, i.e., at the end of the graph a minimum flow d must be maintained (minimum acceptable flow).

The values of flows in all arcs of the network are determined by optimization of this model of WRS satisfying all the boundary constraints. The determination of the optimal flow in arcs of the network is solved by linear programming with minimization of the objective function (minimum costs of water treatment, waste water treatment, pumping, etc.).

The linear optimization model can be formulated mathematically as a linear function

$$z = \sum_{i,j} c_{i,j} x_{i,j} \tag{9.36}$$

where $c_{i,j}$ are the unit costs.

This function is minimized under constraints a) to e) and under the non-negativity constraint $x_{i,j} \geqq 0$.

V. Štichová, 1972 used a simple network model for optimization of the WRS in the basin of the Blanice River with a single reservoir for municipal and industrial water supply. The water demands, capacities of water treatment stations, the estimation of losses in reservoirs and the network were given together with the costs of water treatment, waste water treatment, pumping into water works reservoirs, etc. The linear model was optimized by a standard linear program "MOSI 3" on computer NE4100.

In summary, the static model is criticised as it neglects the time dimension and yields an unrealistic representation of actual tasks. The random variation of the model input of the WRS can be used to give a dynamic model of the network. However, the dimension of the model with the time change of only one parameter of the system (reservoir inflow) increases linearly (Fig. 9.11.). This idea of the model dynamization cannot reflect the process of storage of water in reservoirs, which is a complex inventory problem and its optimization is assumed by methods of linear programming.

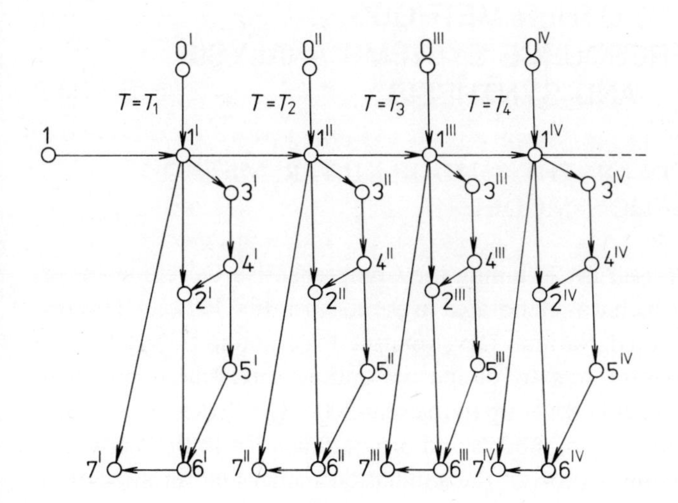

Fig. 9.11 Scheme of the model network with time dimension on input

The application of the graph theory in mathematical modelling of WRS offers an interesting impetus. However, the advantages of this method (good insight into the problems, easy algorithmization with representation of WRS as a graph and in matrix analysis) seem to diminish with the growing complexity of WRS, with the inclusion of the time changes of all the important input parameters and with optimization reflecting the actual properties of the process.

10 OTHER METHODS OF WATER RESOURCE SYSTEMS ANALYSIS AND SYNTHESIS

10.1 THE COMBINATION OF THE OUT-OF-KILTER METHOD AND THE SIMULATION MODEL

A special method was devised in "Planning the Comprehensive Development of the Vistula River System", which was elaborated in the co-operation between Hydro-project Warsaw and the United Nations Development Programme (UNDP). The main objective of the program was to design an optimal investment program for WRS in the basin of the Vistula River up to the year 2000. The Project was multi-purpose, and it included a number of models and programmes. From the methodological point of view, the combination of the simulation model and an allocation optimization model, using the Ford-Fulkerson "*out-of-kilter*" method, was the most interesting.

The WRS included water reservoirs, rivers and diversion channels, which served to transport water to the users. The model of this system was set up in the form of a network (graph) with flows in arcs of the network which corresponded to discharges or flows of water towards certain goals. These were, for example, the supply of water for various demands: minimum flow augmentation, or the unused part of flow. Each arc was assigned a coefficient related to the importance of water use. This coefficient was related to the potential losses that might occur due to a reduction in the water supply for this goal, and it was proportional to the unit loss due to the deficit of water for this goal (e.g., municipal water supply).

The allocation model is combined with the simulation model in the following manner. At each time step of the simulation model a vector of optimal flows is determined by the allocation optimization model. The water resources are allocated to the users in such a way as to minimize the weighted sum of deficits. The unit loss coefficients were used as weights.

10.1.1 The Out-of-Kilter Method

The principles of a special technique – an out-of-kilter optimization algorithm can be described briefly. The method starts by *labelling the nodes* with accompanied correction of flows in arcs of graph (network) to reduce the value of the objective function. The method is iterative and every feasible flow can be used as the starting state. Each arc of the graph has a limited capacity.

For clarifying the method and its assumptions let us consider the following linear problem in this form:

$$a_{i,1}x_1 + a_{i,2}x_2 + \dots + a_{i,n}x_n = b_i \quad \text{for } i = 1, 2, \dots, m \tag{10.1}$$

$$d_j \leqq x_j \leqq h_j \quad \text{for } j = 1, 2, \dots, n \tag{10.2}$$

$$c_1x_1 + c_2x_2 + \dots c_nx_n = \min \tag{10.3}$$

where $a_{i,j}$, b_i, d_j, h_j, c_j are components of a vector of constants, the equation (10.1) describes the form of the graph, the equation (10.2) gives the lower and upper bounds of the arc flow and the equation (10.3) is the objective function. Assume that $\boldsymbol{x} = (x_1, x_2, \dots, x_n)$ is the vector of variables that satisfy the equations (10.1) and (10.2), i.e., vector $\boldsymbol{x}$ is a feasible solution and assume that $\boldsymbol{z} = (z_1, z_2, \dots, z_m)$ is the vector of dual prices that meet the implications

$$c_j + (z_1a_{1,j} + \dots + z_ma_{m,j}) > 0 \rightarrow x_j = d_j \tag{10.4}$$

$$c_j + (z_1a_{1,j} + \dots + z_ma_{m,j}) < 0 \rightarrow x_j = h_j \tag{10.5}$$

for all j. Using the dual theorem of linear programming and further relationships it can be shown[1]) that vector $\boldsymbol{x}$ is the minimizing solution and that equations (10.4) and (10.5) can be used as the optimality criteria.

If vector $\boldsymbol{x}$ satisfies condition (10.1) and if:

$$s_j = z_1a_{1,j} + z_2a_{2,j} + \dots + z_ma_{m,j} \tag{10.6}$$

then for j-th component of vector $\boldsymbol{x}$, i.e. x_j, the following combinations of relationships among c_j and s_j, on one hand, and d_j, x_j, h_j, on the other, are possible. These combinations classify x_j in an exhaustive manner:

$$c_j + s_j > 0;\ x_j = d_j \tag{10.7}$$

$$c_j + s_j = 0;\ d_j \leqq x_j \leqq h_j \tag{10.8}$$

$$c_j + s_j < 0;\ x_j = h_j \tag{10.9}$$

$$c_j + s_j > 0;\ x_j < d_j \tag{10.10}$$

$$c_j + s_j = 0;\ x_j < d_j \tag{10.11}$$

$$c_j + s_j < 0;\ x_j < h_j \tag{10.12}$$

$$c_j + s_j > 0;\ x_j > d_j \tag{10.13}$$

$$c_j + s_j = 0;\ x_j > h_j \tag{10.14}$$

$$c_j + s_j < 0;\ x_j > h_j \tag{10.15}$$

---

[1]) See Ford and Fulkerson, 1962; Yermolyev and Melnik, 1968; Busacker and Saaty, 1968; Harrary, 1969.

If all components of vector $\boldsymbol{x}$ satisfy equations (10.7), (10.8) or (10.9), then the solution if *feasible and optimal.* If any flow in arc x_j satisfies any one of the three conditions (10.7), (10.8) or (10.9), we say that the arc is in-kilter; otherwise, the arc is out-of-kilter.

The algorithm looks for some component that is out-of-kilter and step by step it is transformed to the in-kilter state, all the remaining components staying in-kilter. Other components that are out-of-kilter do not change, or they may be improved. The algorithm is used and converges for integer numbers. All the input values are given as integers.

The claim for a feasible starting solution of the iterative method does not complicate the solution as a *trivial solution* with all components of $\boldsymbol{x}$ equal to zero will do (with $d_j = 0$ for all j). If a feasible non-zero solution is known, than the time for a solution to be reached is greatly reduced. Therefore, this method is advantageous in a set of tasks where only certain parameters change.

The graph is described by the *set of arcs and nodes.* Each arc is directed and determined by starting node i and terminal node j[1]).

Three numbers are assigned for every arc (i, j): $d_{i,j}$ (lower bound of flow), $h_{i,j}$ (upper bound of flow or capacity) and $c_{i,j}$ (costs or cost values); the following constraints must be satisfied

$$0 \leqq d_{i,j} \leqq h_{i,j}; \qquad 0 \leqq c_{i,j} \tag{10.16}$$

For the description of the graph and solution of optimal flow the term *circulation* is introduced, which simplifies the model. Circulation is a non-negative integer vector $\boldsymbol{x} = (x_{i,j})$ having for each arc (i, j) one component and meeting the condition of mass balance for each node i, i.e.

$$\sum_{j=1}^{m} (x_{i,j} - x_{j,i}) = 0 \quad \text{for} \quad i = 1,2, \ldots, n \tag{10.17}$$

If this circulation satisfies the condition

$$d_{i,j} \leqq x_{i,j} \leqq h_{i,j} \quad \text{for all arcs } (i, j) \tag{10.18}$$

than $\boldsymbol{x}$ is feasible circulation. The component of circulation $x_{i,j}$ is the flow in arc (i, j).

A feasible circulation that gives a minimum sum is

$$\sum_{i=1}^{n} \sum_{j=1}^{m} c_{i,j} x_{i,j} \rightarrow \min \tag{10.19}$$

---

[1]) In equations (10.1 – 10.15) index j denotes the number of arc. In the following text i and j denote the number of nodes, therefore, the data for arcs must be assigned by two indices i and j. This change of notation is necessary for the introduction of the notion of circulation.

and it is the *optimal solution* with minimum losses. Therefore the optimum circulation is sought. First of all it is necessary to find out if there is a non-trivial solution. This non-trivial solution exists if, and only if, the inequalities

$$\sum_{\Omega} h_{i,j} \geqq \sum_{\Omega} d_{i,j} \tag{10.20}$$

are satisfied. The sum in inequality (10.20) concerns set Ω, i.e., the set of arcs (i, j). where i is the element of set L and j is the element of its supplement $\bar{L}$; L is the subset of the set of all nodes. Condition (10.20) is to be satisfied for all subsets L.

Using the circulation model, the optimal conditions are equivalent to conditions (for all arcs (i, j))

$$c_{i,j} + z_i - z_j > 0 \rightarrow x_{i,j} = d_{i,j} \tag{10.21}$$

$$c_{i,j} + z_i - z_j < 0 \rightarrow x_{i,j} = h_{i,j} \tag{10.22}$$

where $z = (z_i)$ is the vector of integer numbers with one component for each node i, called node costs (Ford and Fulkerson, 1962).

If $\boldsymbol{x}$ is a feasible circulation and if z exists and satisfies implications (10.21) and (10.22), then the circulation $\boldsymbol{x}$ is optimal.

Denoting

$$f_{i,j} = c_{i,j} + z_i - z_j \tag{10.23}$$

and introducing the *out-of-kilter number*, which equals zero for in-kilter states and is non-negative for out-of-kilter states, the following list of states and out-of-kilter numbers (state numbers) is obtained:

	State		State number
(a)	$f_{i,j} > 0$	$x_{i,j} = d_{i,j}$	0
(b)	$f_{i,j} = 0$	$d_{i,j} \leqq x_{i,j} \leqq h_{i,j}$	0
(c)	$f_{i,j} < 0$	$x_{i,j} = h_{i,j}$	0
(a_1)	$f_{i,j} > 0$	$x_{i,j} < d_{i,j}$	$d_{i,j} - x_{i,j}$
(b_1)	$f_{i,j} = 0$	$x_{i,j} < d_{i,j}$	$d_{i,j} - x_{i,j}$
(c_1)	$f_{i,j} < 0$	$x_{i,j} < h_{i,j}$	$f_{i,j}(x_{i,j} - h_{i,j})$
(a_2)	$f_{i,j} > 0$	$x_{i,j} > d_{i,j}$	$f_{i,j}(x_{i,j} - d_{i,j})$
(b_2)	$f_{i,j} = 0$	$x_{i,j} > h_{i,j}$	$x_{i,j} - h_{i,j}$
(c_2)	$f_{i,j} < 0$	$x_{i,j} > h_{i,j}$	$x_{i,j} - h_{i,j}$

(10.24)

The arc (i, j) is in-kilter if it is in one of the first three states, i.e., (a), (b), or (c). For arcs out-of-kilter, the out-of-kilter number for states (a_1), (b_1), (b_2), (c_2) gives the degree of deviation of the arc flow $x_{i,j}$ from the optimum, the out-of-kilter number for states (a_2) and (c_1) measures the deviation from conditions (10.21) and (10.22). The out-of-kilter algorithm ensures that all the out-of-kilter numbers do not increase, i.e., they remain the same or decrease. This was developed by Ford and Fulkerson, 1962 and a program in FORTRAN is included in CMEA program 1A.301. The description of this algorithm and an example of its numerical application is given in section 10.4.

10.1.2 Application of Out-of-Kilter Algorithm

The algorithm described is used for optimal allocation of the amount of water that is available in a time period t. If there are no reservoirs in the system, the decisions on optimal allocation in particular periods are mutually independent, and the total optimum is the sum of optimal solutions for individual periods. If some reservoirs are included in the system, some amount of water inflow inperiod t can be maintained for future periods. Then the total optimum is no longer the sum of optimal allocations in individual periods. According to the fact that users with different costs values (unit losses due to deficits) are in the system, it may be reasonable to admit small deficits in period t for users with low cost values in order to reduce the losses in periods $t + n$ for users with high cost values. In simulation models in the General Water Plan of Czechoslovakia this problem was solved heuristically preventing the losses due to deficits for users with high cost values in a deterministic simulation model.

In WRS in the basin of the Vistula River at first a solution on the assumption of known future flows was obtained. The *length of period* that should be taken into account in the optimization procedure was investigated. It is apparent that this length will be related to the operation cycles of reservoirs, and that it can include 12 months or more. The practical experience gained in the investigation of WRS in the basin of the Vistula River have shown that six months are sufficient.

The allocation problem in this model was solved not only for the current month, but also for the following months, six in total. Then changes in the system (the changes of storage of reservoirs included) were carried out using the simulation model, and the allocation problem was solved for the subsequent period.

10.1.3 Operating Rules Determination by a Stochastic Simulation Model

In simulation runs with six months in out-of-kilter procedure the value of the objective function (giving the sum of losses) was significantly lower than in simulation runs with only one period in out-of-kilter procedure. The former value of objective function is probably very far from the value for actual operation when the estimate of future flows is uncertain. Therefore, other methods were tested that were more realistic in system operation. First, it was assumed that the value of flow was known for the current period, and for future values *modal values* were used. If an observation for a longer period was available, a more reliable forecast could be formulated based on regression to past flows[1]).

For the computation of regression relationships, Kindler (1975) developed a method using the *stochastic simulation model.* As input it uses a multi-dimensional model, derived by Matalas for a first-order Markov chain and log-normal transformation. In the stochastic simulation model, the allocation is performed by the out-of-kilter method in each period, using the known flows for the future months, i.e., using a total of six months. For each month the storage at the end of the monthly period is related to the storage at the beginning of this period and to inflows in the system. Dropping the index m indicating that all values in equation (10.25) relate to month m, the regression equation has the following form:

$$V_j' = b_{0,j} + b_{1,j}Q_1 + b_{2,j}Q_2 + \dots + b_{n,j}Q_n + b_{n+1,j}V_1 + \dots + b_{2n,j}V_n \qquad (10.25)$$

where $b_{i,j}$ are regression coefficients, Q_i is the inflow in system for reservoirs 1 to n, V_i is the initial storage of reservoirs 1 to n (i.e., at the beginning of month m), and V_j' is the final storage of reservoir j (i.e., at the end of month m; n equation for each reservoir $j = 1$ to n).

In the actual operation of reservoirs the inflows for the current period are presumed to be known, however, the operation rule in WRS is defined by equation (10.25) for each reservoir $j = 1, 2, \dots, n$ and for each month of the year m. The resulting value of the objective function is greater than in the case of an ideal forecast, however, it is better (by approx. 30% over the example given by Kindler, 1975) than in calculations without any forecast.

10.1.4 Out-of-Kilter Method Parameters

The basic assumption of the successful application of the out-of-kilter method in practice is the appropriate estimation of *cost values in arcs* of the graph. It means proper estimation of the losses due to deficits for different users. This problem was partly solved in the Vistula River Project using the following approximations:

– the municipal water supply was taken as highest priority withdrawal and the cost values were taken as prohibitive to prevent deficits for this goal,

– the industrial water supply was evaluated by the relationship between the product and the amount of water supplied,

---

[1]) In the WRS Project in the basin of the Vistula River, the observations of flows were only available for the period 1951 – 1965, which was not sufficient for the determination of regression coefficients in forecast.

– the coefficient for water supply for irrigation was determined by the relationship of annual production increased by irrigation with different values for the monthly periods,

– the coefficients for minimum flow augmentation were determined by lowest values in accordance with the overall tendency to minimize the investment costs,

– the coefficients for river flow regulation for water power plants were determined by the value of power production.

It is apparent that these coefficients were used as an ordering principle for priority determination. They are not ideal as they cannot reflect the sensitivity of users in industry to water deficits, the depth of eventual deficits, their relative losses, etc. Nevertheless, the allocation optimization using these approximations meant significant progress for the solution of WRS problems and the combination of simulation models with the out-of-kilter algorithm was successful for the optimization of WRS.

The method was applied to a practical task. However, further work is necessary and some problems need to be solved, mainly in the determination of cost values and in the method of flow forecast for WRS operation policy (Kos, 1979b).

10.1.5 Examples of a Numerical Solution by the Out-of-Kilter Method

As a simple example, illustrating the out-of-kilter method, the flow optimization in the network in Fig. 10.1., is shown. The input and output values of flows and the optimal flows in arcs are given in Table 10.1. For clarification of the term circulation, an example of the circulation in node 6 is given in Fig. 10.2 for the network in Fig. 10.1.

An example of a practical application of this method in WRS is presented by Kindler (1975). This WRS includes three reservoirs with approximative active

Table 10.1 The optimal flows in arcs of the simple network

Number of arc j	Cost value c_j	Optimal flow x_j	Number of node i	Inflow (+) or outflow (–)
1	10	25	1	– 5
2	11	10	2	–15
3	15	15	3	–45
4	20	20	4	+15
5	40	0	5	+20
6	30	30	6	+30
7	40	0		
8	49	0		

storages of 0.346 km³, 0.225 km³ and 0.297 km³. The out-of-kilter method requires that all the data should be expressed in the same units, i.e., the flow in arcs (in this case tens of litres per second in an average month with 2,629,800 seconds). Therefore the reservoir storages were transformed to these flow units, and they are 13 156 (i.e., 132 m³ s⁻¹ per month), 8556 and 11 294, respectively. The initial storage of reservoirs is assumed as the half of active storage.

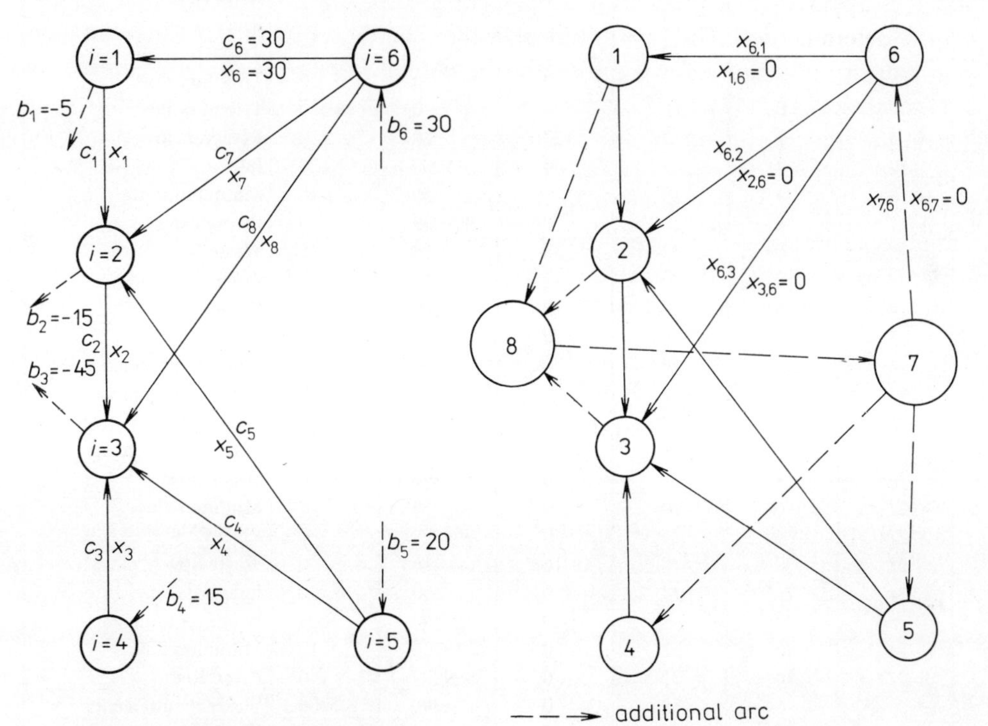

Fig. 10.1 Graph of a simple network
$i = 6$ – number of node, c_j – cost value (for one unit of flow x_j), d_j, h_j – lower and upper bounds of flow in arc j, x_j – flow in arc j
Balance in node 6: $x_6 + x_7 + x_8 = 30$; i.e. $a_{6,j} = 0$ for $1 \leqq j \leqq 5$, $a_{6,j} = 1$ for $6 \leqq j \leqq 8$, $b_6 = 30$

Fig. 10.2 Circulation in node 6 of the simple network
Equation (10.17) for node 6:
$x_{6,1} + x_{6,2} + x_{6,3} + x_{7,6} = 0$
node 7 is the source, node 8 is the sink

The goals of the WRS are the municipal and industrial water supply in five industrial centres ($a-e$), minimum flow augmentation at some points of the system. A schematic representation of WRS is given in Fig. 10.3. The transformed representation in the form of a graph is given in Fig. 10.4. The numbers of nodes in both schematic representations are equal to help a comparison between the two figures.

Table 10.2 Matrix description of the graph

No. of arc	From node	to node	Bounds of flow lower	upper	Unit loss	Remark
1	2	3	4	5	6	7
1	2	6	0	999 999	0	River flow
2	2	1	0	1000	0	Transbasin diversion
3	1	4	0	999999	0	River flow
4	3	14	0	999999	0	River flow
5	0	10	0	0	0	Tributary flow
6	0	15	0	0	0	Tributary flow
7	0	17	0	0	0	Tributary flow
8	1	0	120	120	0	Reservoir losses
9	4	5	0	1000	0	Inflow
10	4	12	0	50	– 10	Minimum flow
11	4	12	0	999999	0	River flow
12	0	12	58	58	5	Waste flow
13	2	0	41	41	0	Reservoir losses
14	6	7	0	900	0	Inflow
15	7	5	0	550	0	Inflow
16	7	9	0	250	– 1000	Water requirements
17	4	9	0	90	– 1500	Water requirements
18	9	12	0	350	0	Waste inflow
19	7	8	0	75	– 1500	Water requirements
20	8	12	0	390	0	Waste flow
21	6	10	0	230	– 10	Minimum flow
22	6	10	0	999999	0	River flow
23	0	10	16	16	0	Waste flow
24	10	11	0	590	0	Inflow
25	11	8	0	590	– 1000	Water requirements
26	10	12	0	300	– 10	Minimum flow
27	10	12	0	999999	0	River flow
28	5	13	0	1390	– 5000	Water requirements
29	13	12	0	910	0	Waste flow
30	12	16	0	999999	0	River flow
31	3	0	59	59	0	Reservoir losses
32	14	15	0	190	– 10	Minimum flow
33	14	15	0	999999	0	River flow
34	0	15	25	25	0	Waste flow
35	15	11	0	600	0	Inflow
36	15	16	0	210	– 10	Minimum flow
37	15	16	0	999999	0	River flow
38	15	16	0	100	– 1500	Water requirements
39	16	17	0	999999	0	River flow
40	15	0	0	600	– 500	Water requirements
41	17	0	0	999999	0	River flow
42	17	0	0	1900	– 10	Minimum flow

The river flow is reduced by the value of minimum flow if the node is connected by an additional arc for minimum flow. For arcs 5, 6, 7 the lower and upper bounds are given for each period.

The definition of the graph with a description of the flows in the arcs is given in Table 10.2. The arcs are defined by the starting and terminal nodes, lower and upper boundaries and cost value, which equals the unit loss. In each node of thc graph (network) the circulation constraints must be satisfied, i.e., mass balance in the node. In nodes where water is consumed (the mass balance would not be met if the surface flow alone is assumed) the evaporation, infiltration or transbasin transfer

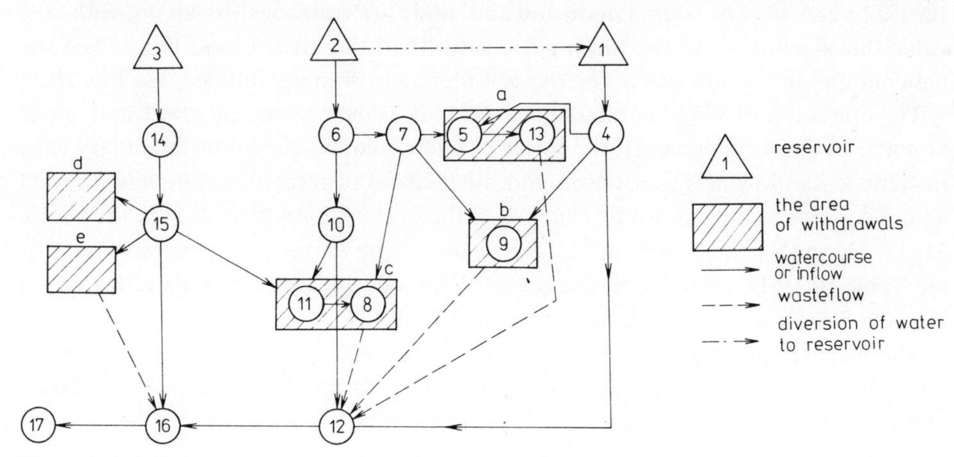

Fig. 10.3 Schematic representation of WRS for the simulation model used in combination with the out-of-kilter algorithm

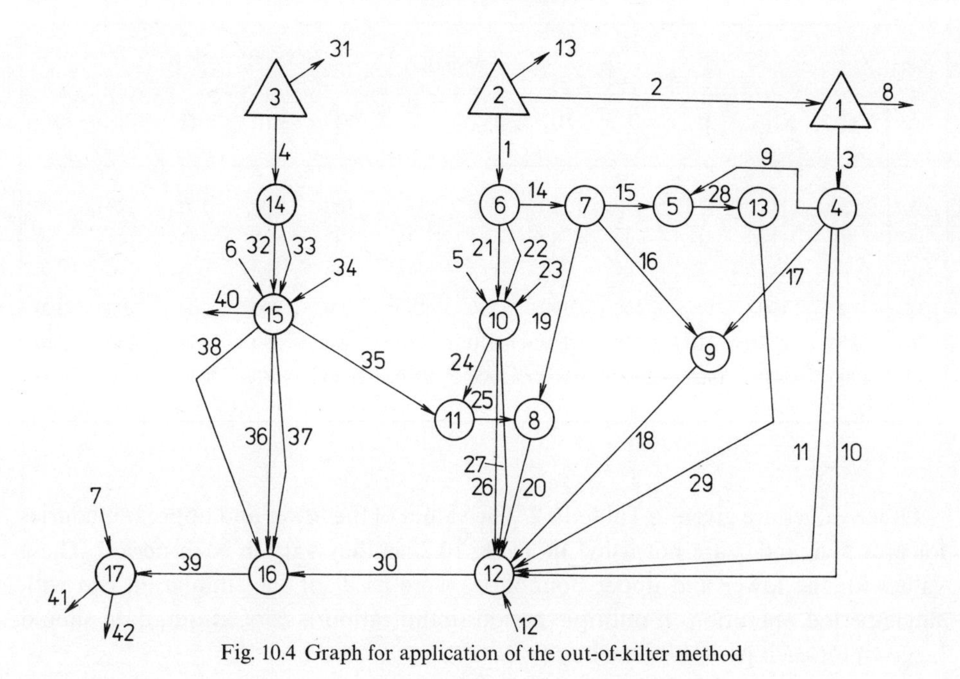

Fig. 10.4 Graph for application of the out-of-kilter method

should be taken into account by connecting these nodes to a hypothetical node called a sink (or super sink) to secure zero circulation, i.e., balance of inflows and outflows in these nodes (in the graph in Fig. 10.4, nodes 13, 8 and 9). This sink node is connected with further nodes by arcs: 8, 13, 31, 40, 41 and 42.

A further hypothetical node called source (or super source) is connected with the graph nodes receiving water. In this way zero circulation is secured in nodes 1, 2, 3, 10, 12, 15 and 17. The source node and sink node are connected by an arc with cost value that is approximately 100 times greater than the other costs. This closes the network and the conditions of the out-of-kilter algorithm are fulfilled (see Fig. 10.2).

The operation of WRS with a defined network is determined by lower and upper boundaries and cost values. If the lower boundary equals the upper boundary (e.g., in some waste flow, in evaporation and other losses of reservoirs), the flow in such an arc is given, and it cannot be changed in the optimization process (e.g., arcs 8, 13, 31, 12, 23, and 34; the values in arcs 8, 13, and 31 are variable for different months; see Table 10.3). The hydrological input of WRS and the active storage of reservoirs 1, 2, and 3 are treated in the same way (arcs 5, 6, and 7).

The water requirements in WRS are given as the upper boundaries of flows in arcs 10, 16, 17, 18, 19, 20, 21, 25, 26, 28, 29, 32, 36, 40 and 42, the upper boundaries of arcs 32, 36, and 42 are variable for different months as shown in Table 10.3.

Table 10.3 Variable requirements in system

No. of arc	Month											
	XI	XII	I	II	III	IV	V	VI	VII	VIII	IX	X
8	120	118	116	118	120	130	116	100	100	100	120	116
13	41	39	38	40	42	47	38	32	33	33	41	39
31	59	58	56	58	60	66	56	49	50	50	59	57
32	190	190	190	190	190	190	190	200	200	200	200	190
36	210	210	210	210	210	210	210	290	290	290	290	210
42	1900	1900	1900	1900	1900	1900	1900	2040	2040	2040	2040	1900

Other values are given in Table 10.2. The values of the lower and upper boundaries for arcs 5, 6, and 7 are not listed in Table 10.2, as they vary in each period. These values for the lower and upper boundaries were used for the simulation run with single-period operation. If multiple-period optimization is carried out, data should be given for each period for each arc.

The output of the out-of-kilter optimization is the allocation of water in investigated period i. In the simulation model, changes due to this optimized allocation are carried out, the model proceeds to step $i + 1$ and the calculation is repeated for the new values of flows and reservoir storage volumes. The main output of this case study is the determination of coefficients in equation (10.25). An example these coefficients is given in Table 10.4 for reservoir 1 for sample months November, February, April and August.

Table 10.4 Regression coefficients in the equation (10.25)

Month	b_0	b_1	b_2	b_3	b_4	b_5	b_6
November	−560	0.299	0.087	0.301	0.966	0.063	−0.041
February	−343	1.083	0	0	1.003	−0.057	0
April	0	0.531	0	0.110	0.967	0	0
August	−730	0.149	0.197	0.165	0.983	0.059	−0.038

The multiple correlation coefficients exceeded 0.98

10.2 THE COMBINATION OF THE CHANCE-CONSTRAINED MODEL AND THE SIMULATION MODEL

Many methods of WRS operation require complicated calculations which make heavy demands on computer time without reflecting the stochastic character of the problem. The chance-constrained model developed by ReVelle *et al.* (1967), ReVelle and Kirby (1970), Joeres *et al.* (1971) and completed by Eastman and ReVelle (1973) by direct capacity determination is an adequate method for relatively simple models using a linear programming technique and the solution of a set of linear equations. The resulting model is stochastic.

10.2.1 The Chance-Constrained Model

The chance-constrained model is first described for a single multi-purpose reservoir. The objective is the minimization of its total storage V_c, satisfying the linear constraints that express goals such as flood control, water supply, minimum storage requirement, minimum and maximum flow constraints. These constraints are expressed in the following ways.

a) Flood control (freeboard) constraint

$$P\{V_{z_i} \leqq V_c - V_{r_i}\} \geqq p_1 \tag{10.26}$$

The storage of water V_{z_i} in the reservoir shall be less than, or equal to, the total storage reduced by the flood control storage (freeboard) with at least p_1 reliability for period i

b) Minimum storage (pool) constraint

$$P\{V_{z_i} \geqq V_{m_i}\} \geqq p_2 \tag{10.27}$$

The storage V_{z_i} greater than or equal to, minimum V_{m_i} shall be kept with at least p_2 reliability for period i.

c) Minimum release constraint

$$P\{x_i \geqq y_i\} \geqq p_3 \tag{10.28}$$

Release x_i greater than, or equal to, minimum release y_i must be committed with at least p_3 reliability for each time period i.

d) Maximum release constraint

$$P\{x_i \leqq z_i\} \geqq p_4 \tag{10.29}$$

This constraint that release commitment x_i determined for period i shall be less than or equal to the maximum acceptable commitment z_i with reliability p_4, was not used in the calculation. In case studies, it is often done for monthly periods and for flood control in many WRS in ČSSR the monthly release is not an important parameter as maximum flow is reached in shorter time intervals.

10.2.2 Principle and Application of Linear Decision Rule

The linear decision rule assumes that the amount of water x_i, released from a reservoir in a period i is a linear function of the storage of water at the beginning of period i, i.e., at the end of the period $i - 1$ and of the decision parameter b_i, determined by linear programming to optimize some objective function. The release commitment is

$$x_i = V_{z_{i-1}} - b_i \tag{10.30}$$

Substituting the linear decision rule into the continuity (mass balance) equation

$$V_{z_i} = V_{z_{i-1}} + f_i - x_i \tag{10.31}$$

where f_i is the inflow into reservoir in period i, yields

$$V_{z_i} = f_i + b_i \tag{10.32}$$

If the linear decision rule is followed in period $i - 1$, the similar equation for $V_{z_{i-1}}$ is substituted into (10.30), which yields

$$x_i = f_{i-1} + b_{i-1} - b_i \tag{10.33}$$

Using the linear decision rule, the stochastic variable V_{z_i} in chance constraints (10.26)–(10.28) can be substituted by variable f_i. The probability distribution of variable V_{z_i} is unknown and the probability distribution of variable f_i is known or can be calculated. Substituting equations (10.32) and (10.33) into constraints (10.26) to (10.28) and rearranging, yields the constraints in the following form:

$$P\{V_c - b_i - V_{r_i} \geqq f_i\} \geqq p_1 \quad (10.34)$$

$$P\{V_{m_i} - b_i \leqq f_i\} \geqq p_2 \quad (10.35)$$

$$P\{b_i - b_{i-1} + y_i \leqq f_{i-1}\} \geqq p_3 \quad (10.36)$$

Using the cumulative distribution function of inflows f_i, the constraints can be transformed into their deterministic equivalents. From the cumulative distribution function the value $r_i(p_1)$ is determined: it is the inverse of this function of f_i evaluated at p_1, or it is the value which the flow in period i exceeds on average in $100(1 - p_1)\%$ of time. Similarly, the value $r_i(1 - p_2)$ is exceeded on average in $100\,p_2\%$ time (and similarly for p_3).

Using these transformations, the following linear program can be formulated:

$$V_c \rightarrow \min \quad (10.37)$$

with constraints

$$V_c - b_i - V_{r_i} \geqq r_i(p_1) \quad (10.38)$$

$$V_{m_i} - b_i \leqq r_i(1 - p_2) \quad (10.39)$$

$$b_i - b_{i-1} + y_i \leqq r_{i-1}(1 - p_3) \quad (10.40)$$

The assumption of the effective use of the decision rule in WRS operation requires that the users were able to use the variable release commitments. For many goals, as municipal water supply, the amount of water that exceeds the demand can hardly be utilized. Knowledge of the possibility of water deficits in a future period, however, is important for a reduction of losses due to deficits. The operation policy of reservoirs in practice during drought periods is similar to the linear decision rule, but it is more complicated. However, in many cases the application of the linear decision rule is a better approximation of reality than the operation with a constant release.

Using the linear decision rule separate operational reliability in each month is considered, neglecting the relationship between the probability distributions of flows in adjoining months. The relationship between reliabilities in a chance-constrained model with the linear decision rule and annual occurrence or time-based reliabilities of WRS (see reliabilities R_1 or R_2 in 6.3.7) can hardly be determined generally; they depend on the correlation between monthly flows and other hydrological and operational details. Therefore other methods were sought that could provide relationships, comparable to the overall reliability of WRS operation. More appropriate for this aim is the chance-constrained model with direct release optimization without linear decision rule.

10.2.3 Release Optimization in the Chance-Constrained Model

In the model developed by Curry *et al.* (1973), the releases x_i were directly optimized in period i. The operation of a single multi-purpose reservoir is optimized using the mass balance (continuity) equation and the chance-constraints (10.26) and (10.27), i.e.,

- freeboard constraint,
- minimum storage constraint, with given reliabilities.

The minimum and maximum release constraints are not stochastic, therefore the release must satisfy the minimum and maximum constraint in each case.

In the transformation of the stochastic problem into a deterministic one, the cumulative distribution functions of flows in individual months do not appear, but a cumulation of convoluted distribution functions is necessary (see below). In the continuity equation the evaporation and infiltration losses from the reservoir are added in relation to reservoir storage.

$$V_{z_i} = e_i V_{z_{i-1}} + f_i - x_i \tag{10.41}$$

where e_i is the fraction of water remaining contingent upon losses due to evaporation and infiltration (e.g., for 2% losses in a month i, $e_i = 0.98$). Substituting the equation (10.41) into the chance-constraint (10.26) for $i = 1$ yields

$$P\{e_1 V_{z0} + f_1 - x_1 \leqq V_c - V_{r1}\} \geqq p_1 \tag{10.42}$$

Rearranging and substituting the value $r_1(p_1)$ from the cumulative distribution function such that only 100 $(1 - p_1)$% of random values are greater than r_1 (p_1), the chance-constraint is transformed to

$$V_c - V_{r1} - e_1 V_{z0} + x_1 \geqq r_1(p_1) \tag{10.43}$$

Using equation (10.42) for $i = 2$ and substituting for V_{z1} from equation (10.41) yields

$$P\{e_2 e_1 V_{z0} - (e_2 x_1 + x_2) - V_c + V_{r2} \leqq -(e_2 f_1 + f_2)\} \geqq p_1 \tag{10.44}$$

The *convoluted cumulation distribution function* of the random variables f_1 and f_2 $(e_2 f_1 + f_2)$ evaluated at the point p_1 gives the value $r_2(p_1)$; on the average it will be exceeded in 100 $(1 - p_1)\%$. The chance-constraint is then transformed to the deterministic equivalent:

$$(V_c - V_{r_2} - e_2 e_1 V_{zo}) + (e_2 x_1 + x_2) \geqq r_2(p_1) \tag{10.45}$$

By similar steps the following relationship can be obtained for the n-th period (defining $e_{n+1} = 1$):

$$\left(V_c - V_{r_n} - V_{zo} \prod_{i=1}^{n} e_i\right) + \sum_{i=1}^{n} \left[\left(\prod_{k=i+1}^{n} e_k\right) x_i\right] \geqq r_n(p_1) \tag{10.46}$$

where $r_n(p_1)$ is the value of random variable w_n with probability of exceedance $(1 - p_1)$.

The random variable w_n is defined by

$$w_n = \sum_{i=1}^{n} \left(\prod_{k=i+1}^{n} e_k\right) f_i \tag{10.47}$$

Equation (10.46) is the deterministic equivalent of the chance-constraint (10.26) for the chance-constraint model without the linear decision rule in which the optimal release quantities x_i are the decision variables. In a similar way the deterministic equivalent for the minimum storage chance constraint with a given reliability can be derived. We get

$$V_{m_n} = V_{zo} \prod_{i=1}^{n} e_i + \sum_{i=1}^{n} \left[\left(\prod_{k=i+1}^{n} e_k\right) x_i\right] \leqq r_n(1 - p_2) \tag{10.48}$$

where V_{m_n} is the required minimum storage.

If in operation the withdrawals Q_{d_i}, which are determined for periods i, are additional requirements, then in equations (10.41)–(10.48) instead of x_i the sum $x_i + Q_{d_i}$ is considered. If these water requirements are stochastic, then the value $f_i - Q_{d_i}$ is assumed in all equations instead of flows f_i (Kos *et al.*, 1974).

The model with direct optimization of release x_i can be enlarged for a system of reservoirs interconnected by diversions and pumping of water. The continuity equation for reservoir k in time i is

$$V_{z_{k,i}} = e_{k,i} V_{z_{k,i-1}} + f_{k,i} - x_{k,i} + \sum_{j=1}^{n} (a_{k,j} x_{j,i} - b_{k,j} q_{k,j,i} + b_{j,k} q_{j,k,i}) \tag{10.49}$$

and the optimized variables $q_{k,j,i}$ give the amount of water transferred from reservoir k into reservoir j in time period i. The coefficients $a_{k,j}$, $b_{k,j}$ express the existence of the connection between reservoirs k and j. The coefficients $a_{k,j}$ are equal to one if water is transferred reservoir k to reservoir j; otherwise they are equal to zero. For $k = j$ both the coefficients are equal to zero. The coefficients $a_{k,j}$ are used for water diversion, the coefficients $b_{k,j}$ are used in the same manner for pumping.

Using a special assumption, the solution of equations (10.46) to (10.48) can be simplified (Kos and Zeman, 1976). Assuming $e_i = e$ constant for all periods and assuming the priority of water supply function with constant release $x_i = x$, but with random demands subtracted from inflows, the following simplification is possible

$$\sum_{i=1}^{n}\left[\left(\prod_{k=i+1}^{n} e_k\right) x_i\right] = \frac{1 - e^{n-1}}{1 - e} x = H_n \tag{10.50}$$

$$e^n = B_n \tag{10.51}$$

The inequality (10.46) is reduced to

$$V_c - V_r - V_{zo} B_n + H_n x = \max_j \left[r_n(p_1)\right] \tag{10.52}$$

and the inequality (10.48) is changed to

$$V_{m_n} - V_{zo} B_n + H_n x = \min_j \left[r_n(1 - p_2)\right] \tag{10.53}$$

In inequalities (10.52) and (10.53) the j-th month is the first month of convolution. For the value $V_{m_n} = 0$ this simplification yields

$$V_z = V_{zo} = \max_n \left[\frac{1}{B_n}\left(H_n x - \min_j \left(r_n(1 - p_2)\right)\right)\right] \tag{10.54}$$

and the total storage of the reservoir is

$$V_c = V_{c_n} = V_r + B_n V_z - H_n x + \max_j \left[r_n(p_1)\right] \tag{10.55}$$

In equation (10.55) the value n is used, which yields a maximum in equation (10.54).

The solution requires the knowledge of the cumulative convoluted distribution function of flows. As theoretical probability distribution function for this purpose the normal distribution, log-normal distribution, and Gumbel's distribution were tested. The parameters were derived by moments and by quantiles. The best fit of empirical and theoretical data and the easiest use on a computer was achieved by *Gumbel's distribution with method of quantiles* (Dupačová and Kos, 1979) (however, the statistical tests of goodness-of-fit were also satisfied by other distributions).

Using the method of quantiles, values such as $r_n(p_2)$ were determined. The quantiles for 5% ($X1$) and 95% ($X2$) were determined from empirical distributions. Quantities A, S, and Z were calculated, using these values:

$$A = \ln\left(-\ln\left(1 - p_2\right)\right) \tag{10.56}$$

$$S = \frac{4.067}{X1 - X2} \tag{10.57}$$

$$Z = 0.27X1 + 0.73X2 \tag{10.58}$$

Then, $r_n(p_2)$ could be calculated

$$r_n(p_2) = \frac{A}{S} + Z \tag{10.59}$$

The calculational and economic constraints limited the maximum values of n to 48, i.e., 4 years.

10.2.4 Chance-Constrained Model and Simulation Model

In the General Water Plan of Czechoslovakia the deterministic simulation model was used in most case studies. In some subsystems of WRS, when a relatively high reliability was required for municipal or industrial water supply with high response to eventual water deficits, the stochastic model was used. The combination of the chance-constrained model with the simulation model was one of the applied methods of stochastic models. The main aim of this combination was the determination of reliability of withdrawals (Kos, 1975).

Table 10.5 Hydrological characteristics of points in WRS

No. of point	1	2	3	4	5	6	7	8	9
Mean annual flow in $m^3 s^{-1}$	0.96	5.56	3.18	1.45	1.45	3.18	5.56	0.96	21.7

Let us consider a system whose schematic representation is in Fig. 10.5. The goal of four reservoirs (Nos. 1 to 4) is the water supply in points 5 and 9. The minimum flows in the rivers at these points are 0.32 and 1.72 $m^3 s^{-1}$, respectively, and the water requirements for consumption (trans-basin diversion) are 0.92 and 7.21 $m^3 s^{-1}$, respectively. In addition, the amount of possible withdrawals without consumption at points 6 to 8 is investigated.

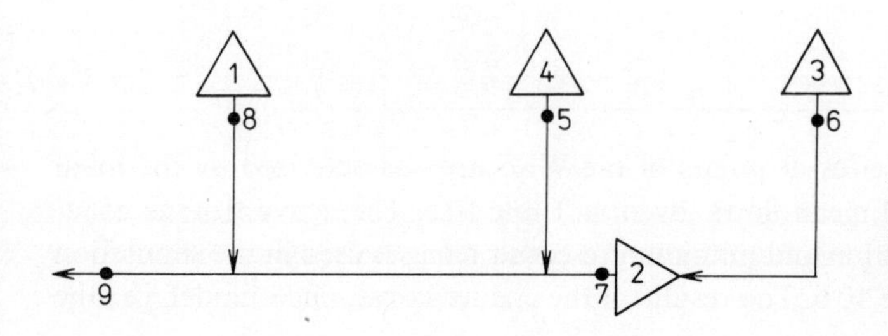

Fig. 10.5 Schematic representation of WRS when the chance-constrained model was applied

The determinstic simulation model was used for the evaluation of variable reservoir releases for river flow regulation at point 9 with different alternatives of operation policy. As the main results of simulation runs with the necessary reservoir capacities, the operation without deficits using the monthly flows in the period of

Table 10.6 The basic input parameters of WRS

No. of reservoir	1	2	3	4
Active storage used in computation (10^6 m^3)	30.5	45.0	68.0	54.6
Minimum release (m^3 s^{-1})	0.12	1.00	0.46	0.32

1931 – 1970 was determined. Then the chance-constrained model was used for the same WRS using the above simplification (e_i = e; $x_i = x$ const) with direct release determination and stochastic release requirements caused in reservoirs due to river flow regulation and given as the output of the deterministic simulation model.

Table 10.7 Main output values of WRS

No. of alternative	No. of point	Release interval n	Active storage [10^6 m^3]	Draft [m^3 s^{-1}]
1	8	33	27.50	0.89
2	8	48	72.32	1.07
3	7	7	23.74	3.00
4	7	20	39.52	3.60
5	7	20	84.73	4.32
6	6	16	49.92	1.00
7	6	16	59.41	1.20
8	6	27	70.92	1.44
9	5	44	64.96	1.27
10	5	42	52.23	1.17

The hydrologic time series at points of the WRS are characterized by the main parameter – the annual mean flows given in Table 10.5. The active storage of reservoirs used in computation and minimum reservoir releases used in the simulation model are given in Table 10.6. The results of the chance-constrained model, i.e., the relationship of draft Q_n and the necessary active storage V_z for the reservoir operation for river flow regulation at points 5 and 9, for 97% reliability, and the optimal length n of the convolution period used in the equations (10.54) and (10.55) are given in Table 10.7.

By logarithmic interpolation, i.e., using the relation a, b – constants

$$Q_n = a + b \log (V_z) \tag{10.60}$$

Table 10.8 The active storage volumes of reservoirs of WRS for reliability of release 97%

No. of reservoir	1	2	3	4
Active storage volume (10^6 m^3)	32	50	75	61
No. of point	8	7	6	5
Draft (m^3 s^{-1})	0.92	3.96	1.50	1.23

which can be used approximately for the whole range of required release and that fits better than the linear interpolation, the values of the draft in the points of WRS for the given active storage of reservoirs were determined. The active storages V_z are greater than those used for calculations in the simulation model, as they were not reduced by losses. These losses were included in the chance-constrained model in such a manner that the losses in both models were approximately the same[1]).

Table 10.8 indicates that the draft at point 5 (i.e., 0.32 + 0.92 = 1.24 m^3 s^{-1}), which was derived by the deterministic simulation model, has a 97% reliability as derived by the chance-constrained model because in both models approximately the same values of draft appeared. If the 99% reliability were required, differences would occur and a higher active storage would be necessary for draft 1.24 m^3 s^{-1}.

10.2.5 Conclusions of the Chance-Constrained Model

Simulation modelling is a very good technique, if the validity of the model is measured by its ability to approximate reality correctly. However, optimization is not implicitly included in the simulation model and the optimum is reached by many

[1]) Assuming 2% monthly losses of reservoir storage the total loss for 9 months of the drought period will be $0.98^9 = 0.83$ and for 12 months $0.98^{12} = 0.79$, which is approximately 17–21%.

repetitions of calculation runs changing the input parameters. If the stochastic character of the modelled phenomena is to be reflected, i.e., the stochastic simulation model is required, then sophisticated methods are necessary for the generation of hydrologic input data. As optimization of the stochastic simulation model is time-consuming, it is only suitable for the detailed investigation. Therefore, in planning the verification of results of the deterministic simulation model is often done in only few cases.

For these reasons some other methods were tested to reflect the stochastic character in the models. A combination of the deterministic simulation model with the chance-constrained model is an appropriate method for this goal, which can be used for the determination of reliability of a WRS and for its optimization with relatively low computer time requirements.

The use of the linear decision rule was tested. However, the results of the chance-constrained model can hardly be compared with the results of the simulation model with fixed target demands on WRS, i.e., with demands predetermined before the computation (possibly variable during the year). Such fixed demands were typical for municipal and industrial water supply. Water requirements for irrigation are random variables, dependent on meteorological, agricultural and other elements, but related in a small degree to the water commitment as in the linear decision rule. Partly, the linear decision rule can be used for a combination of flood control and water power plant operation where the water commitment could be used for dummy power production.

The direct release optimization without the linear decision rule was more advantageous for the combination of the simulation model with the chance-constraint model. The results can be easily compared with those of the simulation model with occurrence-based reliability.

The chance-constrained model with the linear decision rule has a direct solution or a linear program of a small dimension can be used. If the direct release optimization (without simplification as shown) is used then for longer periods (e.g., for 20 years) extensive linear programs are obtained. For example, for 15 reservoirs with one diversion each, over 20 years, with monthly periods, the number of constraints (without non-negativity constraints) would be 7200 and the number of bounded variables 7200. If water supply is a priority goal, the investigated period can be reduced to 60 months or the described simplification can be used. This simplification is advantageous for withdrawals from reservoirs for the municipal water supply with priority over river flow regulation for the water supply for industry and irrigation.

The operation policy of WRS with priorities does not often yield the overall optimum and the possibilities of the chance-constrained model are investigated especially in combination with the simulation model to utilize this combination for better response to minimum storage and free-board constraints (Kos, 1079a). However, the present application of this combination, which was used for operation reliability determination under conditions given by the simulation model, is a progressive technique for many cases.

10.3 STOCHASTIC ANALYTICAL MODEL

The analytical model has to meet a number of requirements to reflect the modelled reality correctly. It should include the longterm (over-year; carry over) and short-term (seasonal; within year) release operation, storage requirements, and flood control function of reservoirs, optimization of hydroelectric power plant capacities, targets for water supply and for various goals, low flow augmentation, etc. In addition, the analytical model must contain some economic functional relationship as the dependence of capital, operation, maintenance and repair costs on reservoir storage, relationship between reservoir water level and reservoir storage, the water supply gross benefits from hydroelectric power and gross recreational benefits, and loss functions expressing the reduction of benefits due to the deficits in water supply, reduction of recreation possibilities and hydroelectric power generation below the level of firm energy commitment.

If a situation of water resources development is modelled, the model should have dynamic properties.

These conditions can be met by a simulation model, however, this model does not include the system optimization. At present there is no analytical model that can satisfy all the conditions listed. Some requirements have to be weakened and some have to be omitted. The character of the task implies the modelling of the stochastic character of the process. The main function of WRS does not appear during periods with average flows, but during extreme deviations such as drought or flood situations. Therefore a stochastic optimization model that includes the computation of the risk of system failure is necessary. Such a model was developed by Jacoby and Loucks (1972) and is called the *screening model.*

The principle of the model is the probabilistic evaluation of different stages of reservoir storage, uncontrolled and controlled flows, reservoir releases, maximization of the expected benefits depending on these probabilities using the linear programming technique. In this model the discrete values of flows and reservoir storage are used with the corresponding probability of occurrence that are used for the determination of active and flood control storage of reservoirs, parameters of the hydroelectric power plants and parameter for operation of the system.

The main goals are represented by water supply (V), hydroelectric power (H), flood control (O), and recreation (R), respectively. The site or location (point) in the basin is denoted by the subscript g. For example the annual gross flood control benefits are U_g^O. The total annual costs at site g are N_g. The objective function of the model, which is maximized, is formed by the expected (E is the symbol of expectation) total annual net benefits from all sites, i.e.

$$\max E \sum_g (U_g^V + U_g^H + U_g^O + U_g^R - N_g) \tag{10.61}$$

In the simulation model the basic time period is one month. In the analytical model the number of intervals has to be reduced. In the given example the year was divided into two intervals: the spring period with high flows (March, April, May) and the rest of the year.

Table 10.9 Conditional and unconditional probabilities at control point Vilémov on the Jizera River

Conditional probabilities $P_{i,j}^t$ for $t = 1$			
	j		
i	1	2	3
1	0.533	0.592	0.452
2	0.167	0.176	0.170
3	0.300	0.232	0.378

Conditional probabilities $P_{i,j}^t$ for $t = 2$			
	j		
i	1	2	3
1	0.605	0.616	0.310
2	0.184	0.151	0.304
3	0.211	0.233	0.386

Unconditional probabilities $P_{i,t}$			
	i		
t	1	2	3
1	0.517	0.215	0.268
2	0.524	0.170	0.306

In each of these time intervals some streamflow intervals are determined. In the example discussed three intervals (states) were used, describing low, mean and high streamflows. The range of possible flows is divided into three consecutive intervals, i.e. i or j (i.e., $i = 1, 2$, or 3; $j = 1, 2$, or 3), in such a way that for each period t the conditional probabilities $p^t_{i,j}$ are the same for all sites. This conditional probability expresses the probability that the unregulated flow will be in interval j in period $t + 1$, given an unregulated flow in interval i in period t. The choice of the same probabilities simplifies the formulation of the linear program.

The unconditional steady state probabilities $P_{i,t}$ expressing the probability of each flow interval i in each period t, are calculated from conditional probabilities $p^t_{i,j}$ by solving the following set of simultaneous linear equations

$$P_{j,t+1} = \sum_{i=1}^{n} P_{i,t} P^t_{i,j} \qquad \text{for } j = 1, 2, \ldots, n - 1 \quad t = 1, \ldots, T \tag{10.62}$$

$$\sum_i P_{i,t} = 1 \qquad \text{for } t = 1, \ldots, T \tag{10.63}$$

In this example, $n = 3$ and $T = 2$. For the values of conditional probabilities given in the Table 10.9 for the gauging station Vilémov, the following set of equations will be obtained

$$\begin{aligned}
P_{1,2} &= P_{1,1} \cdot 0.533 + P_{2,1} \cdot 0.592 + P_{3,1} \cdot 0.452 \\
P_{2,2} &= P_{1,1} \cdot 0.167 + P_{2,1} \cdot 0.176 + P_{3,1} \cdot 0.170 \\
1 &= P_{1,1} + P_{2,1} + P_{3,1} \\
P_{1,1} &= P_{1,2} \cdot 0.605 + P_{2,2} \cdot 0.616 + P_{3,2} \cdot 0.310 \\
P_{2,1} &= P_{1,2} \cdot 0.184 + P_{2,2} \cdot 0.151 + P_{3,2} \cdot 0.304 \\
1 &= P_{1,2} + P_{2,2} + P_{3,2}
\end{aligned} \tag{10.64}$$

By solving these equations the values of the unconditional probabilities in Table 10.9 are obtained. In equations (10.62) to (10.64) a cycle was assumed; then the period $t = 1$ is followed by $t = 2$ and this period is followed again by $t = 1$, etc.

In (10.64), the equation for $j = 3$ cannot be used because then the sum of the first equations (i.e., for $t = 1$ and $j = 1, 2, 3$) would equal the fourth equation (expressing

Table 10.10 Discrete values of uncontrolled flows $G_{i,t,1}$ (10^6 m^3)

	i		
t	1	2	3
1	41.52	70.19	83.00
2	53.25	88.76	108.48

that the sum of probabilities $P_{i,t}$ is equal to one for $i = 1, 2, 3$). The sum of conditional probabilities $P^t_{i,j}$ for the given i, t and $j = 1, 2, 3$ is equal to one and the equations would be linearly dependent.

In further calculation the streamflow values are replaced in each interval i or j by one discrete value $G_{i,t,h}$, where h denotes the gauging station, t is the time period, in such a way as to maintain the first two moments of the probability distribution of uncontrolled flows. The third condition is maintaining $G_{2,t,h}$ in the middle of the medium interval for $i = 2$. The values of $G_{i,t,h}$ for one station are given in Table 10.10. The values $I_{i,t,g}$, the discrete uncontrolled flows at each site g within the basin, are computed from values $G_{i,t,h}$ by reduction coefficients $k_{h,g}$:

$$I_{i,t,g} = k_{h,g} G_{i,t,h} \tag{10.65}$$

$$k_{h,g} = \frac{Q_a^g}{Q_a^h} \tag{10.66}$$

where Q_a^g is the mean annual flow in site g and Q_a^h is the mean annual flow in site h (gauging station).

The flows controlled by a subsystem of WRS $Q_{i,t,g}$ in interval i, in period t and site g are determined by the mass balance equation (10.67) as the uncontrolled flows reduced by the change of flows immediately upstream from site g and consumption

$$Q_{i,t,g} = I_{i,t,g} - \sum_A (I_{i,t,a} - R_{i,t,a}) - \sum_B Z_{i,t,b} \quad \text{for all } i, t \tag{10.67}$$

where $a \in A$ (A is the set of all control sites immediately upstream from site g) and $b \in B$ (B is the set of all withdrawals sites between those control sites and site g), $Z_{i,t,b}$ is the consumption in sites b, $R_{i,t,a}$ is the expected release in sites a; both in period t and interval i. All values in equation (10.65) and following equations are given in mil. m^3 in period t.

In equation (10.67), the occurrence of all values is assumed within the same state interval i. It is a simplifying assumption, which is realistic in flow regimes with similar time patterns. For this analytical model this assumption can be used for all rivers in Czechoslovakia, with the exception of the Danube.

In further investigations, the reservoir storage volumes are divided into three discrete state intervals denoted as k or m (i.e., $k = 1, 2$, or 3; $m = 1, 2$, or 3). Then states k or m correspond to intervals just as the states i or j designate the intervals of streamflows. However, the storage volume intervals k and m are unknown, therefore the reservoir storage volumes $S_{k,t,g}$ or $S_{m,t+1,g}$ are also unknown. The reservoir releases depend on these values and they can therefore be denoted as $R_{k,i,m,t,g}$; using the mass balance equation we get

$$R_{k,i,m,t,g} = S_{k,t,g} + Q_{i,t,g} - S_{m,t+1,g} \tag{10.68}$$

for all k, i, m, t.

To secure a uniform filling and releasing of reservoirs in the whole WRS and to satisfy the assumption in equation (10.67) the relationship for the determination of state m is the following[1]):

$$m = \text{entier}\ \frac{k + i}{2} \tag{10.69}$$

As m is uniquely defined by k, i, the subscript m can be omitted.

The reservoir release depends on the reservoir state k and therefore it is denoted as $R_{k,i,t,g}$; similarly the controlled flow $Q_{k,i,t,g}$ and consumption $Z_{k,i,t,g}$ depend on state k and the subscript k is used. Therefore the equation (10.67) becomes (10.70)

$$Q_{k,i,t,g} = I_{i,t,g} - \sum_A (I_{i,t,a} - R_{k,i,t,a}) - \sum_B Z_{k,i,t,b} \tag{10.70}$$

for all k, i, t.

The change of the equation (10.68) consists in the omission of the subscript m in release R.

The conditional probability for different states of flows and reservoir storage volumes $P_{k,i,m,j,t}$ is the probability of having an initial storage volume within interval m and a flow within interval j in period $t + 1$ given an initial storage volume within interval k and a flow within interval i in period t. Since the conditional probabilities $P^t_{i,j}$ are the same at each reservoir site and the relationship of m to k, i is given by the equation (10.69), the conditional probabilities are:

$$\begin{aligned} P_{k,i,m,j,t} &= P^t_{i,j} \ \text{if}\ m = \text{entier}\ \frac{k+1}{2} \\ P_{k,i,m,j,t} &= 0 \quad \text{otherwise} \end{aligned} \tag{10.71}$$

Knowing these conditional probabilities the steady state joint probabilities $P'_{k,i,t}$ can be calculated in the same way as in the equation (10.62):

$$\begin{aligned} P'_{m,j,t+1} &= \sum_k \sum_i P'_{k,i,t} P_{k,i,m,j,t} \\ &\text{for } j = 1, 2, \ldots, n-1 \\ &\quad\ \ t = 1, 2, \ldots, T \\ &\quad\ m = 1, 2, \ldots, M \end{aligned} \tag{10.72}$$

$$\sum_k \sum_i P'_{k,i,t} = 1 \quad \text{for } t = 1, 2, \ldots, T \tag{10.73}$$

---

[1]) The function entier gives the maximum integer portion of the argument, i.e., of $(k + i)/2$, e.g., entier $(2.5) \doteq 2$.

For the site Vilémov on the Jizera River the probabilities $P'_{k,i,t}$ are given in Table 10.11. The probabilities $P'_{k,i,t}$ are not determined by the choice of intervals i, j, k and m uniquely. The reservoir storage volumes $S_{k,t,g}$ and the release $R_{k,i,t,g}$ are

Table 10.11 Unconditional probabilities $P'_{k,i,t}$ in site Vilémov on the Jizera River

In period $t = 1$			
	k		
i	1	2	3
1	0.382	0.112	0.135
2	0.134	0.103	0.133
3	0	0	0.001

In period $t = 2$			
	k		
i	1	2	3
1	0.342	0.106	0.181
2	0.182	0.063	0.125
3	0	0	0.001

influenced by the joint probabilities $P'_{k,i,t}$, but not determined by them. Therefore a set of constraints can be used and the solution optimized. These constraints are given by the goals of the system, e.g., by water supply, hydroelectric power, flood control, recreation, etc.:

a) The water supply constraints consider the annual water supply targets T_g^V that are allocated to period t and site g using the variables $a_{g,t}$. If the target is not met, deficits $D_{k,i,t,g}^V$ occur. The total water supply available in site g in period t is limited by $Q_{k,j,t,g}$. Then

$$a_{g,t} T_g^V - D_{k,i,t,g}^V \leqq Q_{k,i,t,g} \quad \text{for all } k, i, t \tag{10.74}$$

The consumption $Z_{k,i,t,g}$ in site g is assumed to be a portion $c_{g,t}$ of the actual withdrawal, i.e.

$$Z_{k,i,t,g} = c_{g,t}(a_{g,t}\, T_g^V - D_{k,i,t,g}^V) \tag{10.75}$$

for all i, t.

If the benefits function associated with the water supply target is f_g^V and the loss function is $F_{t,g}^V$, then the annual expected water supply benefits are

$$E(U_g^V) = f_g^V(T_g^V) - \sum_t \sum_k \sum_i P'_{k,i,t} F_{t,g}^V(D_{k,i,t,g}^V, a_{g,t} T_g^V) \tag{10.76}$$

b) The production of hydroelectric power at each reservoir site is dependent on a number of factors: the installed generating capacity of the power plant H_g, the flow through turbines, the mean storage head $H_{k,m,t,g}$, which depends on the initial storage volumes in each period, the number of hours in each period h_t, and the plant factor K_t indicating the position of the hydropower production in period t in the electric grid. The total annual firm power target T_g^H is apportioned to each period by the fraction coefficient $b_{t,g}$. Then the firm energy requirement in site g and period t will be $h_t K_t b_{t,g} T_g^H$. If this requirement is not met, deficits $D_{k,i,t,g}^H$ may occur; otherwise "dump" energy $E_{k,i,t,g}^H$ may be generated. If k_g denotes the constant for converting the product of flow, head and plant efficiency into energy, then the reservoir release $R_{k,i,t,g}$ provides the following maximum of energy produced:

$$h_t K_t b_{t,g} T_g^H - D_{k,i,t,g}^H + E_{k,i,t,g}^H \leqq k_g R_{k,i,t,g} H_{k,m,t,g} \tag{10.77}$$

for all k, i, t and m = entier $[(k + 1)/2]$. The total energy is constrained by the installed generating capacity H_g. Then

$$h_t K_t b_{t,g} T_g^H + E_{k,i,t,g}^H \leqq h_t H_g \tag{10.78}$$

for all k, i, t.

The energy produced has to be positive, i.e., the deficit cannot surpass the requirements

$$h_t K_t b_{t,g} T_g^H - D_{k,i,t,g}^H \geqq 0 \tag{10.79}$$

If function $f_{g,t}^H$ expresses the benefit function for hydroelectric power and the loss function is $F_{g,t}^H$, including not only the deficits of the firm energy, but also the dump energy (as a negative loss), then the expected annual power benefits are

$$E(U_g^H) = \sum_t f_{g,t}^H(b_{t,g} T_g^H) - \sum_t \sum_k \sum_i P'_{k,i,t} F_{g,t}^H(D_{k,i,t,g}^H, E_{k,i,t,g}^H, T_g^H) \tag{10.80}$$

c) The reservoir storage volumes A_g are determined in existing reservoirs; for other reservoirs the following constraint must be satisfied

$$A_{\min,g} \leqq A_g \leqq A_{\max,g} \tag{10.81}$$

The maximum is determined by hydrological, morphological and economical conditions. A minimum value $A_{\min,g}$ is necessary for the reservoir function to be important for WRS. The sum of the maximum active storage $V_{t,g}$ and of the flood control storage in period t, $V_{t,g}^O$ must be less than, or equal to, the total storage volume A_g, i.e.

$$\max_t V_{t,g} + \max_t V_{t,p}^O \leqq A_g \quad \text{for all } t. \tag{10.82}$$

Assuming the relationship in the flood control storage volume in WRS, the flood damage reduction cannot be expressed directly as a function of the values $V^{\mathrm{O}}_{t,g}$, but as a function of some equivalent values $V^{\mathrm{e}}_{t,g}$ that are expressed as the actual flood storage capacity $V^{\mathrm{O}}_{t,g}$ plus the values $V^{\mathrm{h}}_{t,g}$ that include the effect of the upstream reservoir on reducing the peak of a standard project flood, then

$$V^{\mathrm{e}}_{t,g} = V^{\mathrm{O}}_{t,g} + V^{\mathrm{h}}_{t,g} \tag{10.83}$$

and the annual expected flood control benefits are:

$$E(U^{\mathrm{O}}_{g}) = \sum_{t} f^{\mathrm{O}}_{g,t}(V^{\mathrm{e}}_{t,g}) \tag{10.84}$$

where the benefit function $f^{\mathrm{O}}_{g,t}$ was used.

d) The recreation benefits can be assumed to depend on the stability of the reservoir pool and storage volume $V_{k,t,g}$. These volumes are partitioned into the seasonal target storage volume T^{R}_{g}, deficit $D^{\mathrm{R}}_{k,t,g}$ and excess $E^{\mathrm{R}}_{k,t,g}$, then

$$V_{k,t,g} = T^{\mathrm{R}}_{g} - D^{\mathrm{R}}_{k,t,g} + E^{\mathrm{R}}_{k,t,g} \tag{10.85}$$

The number of visitors may vary at different periods of the year. This cyclic variation is expressed by coefficients $r_{g,t}$, the sum of which for all periods t in each site g is equal to one. Using the value of one visitor day and the number of expected visitors, the recreation benefit function f^{R}_{g} and the recreation loss function $F^{\mathrm{R}}_{t,g}$ can be derived. The annual expected recreation benefits are

$$E(U^{\mathrm{R}}_{g}) = f^{\mathrm{R}}_{g}(T^{\mathrm{R}}_{g}) - \sum_{t}\sum_{k}\sum_{i} P'_{k,i,t} r_{g,t} F^{\mathrm{R}}_{t,g}(D^{\mathrm{R}}_{k,t,g}, E^{\mathrm{R}}_{k,t,g}, T^{\mathrm{R}}_{g}) \tag{10.86}$$

In order to enable the use of the linear programming technique all benefit and loss functions are piecewise linear.

e) The computation of expected net benefits requires the values of annual costs in equation (10.61) associated with the construction, the operation, the maintenance and the repair of reservoirs, power plants and recreation facilities. It is assumed that all costs are functions of capital costs and that these costs are defined for the total reservoir storage capacity A_{g}, the power plant capacity H_{g} and recreation facilities T^{R}_{g}. Thus all the components of the objective function (10.61) have been defined. The optimal alternative is then calculated using a linear program under the given constraints. It is convenient to use such a program where the boundaries of the variables are included in a special boundary section. The problem is simplified and the number of constraints is reduced. Using a heuristic approach many constraints can be omitted as they are satisfied by the fulfilment of other conditions.

10.4 THE DESCRIPTION OF SIMPLIFIED OUT-OF-KILTER ALGORITHM APPLICATION IN AN EXAMPLE

In the description of the algorithm, the following notation us used: $c_{i,j}$ is the cost value in arc (i, j), $h_{i,j}$ is the upper boundary of flow in arc (i, j), y_i is the circulation in node i, and $x_{i,j}$ is the flow along arc (i, j).

The circulation y_i in node i is determined by

$$y_i = \sum_i x_{i,j} - \sum_j x_{j,i} \tag{10.87}$$

The corresponding values that change during calculation are denoted by capital letters (e.g., $C_{i,j}$ correspond to $c_{i,j}$). In this simplified algorithm the lower boundary of the arc flow is assumed to be zero.

The algorithm includes the following five steps:

1. The initial values are $C_{i,j} = c_{i,j}$; $H_{i,j} = h_{i,j}$; $X_{i,j} = 0$; $Y_i = y_i$,

2. The values $c_{i,j}$ and $C_{i,j}$ are transformed to non-negative values. *If* $C_{i,j} \leqq 0$ then a further transformation is carried out for each node-source (i.e., where $y_i > 0$) all the values $C_{i,j}$ are reduced by the quantity $\min_j C_{i,j}$ and all the values $C_{j,i}$ are increased by the same quantity.

3. The maximum flow is calculated: starting in a node-source, i.e., where $Y_i > 0$, the flow is allocated to arcs where $C_{i,j} = 0$ labelling the nodes of the graph in the following manner: all nodes i, where $Y_i > 0$ are labelled by a couple (N_i, T_i) defining $N_i = 0, T_i = Y_i$.

In addition, the nodes to which the flow can be directed are also labelled. Assume that node k has been labelled. Then all the unlabelled nodes j are labelled by numbers $N_j = -k$ and $T_j = \min(T_k, X_{j,k})$ if $X_{j,k} > 0$ and $C_{j,k} = 0$. Then the nodes will be labelled (using node m as the starting point) by the numbers $N_m = k$; $T_m = \min(T_k, H_{k,m})$ for all unlabelled nodes m where $C_{k,m} = 0$, $H_{k,m} > 0$.

Having treated node k in this way, the following labelled node is taken and using this labelling procedure the adjoining nodes are labelled. The labelling procedure proceeds till some node with $Y_n < 0$ is reached or as long as it is possible. In the former case flow n can be increased, therefore, go to step 4. In the latter case it is not possible, therefore go to step 5 for potential improvement.

4. Partition of flows: Assume $T = \min(T_n, |Y_n|)$; add T to Y_n and $X_{i1,n}$ where $i1 = N_n$ and subtract from $H_{i1,n}$. Go to node $i1$, if $N_{i1} < 0$; subtract T from $X_{i1,i2}$, $i2 = |N_{i1}|$ and add to $H_{i1,i2}$. If $N_{i1} > 0$ then add T to $X_{i2,i1}$ and subtract from $H_{i2,i1}$. Go to node $i2$ and repeat this procedure until node ik is reached where $N_{ik} = 0$. Subtract T from Y_{ik}. If $\sum Y_i = 0$ for all i there is $Y_i \geqq 0$, then the flow is optimal.

5. Change the potentials: assume A is the set of labelled nodes. For each i that is the element of A $(i \in A)$, the value F_i is determined.

$$F_i = \min\left(\min_{j \in W_i} C_{i,j}, \min_{j \in U_i} |C_{j,i}|\right) \tag{10.88}$$

where W_i is the set of nodes where for all j that are not elements of A are $C_{i,j} > 0$, and U_i is the set of nodes where for all j that are not elements of A are $C_{j,i} < 0$.

Then all values $C_{i,j}$ are reduced and $C_{j,i}$ are increased for each $i \in A$ by the value $F = \min_i F_i$. Go to step 3.

The correctness of this algorithm can be proved by comparison with the equations and procedures described in section 10.1. It is apparent that the algorithm uses

$$\begin{aligned} C_{i,j} &\geqq 0 \quad \text{for } X_{i,j} = 0 \\ C_{i,j} &= 0 \quad \text{for } 0 < X_{i,j} < h_{i,j} \\ C_{i,j} &\leqq 0 \quad \text{for } X_{i,j} = h_{i,j} \end{aligned} \tag{10.89}$$

$C_{i,j} = c_{i,j} - Z_i + Z_j$ where Z_i and Z_j are the final quantities of value F that decreases (Z_i) or increases (Z_j) the cost $c_{i,j}$. By comparing it with the out-of-kilter numbers, assuming $d_{i,j} = 0$ (the lower boundary is zero and $C_{i,j} = f_{i,j}$ in equation (10.24)) the corectness of the described algorithm and its convergence to optimal flow is clear.

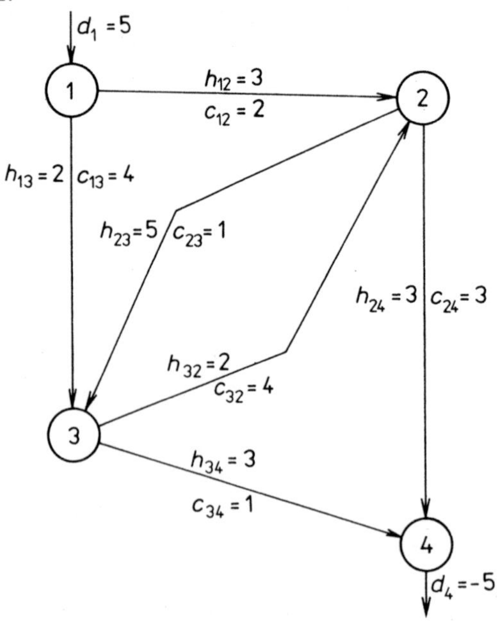

Fig. 10.6 An example of the algorithm out-of-kilter application

The method described is illustrated in a numerical example. The optimal flow in the network in Fig. 10.6 is sought. In step 1 we have $X_{i,j} = 0$. In step 2 the costs of arcs starting in source are changed and we have $C_{1,2} = 0$, $C_{1,3} = 2$. In step 3 the nodes $i = 1, 2$ are labelled by couples $N_1 = 0$; $T_1 = 5$; $N_2 = 1$; $T_2 = 3$. Further labelling is not possible, go to step 5. We have $F = 1$. The costs in arcs are changed: $C_{1,2} = 0$, $C_{1,3} = 1$, $C_{2,3} = 0$, $C_{3,2} = 5$, $C_{2,4} = 2$, $C_{3,4} = 1$[1]).

[1]) $C_{1,2}$ does not change, as it is increased and decreased by the same value ($F = 1$), as $1 \in A$ and $2 \in A$.

In step 3 let us label node $i = 3$; $N_3 = 2$, $T_3 = 3$, go to step 5. We have $F = 1$. If the costs in arcs are changed, then $C_{1,2} = 0$, $C_{1,3} = 1$, $C_{2,3} = 0$, $C_{3,2} = 5$, $C_{2,4} = 1$, $C_{3,4} = 0$, node $i = 4$ can be labelled $N_4 = 3$, $T_4 = 3$, go to step 4. The initial zero flow is increased by $T = \min(3,5) = 3$, then we have $X_{3,4} = 3$, $X_{2,3} = 3$, $X_{1,2} = 3$ and the values $H_{i,j}$ are changed: $H_{3,4} = 0$, $H_{2,3} = 2$, $H_{1,2} = 0$. Then it is possible to label node $i = 1$ as $N_1 = 0$, $T_1 = 2$. Further $F = 1$. The nodes $i = 3,2$ are labelled $N_3 = 1$, $T_3 = 2$, $N_2 = -3$, $T_2 = 2$. Again $F = 1$. If node $i = 4$ is labelled $N_4 = 2$, $T_4 = 2$, the flow can be changed by 2 and the optimal flow will result: $X_{1,2} = 3$, $X_{1,3} = 2$, $X_{2,3} = 1$, $X_{3,4} = 3$, $X_{2,4} = 2$.

11 INFORMATION AND INFORMATION SYSTEMS IN WATER MANAGEMENT

The present information problem, i.e., generation, acquisition, storage, effective processing and retrieval of information, cannot be solved by conventional facilities and procedures. *Automation of information systems* using system approach, computers, modern methods of message transmission, graphic storage devices, data communication equipment, and in addition, various activities of information systems, are necessary to cope with the information explosion.

The information systems include elements from conventional branches (libraries, bibliography, document processing, etc.) and the elements inherent in automation such as programming of computers, cybernetics, logics, linguistics, semionics, statistics, mathematics, and system sciences.

The automation of information processes requires new classification and processing of knowledge in new adequate forms. In many cases an algorithmization of procedures is necessary for computer coding.

The systems approach and application of systems sciences seems to be the most effective method for the identification and analysis of information and for the automation of information processes.

11.1 INFORMATION AND ENTROPY

Information can be defined as a measure of freedom of choice in selecting a message; it is a negenergetic value proportional to the decrease of entropy (disorganization) of a system. This definition is rather broad and further definitions using different aspects of information can be used (syntactic, semantic and pragmatical information).

As the computer is the basic tool in the automation of information systems, the term information will be used to mean syntactic information with the exception of section 11.1 wherein the definition of information is used in relation to entropy. As far as the general meaning is concerned, the term information will be used as a *message about reality* that was accepted as information and can influence the behaviour of the accepting body. The *information system* is a set of subsystems, facilities and persons involved in the acquisition, storage, processing, transmission, retrieval and dissemination of information. In the automation of information systems certain functions of the live part of information systems are allocated to the "non-live" part.

The flow of information is a continuous process of information processing in sequences, determined by information inputs that are often the information outputs of previous information processing.

The set of facilities over which the signal is sent is called the communication *channel*. This notion of a channel includes (1) all the technical facilities which transform the signal before transmitting, (2) transmitting and receiving, (3) the transformation of the received signal, and (4) all the space used for transmitting the signal from transmitter to receiver.

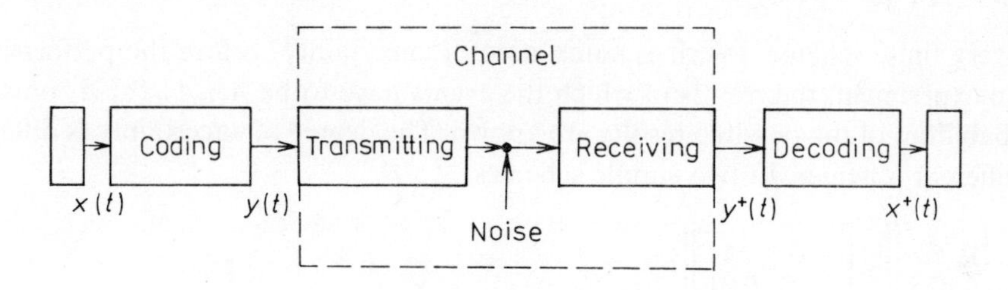

Fig. 11.1 Scheme of a communication system

In Fig. 11.1 there is a schema of a communication system and its elements. The symbols are: $x(t)$ = information source, $y(t)$ = the signal on the channel input (coded message), $y^+(t)$ = the signal on the channel output, and $x^+(t)$ = the decoded (reconstructed from signal) message in the receiving end of the communication system.

The procedure of coding and decoding, i.e., the transformation from message $x(t)$ to signal $y(t)$ and from signal $y^+(t)$ to $x^+(t)$ is expressed mathematically in the form of a functional relationship $y = f(x)$ and $x^+ = \varphi(y^+)$. A general case where the relationships between x and y and between y^+ and x^+ are stochastic was investigated by Kolmogorov (1956).

If the same output signal corresponds to a given input signal, the channel is called noiseless. In such a communication process the transmitted signal is identical with the received one and the input and output messages are two forms of realization of the same message.

The presence of *noise* disturbs this relationship between the transmitted and the received signal. Then both messages and noise form stochastic processes, and in this situation it is possible for different output signals to correspond to one message, or the received signal can be identified with different message realizations. In principle, two basic tasks are performed – the determination and separation (filtration) of signals hidden in noise. Therefore the theory of information can use the methodology of the probability theory and the theory of stochastic processes.

The theory of *entropy* is one of the important parts of the probability theory.

A sample space R partitioned into a set of mutually exclusive and exhaustive random events $A_1, A_2, \ldots, A_n$ will be the starting point. In each experiment only one event can take place. For $n = 2$ a simple alternative exists – a couple of contradictory events. If the events $A_1, A_2, \ldots, A_n$ together with their probabilities $p_1, p_2, \ldots, p_n$ ($p_i \geqq 0$, $\sum_{i=1}^{n} p_i = 1$) are given, a finite schema is thus defined (in the case of $n = 2$ it is known as the Bernoulli scheme):

$$A = \left\| \begin{matrix} A_1, A_2, \ldots, A_n \\ p_1, p_2, \ldots, p_n \end{matrix} \right\| \tag{11.1}$$

Every finite scheme describes some state of uncertainty: before the performance of an experiment, the results of which the events have to be $A_1, A_2, \ldots, A_n$ only the probabilities of the possible results are known. The degree of uncertainty is different in different schemes. In two simple schemes:

$$\left\| \begin{matrix} A_1, A_2 \\ 0.5, 0.5 \end{matrix} \right\| \quad \left\| \begin{matrix} A_1, \quad A_2 \\ 0.99, 0.01 \end{matrix} \right\|$$

the first definitely has more uncertainty than the second one, where the result will "almost certainly" be A_1. In the first example no prediction is possible.

In applications it is desirable to introduce a degree of uncertainty of a finite scheme. This measure was defined by Shannon (1948) as

$$H(p_1, p_2, \ldots, p_n) = -\sum_{k=1}^{n} p_k \lg(p_k) \tag{11.2}$$

All logarithms in (11.2) have an arbitrary, but common base. For $p_k = 0$ by definition $p_k \lg(p_k) = 0$ is used. The quantity $H(p_1, p_2, \ldots, p_n)$ is called entropy of a finite scheme (11.1).

The properties of entropy:

1. The necessary and sufficient condition for $H(p_1, p_2, \ldots, p_n) = 0$ is that one of probabilities $p_1, p_2, \ldots, p_n$ is unity, and all the others are zero. It is the case when the result of the experiment can be predicted with a total certainty – i.e., no uncertainty is present. In all other cases entropy is positive.

2. The scheme when all the results are equally probable, i.e., $p_k = 1/n \, (k = 1, 2, \ldots, n)$ has the highest uncertainty.

3. Suppose A, B are two finite schemes

$$A = \left\| \begin{matrix} A_1, A_2, \ldots, A_n \\ p_1, p_2, \ldots, p_n \end{matrix} \right\| \quad B = \left\| \begin{matrix} B_1, B_2, \ldots, B_m \\ q_1, q_2, \ldots, q_m \end{matrix} \right\|$$

and suppose these schemes are independent, i.e., the probability that both events A_k and B_i occur simultaneously is $p_k q_i$. The set of events $A_k B_i$ ($1 \leqq k \leqq n; 1 \leqq i \leqq m$)

with probabilities π_{ki} forms a new finite scheme that is called the fusion of schemes A and B and denoted AB. Suppose $H(A)$, $H(B)$, $H(AB)$ are entropies of schemes A, B, and AB, respectively, then:

$$H(AB) = H(A) + H(B) \tag{11.3}$$

It follows from:

$$-H(AB) = \sum_k \sum_i \pi_{ki} \lg (\pi_{ki}) = \sum_k \sum_i p_k q_i (\lg (p_k) + \lg (q_i)) =$$

$$= \sum_k p_k \lg (p_k) \sum_i q_i + \sum_i q_i \lg (q_i) \sum_k p_k = -H(A) - H(B)$$

4. Let us investigate the case where schemes A and B are dependent. Denoting q_{ki} as the probability that in scheme B event B_i will occur on the condition that event A_k has occurred in scheme A, we get:

$$\pi_{ki} = p_k q_{ki} \; (1 \leqq k \leqq n, 1 \leqq i \leqq m)$$

Then $-H(AB) = \sum_k \sum_i p_k q_{ki} (\lg (p_k) + \lg (q_{ki})) =$

$$= \sum_k p_k \lg (p_k) \sum_i q_{ki} + \sum_k p_k \sum_i q_{ki} \lg (q_{ki})$$

For each k, $\sum_i q_{ki} = 1$ and the sum $-\sum_i q_{ki} \lg (q_{ki})$ is the conditional entropy $H_k(B)$ of scheme B computed on the assumption that event A_k has occurred in scheme A. We get

$$H(AB) = H(A) + \sum_k p_k H_k(B)$$

The conditional entropy $H_k(B)$ is a random variable in scheme A: its values are determined by the occurrence of events A_k in scheme A. Therefore, the last right-hand-side term is a mean value of the random variable $H(B)$ in scheme A that is denoted as $H_A(B)$. We then get

$$H(AB) = H(A) + H_A(B) \tag{11.4}$$

In a special case where schemes A and B are independent, equation (11.4) will be identical with equation (11.3). In all cases, the inequality $H_A(B) \leqq H(B)$ holds true. This inequality can be interpreted so that knowledge of the results in scheme A can, on average, decrease the uncertainty in scheme B.

5. The required property of entropy is

$$H(p_1, p_2, \ldots, p_n, 0) = H(p_1, p_2, \ldots, p_n)$$

i.e., addition of an impossible event to the given scheme or addition of an arbitrary number of such events does not change its entropy.

If an experiment is carried out, the results of which are shown in the given scheme, some information is obtained (we know then which of the events A_k has occurred) and the uncertainty of the given scheme is totally cancelled. Information given by the results of an experiment rely on the abolishment of some uncertainty which existed before the experiment took place. The higher the uncertainty, the higher the value of the information gained by abolishing its uncertainty. As for the measure of uncertainty of a finite scheme A, its entropy $H(A)$ was chosen, and the amount of information gained by removing the uncertainty can be measured by an increasing function of the variable $H(A)$. The choice of this function means the choice of a scale for the amount of information, and, in principle, it is arbitrary.

However, the properties of entropy show that it is usually adavantageous to consider an amount of information proportional to the entropy.

One unit of information or entropy has the form:

$$\begin{Vmatrix} A_1 & A_2 \\ 0.5 & 0.5 \end{Vmatrix}$$

For the basis of logarithms equal to 2 it is the amount of information of choice of one from two equally probable possibilities. The entropy of this scheme is equal to

$$H\left(\frac{1}{2}, \frac{1}{2}\right) = -\frac{1}{2}\log_2\left(\frac{1}{2}\right) - \frac{1}{2}\log_2\left(\frac{1}{2}\right) = 1$$

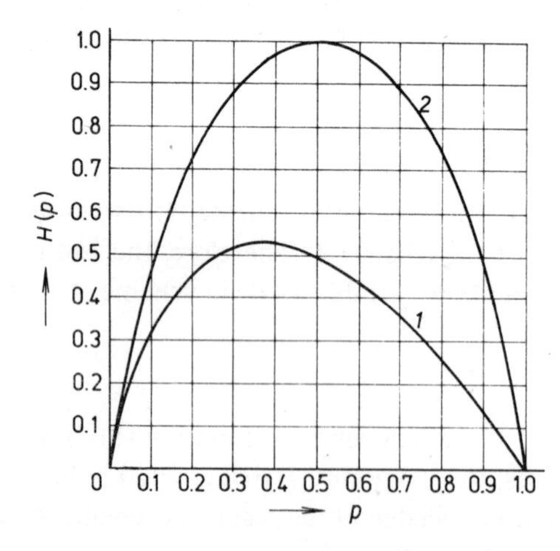

Fig. 11.2 Graph of entropy values

This unit is called *bit* (abbrev. binary digit). The entropy functions $H(p)$ in bits for some basic examples are shown in Fig. 11.2.

For example, entropy of a source that is transmitting a message composed of letters from an alphabet with 32 symbols with an equally probable occurrence is equal

$$H = -\sum_{1}^{32} \frac{1}{2^5}\log_2\left(\frac{1}{2^5}\right) = -\log_2\left(\frac{1}{2^5}\right) = 5$$

Entropy of the source is then 5 bits per 1 symbol. If the source is transmitting, say 40 symbols per minute, then the entropy of the source is 200 bits per minute.

The letters of texts in any one language are not, in fact, independent, their frequencies depend on previous letters of the text. Therefore, the conditional probabilities of each letter can be investigated in a set of experiments, i.e., sample texts; the mean values can be used for the estimation of conditional information (Dupač and Hájek, 1970).

The words in a sentence or in the whole text are not independent. There is some a priori probability of the occurrence of words an word associations. The deviations of observed frequency from this probability are important for the system of automatic indexation (see section 11.6).

Redundancy of information: a scheme in which all elements are equally probable has maximum entropy. Generally, however, the elements have a different probability of occurrence and the entropy of their scheme will be lower. The ratio of the actual entropy H of a given source to the maximum possible entropy H_{max}

$$h = \frac{H}{H_{max}} \tag{11.5}$$

is called the *relative entropy*. The difference $H_{max} - H$ is called *inner* (or redundant) *information* and the ratio

$$R = \frac{H_{max} - H}{H_{max}} = 1 - h \tag{11.6}$$

is a *coefficient of redundacy* or redundancy of source.

The redundancy of information in languages has been much investigated. It facilitates the abbreviation of texts (in a cable, by omitting some words), or it makes possible the increase of reliability of correct acceptance of a message (to estimate and correct the errors in the text). It is important in coding where an optimal abbreviation of the original message is desirable. Khinchin (1954) has shown that for entropy of a message equal to H the minimum value of the coefficient of abbreviation $1 - R$ is

$$1 - R = \frac{H}{H_{max}} \tag{11.7}$$

Using this relationship, the lower boundary of message abbreviation can be found and used for optimal coding. The methods of coding used in practice can be a evaluated by comparing them with this optimal coding.

The notion of entropy can be generalized for continuous random variables in relation to their probability distribution types for one-dimensional and k-dimensional distributions (Lapa, 1971).

11.2 BASIC ACTIVITIES AND FUNCTIONS OF INFORMATION SYSTEMS

The basic activities of information systems are derived from their representation as *communication systems* of scientific information starting from the document and ending at its destination: the user of the information. These activities can be classified into three main groups: generation, processing and use of information.

The generation of information involves activities that are necessary if the document is to be published. This phase, comprising contacts with authors, publishers, reviewers, editors, consultants, etc., is not formalized and in the automation of information systems it forms the environment of the system. The processing of information comprises the acquisition, classification, indexing, organisation of documents and their indices into document files and index files; secondary information is treated in processing, i.e., information of information.

Communication with users requires communication links between the source of information and its user via the information channels and information centres that act as switching centres in the process of information retrieval.

The main activity in information processing is information retrieval, based on the content of the document. Such activity is very complex and deals with a large number of items. Therefore, this is the first stage where automation could be used, e.g., automation of selective dissemination of information.

11.3 THE ORGANIZATION AND DEVELOPMENT OF INFORMATION SYSTEMS

The following aspects are taken into account in the organization and development of information systems: a comprehensive plan for automating management information systems, the functions of computers in information systems, data communication and data transmission systems, modern graphic devices, bibliographical document processing, organization of scientific document files, communication problems of relevant information retrieval, etc.

The production of primary information is influenced by the information systems. In a creative process the question arises, as to whether the problem has already been investigated and, if it has, then the necessary information has to be delivered to the user. In view of the information explosion it is becoming progressively more tedious and complicated to find a solution to this basic information problem.

Automated information systems enable information to be processed by *synthetic intelligence*, which use secondary information stored in computer memories and on the selection and retrieval of information by algorithms and sub-systems based on synthetic intelligence.

Information files are based on national reference services, which are interlinked and form international and world-wide information systems with libraries and their services.

Information centres are a modern branch of information systems, which serve selected groups of users and cater to their requirements.

The development of information systems requires a higher type of centre: analytical information centres which carry out the active acquisition of documents, and produce the criteria for a selective dissemination of information. If these centres are decentralized, then they can promptly react to users' requirements, and provide information feedback.

As water resources systems and water management have an inter-disciplinary character, the selection of documents is a complex problem, and an analytical approach is recommended. Analytical information centres in water management are therefore fully justified.

11.4 THE SYNTHESIS AND ANALYSIS OF INFORMATION SYSTEMS

In the application of the systems approach to the information problem, the starting point is systems analysis, i.e., the analysis of the structure and behaviour of information systems. In this process the whole system is investigated, or this investigation

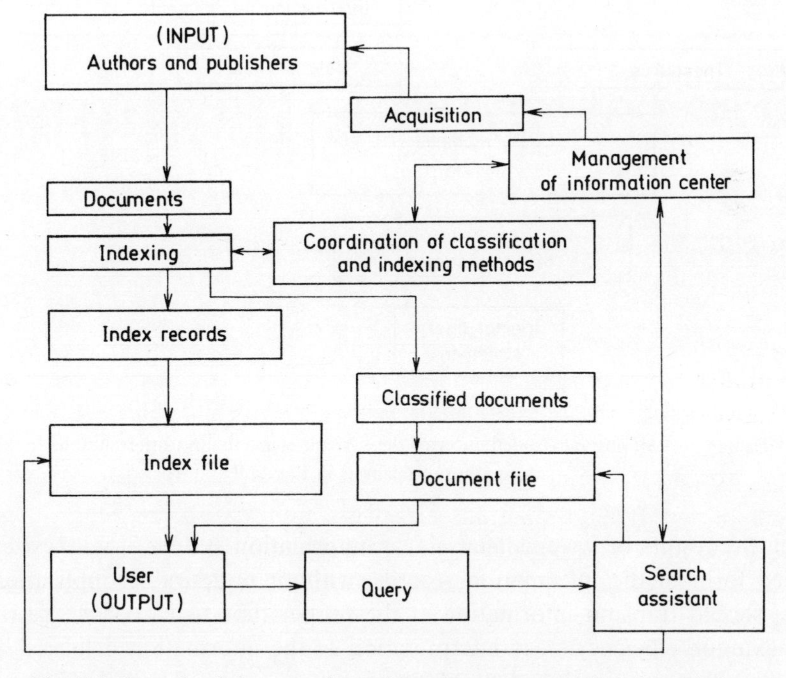

Fig. 11.3 Example of elements of an information system and their interrelations

is focused on the structure or behaviour of the system. If the information system does not yet exist, systems analysis is performed on a simulation model of an information system with the known properties of certain components. Simulation helps to determine the properties of other components and thus the whole structure can be modelled.

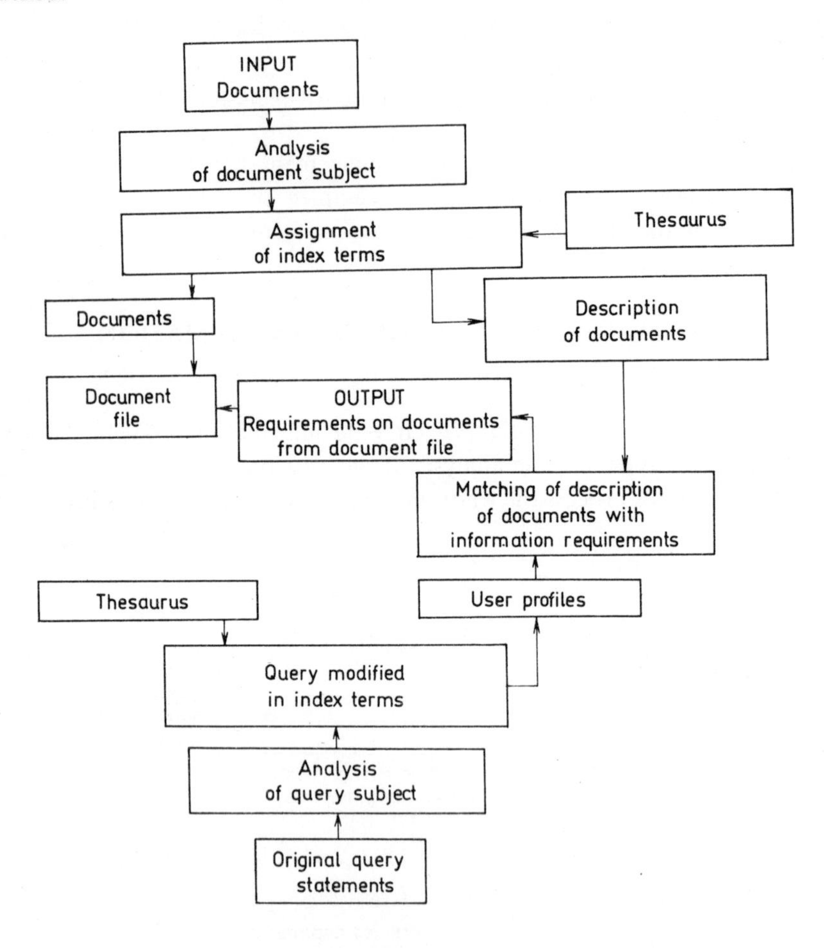

Fig. 11.4 Example of an alternative definition of the system, elements and interrelations for the same information problem as in Fig. 11.3

Using the results of systems analysis, an information system is synthesized with a unified form of the information records, without undesirable duplication. The services should transmit information at the proper time to its correct destination with maximum efficiency, easy interpretation of the information delivered and an information system must be adaptable to changes.

Systems analysis comprises three main steps: the definition of the system, the analysis of the properties of the system and the correction of the system. The definition of an information system includes the choice of elements of the system and corresponding relationships, and the acquisition, organisation and storing of data in a proper form.

Models of recording and data processing algorithms are determined for the elements chosen as shown schematically in Fig. 11.3 for an information retrieval system.

In Figs. 11.3 and 11.4, the systems are defined on a rough discriminating level; they serve as examples of information retrieval and communication of information.

In a more detailed description, the names of the elements are given together with all the important input and output operations, coding, transformations of information in the elements, which are listed together with activities performed in the element, activities for input errors detection and processing, with rules for activities on auxiliary conditions (e.g., if input errors hinder further processing of information), space and time capacities of elements, etc.

Once they have been described, the elements and relationships can be represented graphically in *flow charts*. In information systems, the flow of information plays a decisive role in this mapping of the system. In the automation of information systems, the structure of the system can be represented in matrix form. Systems analysis often reveals that some elements are connected to the system by a very weak link or that the amount of the flow is not important for the objective of the investigation. Such an element can be omitted or, be coupled to another element, or the whole system can be divided into several subsystems which can be investigated separately in further analysis. This disintegration of the system is the main method of investigation of large and complex systems.

Having defined the elements of the system, their interrelationships and relationships to the environment of the system, researchers then attempt to obtain a better definition of the interior of the system, especially of the transformations that take place in flows in and out of the elements. The systems variables and their relationships are investigated. At the same time the consequences of the designed changes are considered in the direction from input to output, and back if feedback is present. The description of the transformations is corrected on the basis of knowledge gained from other systems in operation. In new systems, this is followed by an experimental stage.

In information systems the transformation for each element can hardly formally be described, analytically. Therefore, the description of decisions and operations that take place in each element of the system is an acceptable result, suitable for the simulation of relationships between input and output.

In the *analysis of information systems* the goal is often considered as known that the elements of the system have been determined and so have the decisions and transformation processes which are to take place in these elements. Then task of the systems analysis is thus the efficient transportation of information to the proper place, by adequate means and the processing of data by suitable system software.

The indications of a malfunction in an information system are: an insufficient quantity of information (e.g., in the retrieval of documents), undesired multiple processing of the same information, contradictions in data in information files, delayed and incorrect information without necessary details, and/or a large amount of information that requires too much time for users' evaluation, etc.

The analysis of existing or simulated systems helps to reveal these drawbacks. There are inner and outer methods of applied systems analysis. In outer tests the method of the "black box" approach is used, i.e., the reactions on input are investigated without analysis of the structure of the system. In the inner test, on the other hand, the structure of the system is investigated from the viewpoint of correctness, complexity, and dynamic character of the transformations in each element of the system. The most effective seems to be a combination of these two methods.

In outer tests, past records describing the inputs and their corresponding outputs are analysed first. Sometimes the isolation of various impacts and the determination of components is difficult or impossible. The *experiments* are performed when only one input variable (or a few input variables) are changed and the corresponding response is investigated.

The outer test can be used for control of inner activities of the system if there is some information concerning the system structure. A special kind of test is the response to exceptional inputs, outside the designated range. If such tests are not performed in advance, then they should be carried out during the experimental operation.

In the investigation of the system by outer test, not all the possible situations can be analysed; outer tests as the single methods of analysis are restricted to cases where the structure of the system is not accessible. The improvement of information systems requires, in most cases, a change in the system structure or a transformation in its elements. This approach is conditioned by inner tests.

The main *inner tests* are the tests of continuity and completeness. For instance in a system there is no relationship between representation of two elements, and this relationship is important, as indicated by inner tests. Then the structure of the system should be corrected. For testing the sequences of information flow in the system, flow charts are suitable and the tests by incidental matrices are applied using a computer.

In information systems that originated by the aggregation of activities previously performed separately and connected in the process of centralization due to computer data processing, inner *tests of compatibility* are important. The elements of the system have compatible relationships if output of one element is a suitable input into another element. Automation of the information process should respect compatibility not only between elements, but also between subsystems so that a form of output should fit a form of input in further processing. If this condition is not fulfilled, e.g., in the application of the information service on magnetic tapes, a further element must be added to deal with this problem to secure compatibility.

Further tests that serve the correctness and effectiveness of flow of information in a system are called *communication, linkage and transformation tests.* According to the dynamic character of information systems the *loading* of individual elements should be tested. Data concerning the amount of flow, overloading or underloading, the frequency of requirements, their variability in time (continuous or impulse character) are collected and analysed.

The *matrix notation* of information systems structure helps to mechanize these tests. For instance consumptive elements can be picked out (i.e., the elements that receive more information than they transmit), elements that are sources of information, elements with a maximum variability of input and output, elements that are critical for the improvement of information systems, feedback in the system and mainly elements that produce a maximum delay of information.

Testing should start with simple and inexpensive tests and should proceed to more detailed and expensive tests. In processing a large quantity of data or in testing more systems with similar properties the methods of mathematical statistics and the theory of probability can be used.

Testing disturbances in the system is often called *diagnosis of the system.* In the design and building of a new system the knowledge of statistical characteristics helps to estimate the probability of disturbances and to enable proper measures to be taken to prevent the system from failing before it actually does.

The method of *tracing the signals* helps to find void places in information systems. Input standard values are used and the deviations from the standard values are sought throughout the system. For exceptional situations the method of *artificial overloading* is useful, which enables the discovery of the weakest points of information flow in the system.

11.5 THE LANGUAGE OF INFORMATION SYSTEMS

The *information language* is a tool for the description and transformation of information. In automated information systems it is used in the algorithmization of different phases of data processing.

In the description of information the process of abstraction is used mainly, i.e., information is transformed in the direction to greater generality.

Information storing is a component of information processing. In automated systems the form of storing must be suitable for both man and computer. The system should have a dynamic behaviour with a possible modification of information, including its correction, supplementation, displacement and rearrangement. In information systems some part of the information should serve to retrieve information, which is the main aim of information systems. This part of information makes the greatest impact on the information language.

11.5.1 Characteristic Properties of Information Languages

For information retrieval the collection of documents needs to be in a certain order and there should be some means of searching, matching, and recognizing whether or not the document contains the desired information. To avoid reading the whole document file, a *catalogue* is used, i.e., a list of records with a description of the document content and their location in the file.

Each entry in this catalogue is called the *index record*, all index records form the *index file*. A special language is used in index records and index files. The requirements on this language in automated information systems are as follows:

– the language has to be understood by both man and computer; if it is designed for a broader community of water resources researchers, its syntax and semantics should be simple,

– the use of a computer requires more rigid and unambiguous rules as to format, syntax and vocabulary,

– the format of records should be suitable for their organization into files,

– this language should be apt for development and modification in application.

When automating information systems, the syntax and vocabulary of information languages should be investigated thoroughly. A vocabulary is a set of words that is used in a language. As the index language differs from the natural one, its vocabulary is restricted to words serving the description of documents in index records. The syntax is a set of rules for combining elements of vocabulary into language units, thus producing a meaning not expressible by elements of vocabulary.

The basic requirements on an index language are: expressiveness, unambiguity, compactness and low cost. *Expressiveness* measures the ability of the language to describe the subject. The Universal Decimal Classification is not very expressive, but it is not ambiguous; a natural language, on the contrary, is very expressive. *Unambiguity* requires a control of synonyms and homographs. The numeric code is most *compact*, but it does not meet the requirement of being easily understood by man.

The costs include the preparation of the index language, costs of indexing, training of indexers, maintenance and development of the language, costs of indexing errors, etc.

Examples of index languages are: hierarchical classification (e.g., Universal Decimal Classification), subject headings, and key words. The system of key words can be fixed or free. The fixed system uses a *thesaurus*, i.e., a dictionary with interrelations in a fixed content vocabulary.

The syntax of index languages in information systems should have the properties of algorithmic languages for their representation on computers. The simplest syntactical rule is the *joining of terms*. The relationship of combined terms is described precisely. The added term narrows the region described by the first term. A higher type of syntax is, for example, filling out of items on a form.

Another example of syntax is the addition of descriptors or key words into terms to describe the content of a document in its proper context. This method helps in a system of key words with a fixed vocabulary to restrict ambiguity.

In the natural language retrieval is difficult. If for instance, the word "reservoir" is sought in the text, the whole text has to be read. In order to facilitate retrieval, different *systems of ordering* were developed. One such system is KWIC (key-word-in-context), where all significant permutations of the document title are alphabetically ordered.

11.5.2 Comparison of Information Languages

Information languages can be compared from different points of view, such as expressiveness, non-ambiguity, compactness, etc. The decisive aspect in the automation of information systems is their *compatibility* with computer algorithms.

A further requirement is their internationality, flexibility for the investigated subject (ability to express different branches of science) and universality (possibility of expressing different alternatives of information requirements).

No single language can fulfil all these requirements completely. Therefore different languages are constructed with properties suited for the main aim of the information system. Comparison of information languages is useful in the choice of the language for information system in WRS.

Information languages should cover the maximum quantity of subjects. The higher their *expressiveness*, the better their ability to meet these requirements. Classification according to expressiveness can be done in the following way:

1. hierarchical classification,
2. subject headings,
3. keywords with fixed thesaurus,
4. keywords with free thesaurus,
5. tagged descriptors,
6. analytic terms,
7. natural language.

Information systems with subject headings and key words with a fixed thesaurus are often used for those regions in which documents are expected to belong. The flexibility of the thesaurus helps in filling the gap caused by the development of science.

The system of free key words can cover the new subject without difficulties. No synthetic language can provide an ideal degree of precision. Synthetic languages use combinations of descriptors to increase precision of subject description. The number of combinations that can be derived from the basic descriptors is enormous even with a fixed thesaurus. However, the coincidence of user's interpretation and indexer's idea is an issue often investigated.

Information languages are distinguished by their degree of *unambiguity*. By definition the hierarchical classification does not include any synonyms and has no ambiguity. In subject headings and key words with a fixed thesaurus, descriptors with very close meanings can occur. In key words with a free dictionary and further information languages, as discussed, some ambiguity may exist. The control of homograph is almost impossible in these information languages. Ranking information languages according to relative potential ambiguity is the same as ranking according to their expressiveness. Ranking by relative *compactness* is quite the opposite. A hierarchical classification requires fewer symbols, the natural language more of them. The *costs* of information languages, as measured by the selection and use of terms is difficult to determine. The whole process with its complexity should be taken into account, including time requirements of indexing, training of indexers, risk of indexing errors or misinterpretation. Some authors claim that minimum costs are required by information languages using key words and tagged descriptors when hierarchical classification is not necessary, syntax is not complicated and the automation of the information process can be achieved more easily.

11.5.3 Thesaurus

A thesaurus is a dictionary containing words in relation to a precisely defined word content. As defined by UNESCO, it is a controlled dynamic dictionary of terms in semantic and generic relations that covers a certain area of knowledge. This dictionary contains an alphabetically-ordered system of descriptors and indexes of their relationships, a set of expressions with their relationships and rules for their use. The expressions in the thesaurus may be descriptors, uni-terms, key words or expressions from the natural language. It is based on some natural language (e.g., English). The following main types of relationships occur:

– relationships that handle synonyms that specify for each thesaurus entry one or more synonym categories or concept classes (instead of *A* use *X*),

– relationships that handle homographs by an additional descriptor; i.e., in the thesaurus several combinations are used (instead of *A* use *AX*, or *AY* or *AZ*),

– relationship of indication that specifies some different, possibly more accurate descriptor (if *A* compare *X*),

– hierarchical classification – if possible, each descriptor is followed by a broader and a narrower term.

A thesaurus is a grouping of words or of word stems into subject categories (concept classes). It controls the synonyms, gives rules for using standard descriptors (it contains terms whose use is prescribed for content analysis purposes) thus serving automatic retrieval systems.

11.6 MODEL OF AN INFORMATION SYSTEM

The objective of information systems is to predict and control its elements and relationships. It is based on the description and explanation of the reality modelled.

The organizational structure of information elements or the types of activities can be chosen as a basis for model classification. According to the organizational structure the following models are referred to:

– models of library activities, publishing activities, retrieval activities, data management, and catalogues and file organization.

The information process as a communication between the source of information, via information systems, to the user is used in classification by types of activities. The following models can then be distinguished:

– models of publishing primary information,
– models of documents acquisition,
– models of secondary information processing,
– models of information retrieval,
– model of selective dissemination of information,
– models of files organization,
– models of data base set up and management.

In the *model of library activities* there are areas where automation is not desirable (e.g., in the physical manipulation of books); this can be taken into account in world libraries, but not in information centres of water management.

The storage of documents by micro-storage techniques is suitable for automation of their handling and duplication. A highly automated system may be achieved by combining these systems with automatic communication systems and the remote transmission of documents with their print-outs by a user, on request.

Automation of further activities is included in the model of secondary information processing. The main problem – automatic classification and indexing – has not yet been solved. Its main assumption – the automatic coding of texts of documents for computers – is in the development stage. The translation of optical signals gained by reading texts, the recognition of individual symbols exists, and it is commercially available for texts on a particular form, but not for the reading of conventional books, journals and reports. This problem will be solved and a multi-purpose reading apparatus with the necessary storage for this information will be developed.

Progress in computer hardware should be accompanied by the necessary software with systems for the comprehension of texts by computer.

The systems of *automatic indexing* are often based on statistical analysis and the frequency of separate terms in text. Some words often occur in certain associations (e.g., in this chapter "information system"). Thus the associations of this term can be investigated as a whole and the frequency of this association can be determined. The terms or term associations are ordered in descending order of their frequencies and those with the highest frequency are used as descriptors in automatic indexing.

Some authors (e.g., Salton, 1968) criticize this method, as frequency can be a bad indicator of word significance. They claim that a word is significant if its frequency is higher than was expected *a priori*. This method requires knowledge of *a priori* probabilities of terms and term associations for each branch.

Other methods of retrieval of significant words and word associations in the text do exist. However, the problem of automatic indexation has not yet been solved. A promising way probably is in the systems approach with the requirement that the authors of primary information should create the basis for indexing. A summary of papers and books helps very much in indexing, some journals publish not only the primary information, but also the secondary information, including the key words. Then the task of transforming these indices into the form required by the information system and the computer is easier, but still has to be undertaken. These problems are also present in information systems for water management.

The systems approach shows that information systems influence the model of primary information that is separated from information systems, but forms its significant environment.

The model of documents acquisition is connected with the selection of documents and therefore with a model of secondary information processing and a model of information retrieval. In these models there is a centre of automation of information systems for the *selective dissemination of information*, retrospective retrieval, thesaurus operations, automatic classification, data base manipulation, etc. The computer with its peripheries is the main tool of automation (see section 11.7).

In the model of library activities in the conventional form, the advertisement activities and preparation and publication of bibliographic quotations can be automated; full automation is possible in activities using modern materials (microfilm, microfiche, etc.). This process is dynamic with step by step mechanization of the system components and their interconnection in the system.

If secondary information is stored on magnetic tape, then the handling of catalogues, indices volumes, glossaries, directories, etc., is possible using contemporary computer hardware.

The development of computers with big centralized systems and also of decentralized *minicomputers* adds further possibilities to information systems. The minicomputer does not require air-conditioning and uses conventional cassettes, floppy discs, etc., and can be more easily adapted to the user's needs and to the needs of a library. Therefore, besides the centralized information service for water management, decentralized information centres and libraries with close contact with users are progressive forms of information systems.

A further condition of the success of information systems is provided by special purpose equipment such as *copying equipment*, graphic storage devices, printers, etc. The ideal is copying equipment with remote control operated by the user or a group of users, including equipment designed to transform information from one medium such as magnetic tape to another, e.g., microfilm with various kinds of expansion and reduction ratios. However, using the conventional copying equipment the time lag from information requirement to information retrieval can be reduced substantially by the automation of the most critical activities using modern microstorage, copying and conversion devices.

The *model of publishing activities* concerns both primary and secondary information. The automation of primary information concerns printing and editing; creative activity cannot be automated, and experiments with synthetic intelligence are in their initial stages.

Automated information systems help to reduce undesired investigation of the same problem by several researchers and to co-ordinate the methods and forms of publishing. They also help to shorten the time gap between the manuscript and the published book, or journal, by automation of the printing process. If the basis for secondary information (abstracts, key words, etc.) is included in the primary information, then the process of their publication can be fully automated. This procedure can be realized by contemporary computer hardware and equipment and it does exist in some branches to a limited extent.

Models of analytical and research activities cannot be fully automated for all the diverse activities involved. Processing of data on computers is an integral part of these activities; for example: *relevance* judgements, precision determination, questionnaire evaluation, computation of economic effectiveness of systems of selective dissemination of information with various indexing systems and the determination of their advantages for selective dissemination and retrospective retrieval, computation of costs per user, per one profile, per one descriptor, per one relevant document, evaluation of various policies in retrieval based on magnetic tape services, tests of data base organization, estimation of operation of on-line and off-line systems, etc.

Information systems are an important part of *cybernetic control systems* and management information systems. The modelling technique is a basic method of cybernetics. In the cybernetic concept the main stress is given to feedback in models and between models, application of the theory of algorithms for models, especially in quantitative expression and coding for computers, the application of certain parts of automata theory (e.g., finite automata), heuristic programming and approaches to synthetic intelligence, the theory of games and its application to information systems, etc. In the future development of information systems other methods of cybernetics can be applied, e.g., investigation of self organizing systems, systems with automatic pattern recognition using optical or magnetic reading techniques. Self-organization and self-control can be the progressive means of automatic software generation. The process of learning common to living organism's can be utilized for cybernetic automata in information systems, e.g., in automatic indexing, translation from one language to another, etc.

The application of cybernetics (with the exception of systems approach, the use of computers, and modelling) has been used more intensively in control and regulation of technical systems and its use in information systems in water management is in the developmental stage.

11.7 COMPUTER AS THE MAIN INSTRUMENT OF INFORMATION SYSTEMS AUTOMATION

The development of computers made their application to information systems possible. In the automation of complex systems the computer is the main, but not the only tool of automation. Computerized information systems are used for *selective dissemination of information* and information retrieval when the systems are based on magnetic tape services, for the automation of set-up of lists, catalogues, bibliographical references, etc. In these activities problems can occur more in capacities for assembling and coding of input data and necessary software than in methodological issues and computer hardware.

The function of computers in systems of selective dissemination of information and information retrieval is the basic function of modern information systems.

11.8 INPUT OF INFORMATION SYSTEMS IN WATER MANAGEMENT

International co-operation in scientific, technical and economic information, together with the co-ordination of their processing, is the main prerequisite of effective information systems. A basic condition of an effective information system is its data base. The automation of information systems in water management requires the set-up of a data base, its maintenance and replacement. This system is to be co-ordinated with information systems of other branches and with international information systems.

For WRS primary, as well as secondary information is important, especially hydrological data and data concerning the basin where the WRS is located. The data base of such an information system is often developed together with some big project and water plan. An important requirement in this activity is the interface between their information systems and the compatibility of data structure.

Data structure elements are called *bits* and are often organized into bytes (e.g., IBM) or words (e.g., ICL) that can be used as representation of *characters.* A *field* is built up from characters, and a *record* from fields. The field is a set of characters with a defined meaning, a record is a set of fields; the records are descriptive of individuals, the file is a class of individuals. Grouping of records in a file is done according to some organisational structure.

The representation of a character by bits depends on the type of computer. Therefore, the basic level for the information system is the character. In data structure on magnetic tapes the item of structure is one block, on a disc the item is a cylinder. This is a physical structure. It is advantageous if this physical structure can be used as a subclass of the file structure of information systems.

The arrangement of fields in relations in a record, the arrangement of records in a file, files in set of files, etc., is called *data structure.* Data structure can be defined in different ways. The requirements of data structure are similar to the requirements of the index language, as they both serve the same purpose – adequate information storage and retrieval.

The expressiveness of the record structure is the ability to describe information using this structure. If the structure is too simple, it hinders expression of complex information. If some information is sought for in a record, the field of descriptors in this record has to be found. If some difficulties occur in this procedure, the process of retrieval is delayed and if some ambiguity is present, the results may be incorrect. Therefore positional ambiguity must be excluded as far as possible.

Records with high compactness use a relatively small number of characters for coding information. Requirements for storage are then smaller, which is economically better. However, a high level of compactness makes the retrieval procedure more difficult.

Forms of structure: invariant structure, bit coding, (yes-no questions), fixed and repeated fields, tagged fields, adressed fields, and the structure of the natural language.

Invariant structure does not provide any information, and for a given file, it is used for the identification of this file from the rest.

Bit coding is used in systems of fixed fields for identification, to ascertain whether or not a particular index is present. In *fixed fields structure* the position and length of each field is given. However, the content of a field can change. If the length of the content is highly variable, this structure does not make full use of the storage in computers. If the form of the fields and their sequences are given, but their numbers change, this structure is called a *structure of repeated fields.* The *structure of addressed fields* (that are more economical as their length is variable) is organized into two groups of fields – descriptive fields and fields of adresses that delineate the limits of descriptive fields, or their length and the type of their content (heading of fields).

In the application of these structures combinations are preferred which make full use of their advantages. These advantages are derived from their properties such as expresiveness, compactness, etc. For example, the expressiveness of these structures in ascending order is as follows: invariant structure, bit coding, structure of fixed fields, structure of repeated fields, structure of adressed fields and structure of the natural language. In this order the ambiguity, both positional and semantic, grows. Compactness is highest in the structure of fixed and repeated fields, other structures are less compact.

Processing of a file is, in fact, its transformation and involves a search for the parts to be processed. *Searching* is the process of isolating one particular area of the file. The result of this search may be a statement as to whether a given information area is present or not. The processing of files depends on their structure, which is related to their physical storage in the memories of computers.

According to how accessible stored information is, the memories of computers may be classified into *equal*, *direct* (random), and *sequential access memories*. The inner memory of a computer has equal access as every addressable cell of the computer has the same access time. Magnetic discs (and drums) have direct (random) access, other storage devices such as magnetic tape, paper tape, punched cards, etc., have sequential memory access.

The most advantageous is the inner memory that influences the structure of records less. Its capacity is, however, limited and the security of information is low (e.g., it may be destroyed by interruption). Sequential organisation, which is necessary on magnetic tape, is useful in sequential processing only. Duplicating or triplicating of files, which is relatively cheap on magnetic tapes, increases the security of stored information. Magnetic discs provide a lower access time, its security, however, is more complicated. The storage medium is influenced by the frequency of file processing.

The file *structure* is an important factor in the retrieval of information. The simplest organization of records in a file is the *sequential structure* where the records follow each other logically and physically. The addition of records inside the file is, however, more difficult and is not effective.

In further structures the logical and physical structures are different. In a *random structure* the physical structure is random and the ordering principle is given by addresses. In a *chain structure* these addresses are contained in records. A *branching structure* is an extension of the chaining concept. In a *list structure* (or directory structure) addresses are physically separated in a directory. The list structure is the most advantageous for retrieval, where direct access storage devices are necessary.

11.9 EFFECTIVENESS CRITERIA FOR THE PERFORMANCE OF INFORMATION SYSTEMS

Testing is the best way to evaluate the effectiveness of information systems. Its main components should be: the determination of test criteria, design of the test, performance of the test program, results analysis, and design of systems innovation or amendments based on this analysis.

The criteria for the evaluation of systems operations should be contained in answer to the following questions: Are the users satisfied with the system? Does the contemporary system of indexing encompass the main branch problems? Does the delay between publishing the primary document and getting the information to the user influence the performace of the system and diminish its value? Could the user be informed sooner about this document by another means? Are there any substantial differences in indexing methods between the files? Are the descriptors specific enough? Is the applied method of relevance evaluation appropriate? Are the terms of the thesaurus adequate? What are the requirements of users concerning the indices of precision and recall? Are there policies for their increase and can these policies enter the system? How are the outputs of information systems and other services utilized by users? What methods are used by users to increase the indices of precision and recall? Can these methods be automated and used in the information system? What are the most effective methods of interaction between users and the information system? Does the necessity for this interaction influence the system? Do input data of the system (mainly magnetic tape services) include errors, if so what kind of errors? Can the system handle these errors? Are the programs in information systems software flexible enough for possible changes?

The user is interested in *high relevance* and a *short time lag*. The basic problem of the selective dissemination of information and retrospective retrieval is the high degree of relevance. Relevance is often evaluated by *precision and recall indices.* There is an approximately indirect relationship between them. The precision index (PI) is the proportion of relevant documents to retrieved documents, the recall index (RI) is the proportion of relevant documents retrieved by the system to all relevant documents in the file. (For instance 20 documents in the file are relevant. If 10 documents are retrieved and 8 of them are actually relevant, the precision index $PI = 8/10 = 0.8$ and recall index $RI = 8/20 = 0.4$.) If no section is performed and all the documents are retrieved, the recall index will equal one and the precision index will be almost zero. If the retrieval is made by strict requirements a few documents will be retrieved and possibly all the documents will be relevant ($PI = 1.0$), but a relatively high number of relevant documents will not be present in the retrieved part, and the recall index will be low.

In the design of tests for systems evaluation more users should be involved as their evaluation of relevance can differ from that applied in systems software. Some aspects of systems operation policy may be geared to satisfy the users' requirements. The system operation policy is, however, based on the requirements of an average user, i.e., it tries to satisfy the majority of users, but not all of them. It would not be efficient to complicate the system for the specific needs of one, or several users.

Further aspects of systems operation policy are revealed by comparison of *operational effectiveness* and *overall economic efficiency*. Some means serve to increase both forms of efficiency, e.g., automation of information systems. However, in some situations the effort to increase both forms leads to competitive actions and some optimization is necessary to determine the optimal degree of satisfaction of these aspects. This phenomenon can be observed in system hardware elements and in software too, e.g., in indexing methods, information languages, retrieval policy, processing of outputs, methods of their delivery to users, etc.

11.10 SOFTWARE OF INFORMATION SYSTEMS

The software of information systems should be computer-independent. This requirement is, however, seldom fulfilled. Therefore, the software of information systems will differ for individual computers. This problem can be solved by the unification of hardware with centralization of information systems. Beside these tendencies to centralization, the development of information systems with *minicomputers* leads to decentralized systems for special purposes. The software for these systems is often supplied by the producer, and it uses an interactive mode.

Two basic types of information systems can be distinguished – integrated (centralized) systems and distributed (decentralized) systems. If the same data base is used by several subsystems, the integrated information system is often preferred. The key component is the common data base. The basic characteristic are fast response to queries via remote terminals, on line mass storage, continuous up-dating of files, centralized data processing. An integrated information system reduces redundancy, secures more protection of data, allows more than one user to retrieve documents, increases the overall effectiveness due to provision of more timely, relevant and accurate information, reacts better to long-term planning. The integrated system has some disadvantages – special personnel requirements (system analysts), co-operation of subsystems, lower responsiveness to user's needs, the breakdown of the information system may have catastrophic results, high development costs, difficult modifications.

The distributed information system, on the other hand, does not use the method of the universal system (that leads to low efficiency in software overheads and administration), but a modular system. An alternative to the common data base in a distributed system is an aggregation of information systems arranged in such a manner that a set of subsystems is formed and tied together by means of a communication interface. To develop the data base is difficult but not so expensive, and this data base is more suitable to the needs of users. The cost effectiveness with minicomputers is often higher in distributed systems, the overall costs are lower, and the information system can be built in units that are able to function separately.

The system can easily be modified to meet users' requirements. Recovery and control are easily handled, simple software is necessary, the errors in input data are processed in their place of origin, the breakdown of the subsystem can be overcome, new subsystem can be added.

On the other hand, more co-operation is necessary, some redundancy occurs, more communication channels are necessary, the retrieval of data from several subsystems is more complicated, users can gain access easily on one subsystem only, etc.

The effectiveness of computer-based information systems can be measured by the cost-benefit ratio as in other branches of technology. In information systems not only the direct costs and benefits, but also the indirect costs should be considered. The direct benefits involve time reduction in tasks, the direct costs involve costs of computer purchase or rental, costs of computer operation (payroll of employees, costs of supplies, maintenance, power, insurance, etc.).

In information systems many benefits are of intangible nature that can hardly be evaluated in monetary units and stem from better information acquaintance and knowledge of users. It is also important to liberate people from routine work. Indirect costs also include costs of information system reorganisation, personnel training, and transition costs.

The increase of the effectiveness of computer use in information systems is concentrated on its hardware, software, systems approach to co-ordination of software, input data, data base, use of output and actions that are related to this output. The systems approach increases effectiveness by the unification of input and output, thus creating conditions for unique interface.

The software of information systems should be set up on the basis of the project of the information system, containing a preliminary or feasibility study, analysis of the existing system (or simulated system), and the design of the system, system innovation or system amendments. The feasibility study has to state the conditions for further information processing, form a team for project elaboration and find the budgetary constraints. The problems of analysis have already been discussed. The design of the system should be elaborated in alternatives with economic analysis to an extent to satisfy approval by decision-makers.

The computer is the main component of information systems. Therefore the alternatives differ according to which choice is made. For the designated type of computer and the necessary software the design should specify the main activities in the system, the possibility of their integration, a rough flowchart of data processing by computer, define the files and their organisation, estimate the cost and time of their processing and transformation, design the content and form of output, identify the individual runs of data processing, estimate the number of operations in the subsystems of information systems, the periodicity of these runs and operations, list the standards that should be respected, prepare specification for programmers, define methods for data conversion and preparation, etc.

Some of the methods of information system analysis quoted are used not only in system design, but also in the evaluation of its operation. An information system in operation is not a static object. The input data change in content, quantity and form. After exceeding certain limits the combined effect of these changes requires a modification of the system. Further, the requirements of users change with the development of science, technology and their experience with the system. The fast development of computer hardware and software influences information systems by providing new possibilities.

12 WATER RESOURCE SYSTEMS PROJECTS

12.1 THE WATER RESOURCE SYSTEM OF THE HORNAD RIVER BASIN

The area of the Hornád River Basin is 4346.31 km^2 at the Czechoslovak frontier. The important left-bank tributaries are: the Levočský Stream (153.28 km^2), the Svinka Stream (344.56 km^2), the Torysa River (1348.98 km^2) and the Olšava Stream (339.54 km^2). On the right bank the Hnilec River is the most important tributary (654.9 km^2).

The WRS also includes the basin of the Bodva River (890.37 km^2) and the upper basin of the Poprad River (transbasin diversion to the Hornad River basin) as far as Matějovce (311.07 km^2).

Table 12.1 The reservoirs of the WRS in the Hornád River basin

No. of point	Name	River (Stream)	Q_a [m^3 s^{-1}]	MQ [m^3 s^{-1}]	A_z [10^6 m^3]	Capital costs [10^6 Kčs]
1	Hranovnica	Hornád	1.03	0.20	24.4	189
2	Levoča	Levočský	0.43	0.07	10	127
3	Markušovce	Levočský	1.00	0.19	27.6	116
4	Hrabušice	Hornád	2.42	0.46	72.4	330
5	Helcmanovce	Hnilec	5.55	1.09	133.8	926
6	Ružín I	Hornád	16.00	3.00	48.5	existing
7	Ružín II	Hornád	16.00	3.63	2.5	existing
8	Kysak	Hornád	17.50	3.85	10	
9	Obišovce	Svinka	1.91	0.45	40.4	83
10	Krivany	Torysa	1.96	0.32	43.5	292
11	P. N. Ves	Lutinka	0.41	0.06	23.2	355
12	N. Šebastová	Sekčov	1.90	0.32	14.2	94
13	V. Myšla	Olšava	1.62	0.33	29.6	60
14	Bukovec	Ida	0.65	0.11	21.4	190
15	N. Medzev	Bodva	1.26	0.19	24	
16	Jasov	Bodva	1.44	0.21	74.5	422
17	Jablonov	Turňa	0.42	0.06	11.1	108
18	Tichý potok	Torysa	0.885	0.149	27.3	

For a representation of its basic properties and behaviour, the WRS was first of all defined on a rough discriminating (not detailed) level. The goals of the system, its elements and its environment were specified.

Definition of the system on a rough discriminating level (screening model)

The *water reservoirs* are the main elements of the system. From 25 reservoirs that were investigated in this basin by Kos *et al.*, 1979, the following were excluded: Palcmanská Maša (transbasin diversion to the basin of the Slaná Stream – the hydrological influence was considered approximately in hydrological data, and small reservoirs under 10 mil. m^3 of storage with local importance. The list of reservoirs included in the model and their basic parameters are given in Table 12.1.

The model considered the following goals of the WRS (related to control points 20 to 31) that significantly influence the behaviour of the WRS in the Hornád River basin: municipal, industrial and irrigation water supply, dilution of flows for water quality management and hydropower production. Other less significant goals were ignored in the model.

System with detailed discriminating level (simulation model)

Simulation modelling was the basic technique for the system on the detailed discriminating level. Systems analysis proved that for the discriminating level of the General Water Plan the network of points obtained as the output of the system defined on the rough discriminating level was sufficient (Kos and Zeman, 1976), and a more detailed division was not necessary. The schematic representation of the system is shown in Fig. 12.1.

The comparison of the alternatives to WRS was carried out on the basis of a minimum capital costs objective[1]) with some reliability constraints, as the problem of determination of benefits from the multi-purpose WRS has not yet been solved.

In WRS with a priority municipal and industrial water supply, the occurrence-based reliability[2]) of the municipal water supply was 97–99% and of the industrial water supply was 95–97%.

As the period of available monthly flows records was 30 years the values for reliabilities higher than approx. 97% could not be reliably computed by the deterministic simulation model. Therefore, (especially for drafts higher than 70% of the annual mean flow) the generated stochastic monthly time series should be used as the input of the stochastic simulation model. There were few such reliability requirements in WRS of the Hornád River basin and a deterministic simulation model with

[1]) The OMR (operation, maintenance and repair) costs were assumed as a linear function of capital costs.

[2]) The occurrence-based reliability is defined as the ratio of the number of years without water deficits to the total number of years under study plus one (see 6.3.7).

the observed time series for the period 1931 – 1960 were applied. The required operation without deficits in this period offered a reliability greater than 97%, thus satisfying the reliability requirement.

In similar hydrologic conditions where a more detailed investigation was performed by stochastic models, the operation of WRS without deficits in the period

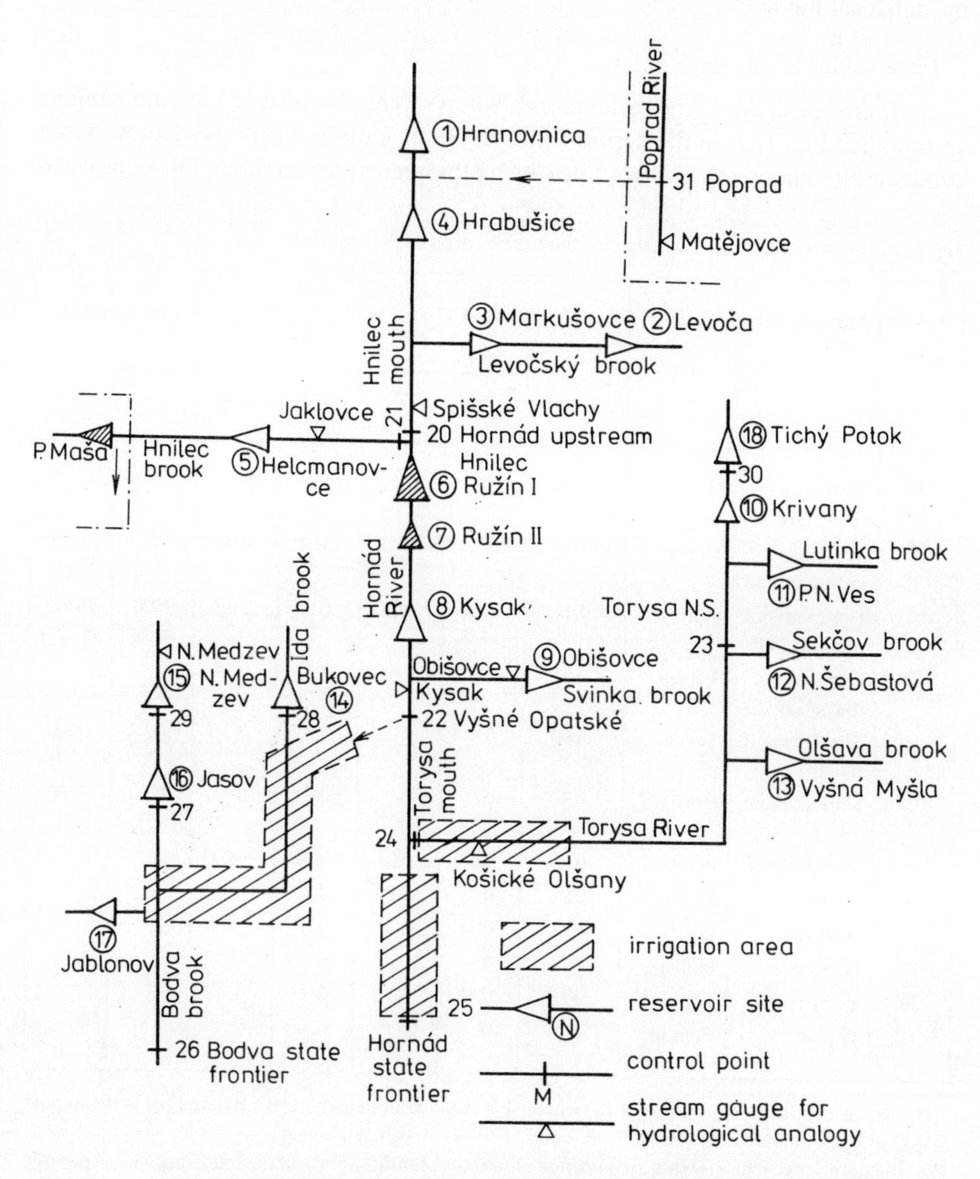

Fig. 12.1 Schematic representation of the WRS in the Hornád River basin

1931 – 1960 offered 95 – 97% reliabilities. If a 99% reliability was required, the draft determined by the deterministic simulation model using the period 1931 to 1960 has to be reduced by 10 – 20% (Kos *et al.*, 1967 – 1970). These estimates were used in the WRS of the Hornád River and the operation of reservoirs with a direct withdrawal from the reservoirs pools for the municipal water supply was analysed more throughly by statistical methods.

Input values of the model

a) *Hydrological data.* The mean monthly flows were observed in 8 stream gauging stations (see Fig. 12.1) in the period 1931 – 1960. In stations where the data were not complete the missing data were filled in by the regression method. The completed

Table 12.2 Estimates of WRS demands for the time horizons 1985 and 2000

No. of point	Name of point	River Stream	MQ $[m^3 s^{-1}]$	C $[m^3 s^{-1}]$		W $[m^3 s^{-1}]$		Irrigation*)	
				1985	2000	1985	2000	1985	2000
20	Upstream of the Hnilec S.	Hornád	1.16	0.02	0.05	0.12	0.14		
21	The Hnilec stream mouth	Hnilec	1.62	0	0	0.05	0.05		
22	Upstream of the Torysa R.	Hornád	4.48	1.04	1.14	1.19	1.33	9300 (18.6)	16 625 (33.25)
23	Upstream of the Sekčov S.	Torysa	0.45	0.07	0.08	0.13	0.14		
24	The Torysa R. mouth	Torysa	0.80	0	0	0	0	0	3700 (7.4)
25	State frontier	Hornád	4.3	0	0	0	0	0	3500 (7)
26	State frontier	Bodva	0.28	0.02	0.03	0.02	0.02		
27	Jasov	Bodva	0.21	0	0.01	0	0		
28	Bukovec	Ida	0.11	0.51	0.51	0	0		
29	N. Medzev	Bodva	0.11	0	0.64	0	0		
30	Tichý potok	Torysa	0.15	0.73	0.73	0	0		
31	Poprad	Poprad	0.98	0	0	0	0		

*) The demands are in the form A (B) where A is the area of irrigation (10^3 ha) and (B) is the annual irrigation water requirement (10^6 m^3/year).
The irrigation water requirements are partitioned into the monthly values in the following way: April 4%, May 17%, June 21%, July 24%, August 22%, September 10%, October 2%.

and adjusted monthly flows in the gauging stations were recorded on magnetic tape. If the gauging stations did not correspond to the dam sites or use points (control points), it was necessary to make adjustments in the simulation model based on the drainage areas or annual mean flows ratios.

b) *Demands*. At the time of analysis (1978) only two reservoirs existed: Ružín I and its short-term release control reservoir Ružín II (with a weekly operating cycle). The purpose of the Ružín I reservoir is to provide an industrial water supply for the Eastern Slovakia Ironworks in Košice and other industries in this area; to dilute streamflows near the mouth of the Torysa River (with a 90% reliability); to use a part of the storage for flood control; hydroelectric power generation and the development of facilities for water-related recreation. The operation depends on the river flow regulation requirements at points Ružín I (3 $m^3 s^{-1}$), Kysak (6 $m^3 s^{-1}$), and the Hornád River downstream of the confluence with the Torysa River (10 $m^3 s^{-1}$).

Meeting the constraints imposed by the last two points depends on reservoir content (and an increase of water supply reliability) thus, a reduction of reliability to 90% is achieved. In this simulation model two alternatives of the operation of the Ružín reservoir for hydroelectric power generation were investigated, i.e., (a) a model with a priority of hydropower to river flow regulation for waste water dilution, and (b) a model without this priority (with some losses in hydropower benefits due to the greater variation of the head). In other reservoirs of the system the primary hydroelectric power generation is not important (i.e., without pumping power plants).

The estimation of steady withdrawals of water for municipalities (sites 27, 28, 29 and 30) and for industry (mainly for the Eastern Slovakian Ironworks and for industries in the towns of Košice and Prešov) for years 1990 and 2000 are given in Table 12.2, where C is the consumption and W is the part withdrawn without consumption (numerically equal to the return flow or waste flow).

At the time of analysis, large-scale irrigation in the basin of the Hornád River did not exist. The plan for its development and the within-year pattern of withdrawals is given in Table 12.2 at points 22, 24 and 25.

The minimum acceptable flows were determined using the principles of the General Water Plan. The values of the minimum acceptable flows MQ for the control points are given in Table 12.2 and for reservoir sites minimum acceptable releases are shown in Table 12.1.

Flood control is not included as a direct goal in the model, however, the flood control storage volumes of reservoirs, determined by the flood control model, are considered.

The *water quality* requirements are included in dilution impact. At points where water pollution needs to be treated, the possibility of enhancing the draft was investigated in relation to an increase of available reservoir storage upstream of those points. If the costs of the additional storage are related to the draft, the cost of dilution can then be calculated.

Evaluation of the basic alternative

The basic alternative of the WRS design and operation is related to the demands estimated for the years 1985 and 2000. A list of the reservoirs that would be necessary to meet these demands is given in Table 12.3.

Table 12.3 Basic alternative to reservoir development until 1985 and 2000

No. of point	Name	River (Stream)	Active storage [10^6 m^3]	
			1985	2000
6	Ružín I	Hornád	48.5	48.5
7	Ružín II	Hornád	2.5	2.5
9	Obišovce	Svinka	0	6.3
12	N. Šebastová	Sekčov	14.2	14.2
14	Bukovec	Ida	21.4	21.4
15	N. Medzev	Bodva	0	24
18	Tichý potok	Torysa	27.3	27.3

Decomposition of system to subsystems

The simulation model of the WRS in the Hornád River basin and the capacity of the computer do not require the decomposition of the system into subsystems. Therefore, the basic model was developed for the whole system. However, the reduction of computer time was achieved by calculations using the following subsystems:

– the subsystem of the Torysa River basin as far as point Prešov,

– the subsystem of the Hornád River upstream of the confluence with the Torysa River, including the transbasin diversion from the Poprad River and to the Slaná Stream and to irrigation in the Bodva River basin,

– subsystem of the Bodva River basin.

Computation of the alternatives of the subsystems and their evaluation

The hydrological water resources potential given by the feasible dam sites in the area investigated is higher than the demands for the system until the year 2000. In subsystems the relationship between reservoir storage and the draft at important control points was calculated (see Fig. 12.2). In this relationship, the demands estimated for the year 2000 and the low flow augmentation at the points investigated were satisfied. The latter requirement can be considered as the river flow regulation for the constant draft. In a configuration of the subsystems a number of reservoirs combinations were assumed. For a reduction in computer time, the reservoirs (without special function in WRS) were classified, using the ratio of capital costs and released draft, and the reservoirs with the lower values of this ratio were included in simulation models.

a) The Subsystem of the Torysa River Basin Closed at Point Prešov

Most economic activity is concentrated in the town of Prešov, and therefore this point (23) was chosen as an important control point for the subsystem. By the year 2000, the capacity of the Tichý Potok and Nižná Šebastová reservoirs will be exhausted. As the Vyšná Myšla reservoir is downstream of point Prešov, the draft at this point can be influenced by the Krivany and P. N. Ves reservoirs only.

The maximum draft at point Prešov is 2.65 $m^3 s^{-1}$ and subtracting the minimum

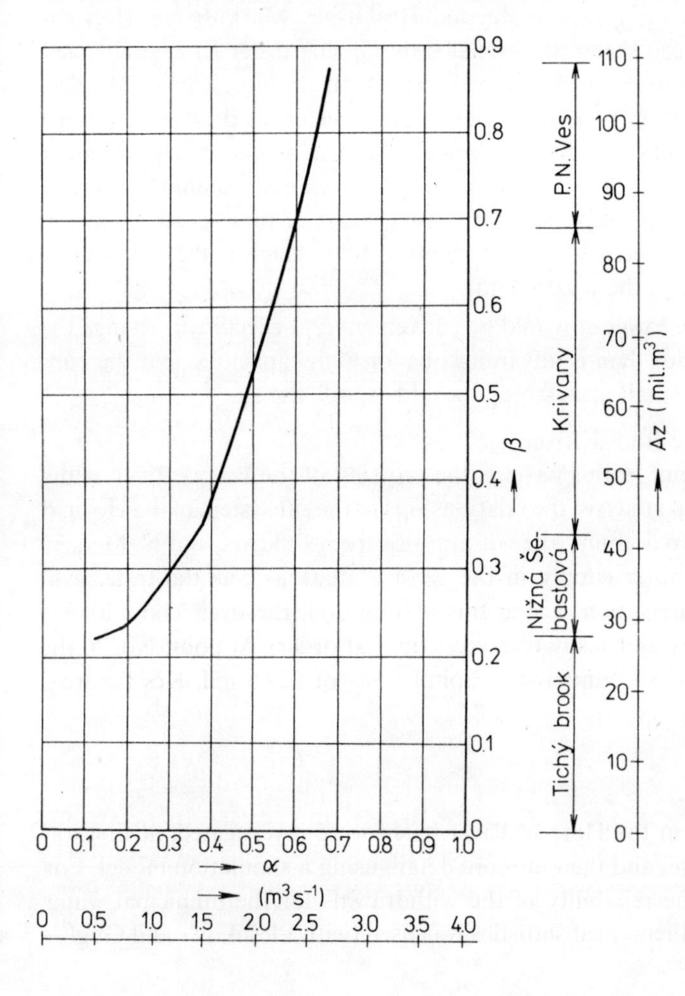

Fig. 12.2 The relation between draft and storage $\alpha = f(\beta)$ at point 23 α = draft/mean annual flow, β = active storage/mean annual flow converted to mil. m^3, A_z = active storage

flow and demands $MQ + C + W$ the remaining "commitment" is 1.98 $m^3 s^{-1}$. This maximum requires the capital costs of 646.6 mil. Kčs (approx. \$ 30 × 10^6) for the construction of the Krivany and P. N. Ves dams. The results in the form of the relationship between draft and storage are presented in Fig. 12.2.

b) The Subsystem of the Hornád River Upstream of the Confluence with the Torysa River

The most important point in this subsystem is the site Vyšné Opatské (No. 22). The model assumes water supply for irrigation below the confluence of the Torysa and the Hornád Rivers.

The transbasin diversion from the Poprad River significantly influences the subsystem and therefore only the Ružín (I and II) and Obišovce reservoirs are necessary to meet the demands. Other reservoirs form an additional commitment function. Using the ratio of capital costs and draft as a minimizing criterion, the reservoirs were placed in the following sequence: Kysak, Hrabušice, Markušovce, Helcmanovce, Hranovnica, Levoča. Using additional criteria, this order can be changed (e.g., Hrabušice has a negative environmental impact on the protected wilderness area "Slovakian Paradise"), but the form of the relationship of draft and storage $\alpha = f(\beta)$ (see Fig. 12.2) would not change significantly. The maximum commitment is $MQ_{com} - MQ_{req} = 11.3 - 4.48 = 6.82\, m^3 s^{-1}$ and the required maximum of added capital costs is 1771 mil. Kčs (approx. \$ 90 × 10^6). Most of these costs go to the Helcmanovce dam (926 mil. Kčs – approx. \$ 50 × 10^6), which is the second key dam in this basin (the first is the Ružín dam).

If the requirements were lower, it would be convenient and effective to change the order of construction as this dam is advantageous for high capacities, and the construction of a dam with a small total storage would "spoil" the site.

c) The Subsystem of the Bodva River

The most important point in this basin is the crossing of the Bodva River at the state frontier (No. 26). In an analysis, the relationships to the subsystem of the Hornád River basin were taken into account (the withdrawals from Bukovec and N. Medzev reservoirs for municipal water supply in the Košice area), as was the transbasin diversion for large-scale irrigation. Using this assumption, the draft commitment can be applied to Jablonov and Jasov reservoirs (in that order). At point No. 26 the additional draft is 2.04 $m^3 s^{-1}$, involving capital costs of 529,6 mil. Kčs (approx. \$ 25 × 10^6).

Conclusions

The problems of WRS in the Hornád River basin were investigated, at first, on a rough discriminating level and then in more detail, using a simulation model. For better determination of the reliability of the withdrawals for the municipal water supply, the method of mathematical statistics was used (with Pleshkov's and Gugly's diagrams). The completion of the hydrological time series was performed by regression analysis, and the validity of this method was verified by Ripple's diagram. A comparison of the results of the different models was used for the verification of the results. This method is often called *multi-modelling*. The application of different methods is really an experiment with models. While experimentation with one model can find the response to input data assuming some given structure of the system, experimentation with several models can test the structural assumptions. The systems approach in combination with the comparison of the results of the solution of one issue by different methods is a progressive way of dealing with the problems of WRS.

12.2 THE WATER RESOURCE SYSTEM OF THE ODRA RIVER BASIN

An example of water resources problem assessment with simulation models and chance-constraint models application is the Project of the Water Resource System of the Odra River Basin (1974).

In this basin, which comprises the industrial centre of Ostrava and its surroundings, the water resources requirements are high as compared with the potential water resources of this area. The isolated water resources facilities cannot meet the demands: therefore a WRS exists, and it is to be developed.

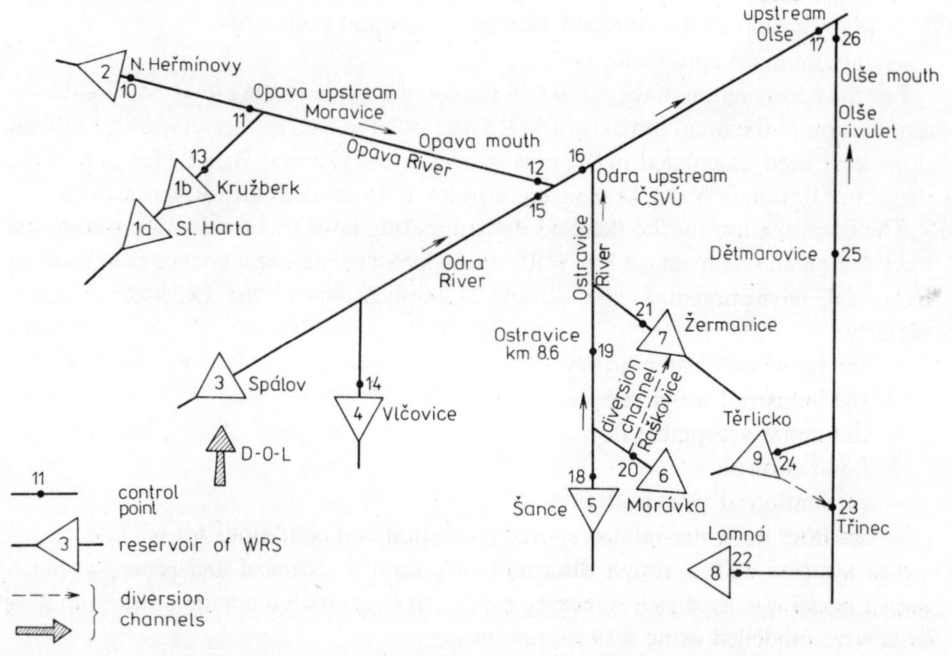

Fig. 12.3 Schematic representation of the WRS in the Odra River basin

The area covers the basin of the Odra River with its tributaries – the Opava, the Ostravice and the Olše Rivers (Fig. 12.3). A characteristic feature is the relatively high concentration of inhabitants and industry in this area. Coal-mining, thermal power plants and ironworks dominate the industrial scene accompanied by the chemical and machinery industries. In view of the climatic and soil conditions, agricultural development is not so important in this area as in others.

The *systems approach* was used at two discriminating levels. First, a model on a *rough discriminating level* was developed, the main interactions were defined, investigated and later used in the simulation model. Some goals were not considered in the simulation model, but they have an impact on input parameters, and some goals were ignored according to the planning level of the General Water Plan.

The WRS was defined on a rough discriminating level to meet the following goals:

– the development of the municipal water supply,
– the development of the industrial and irrigation water supply,
– flood control and its relation to water supply demands,
– water quality control, prevention of pollution, waste water treatment and environmental enhancement,
– hydroelectric power generation,
– navigation, inland waterways and river treatment,
– water-related sports and recreation,
– fisheries,
– stabilization of the riverbed, transport of liquid pollutants,
– minimum acceptable flows.

For the construction of the model on the rough discriminating level and for the determination of the main impacts of WRS, the isolated effects of reservoirs and diversions were used, calculated in the case studies of the General Water Plan and in the Project of Rational Water Management in the Basin of the Odra River in 1972.

The investigation on the detailed discriminating level was focused on issues that have the greatest impact on the WRS and which are the consequence of the social, industrial, environmental and ecological implications in the Odra River basin. Namely:

– the municipal water supply,
– the industrial water supply,
– minimum acceptable flows,
– flood control,
– prevention of river pollution,
– facilities for water-related sports, recreation and conditions for wildlife.

For systems with a rough discriminating level a *technical and economic input-output model* was used as a screening model. Systems with a detailed discriminating level were modelled using a *simulation model.*

The main *goal* of WRS in the Odra River basin was the municipal and industrial water supply. The irrigation water requirements and primary hydroelectrical power generation (i.e., with the exception of pumping power plants) were secondary goals in this basin. Flood control requirements were incorporated as the flood control storage volumes of reservoirs. Recreation influences the operation of WRS by requirements for recreation pool maintenance in the summer months.

Recreational activities are not permitted on the municipal water supply reservoirs. In future developments, some controlled forms of recreation (e.g., swimming, fishing, etc., but not motor boating) are to be permitted in the multi-purpose Slezská Harta reservoir, which has the main purpose the municipal water supply. At present, recreational facilities have been developed and will develop further on the shores of the Těrlicko and Žermanice reservoirs. Due to a concentration of requirements for many purposes in this area, reservoirs cannot be operated giving priority to recreation even in the summer months. The resulting reliability of a recreation pool is,

Table 12.4 Requirements in WRS of the Odra River basin

Control point of WRS, River (Stream)		Time horizon			
		2000		2015	
	M_j	$W_{m,j}$	$C_{m,j}$	$W_{m,j}$	$C_{m,j}$
	$[m^3\ s^{-1}]$				
N. Heřmínovy, the Opava R.	0.57	–	–	0.8	–
Opava R., upstream of the Moravice R.	0.61	–	–	–	0.57
Opava R., mounth	1.46	0.10	0.48	0.14	0.68
Kružberk, the Moravice R.	0.34	–	2.89	–	3.78
Vlčovice, the Lubina Str.	0.11	0.24	–	0.24	–
Odra R., upstream of the Opava R.	0.49	–	1.08	–	1.15
Odra R., upstream of the Ostravice R.	2.21	–	1.46	–	1.47
Odra R., upstream of the Olše R.	3.46	2.5[y])	–	2.5[y])	–
Šance, Ostravice R.	0.30	–	1.6	–	1.6
Ostravice R., km 8.6	0.76	0.2	1.0	0.2	0.95
Raškovice, Morávka R.	0.16	–	0.6	–	0.6
Žermanice, Lučina R.	0.08	–	0.99	–	0.99
H. Lomná, Lomná S.	0.07	–	0.38	–	0.38
Třinec, Olše R.	0.32	0.68	–	0.68	–
Těrlicko, Stonávka S.	0.11	–	0.33	–	0.33
Dětmarovice, Olše R.	0.76	–	1.69	–	1.58
Olše R., mouth	0.97	–[z])	–	–[z])	–

[y]) a parameter indicating low flow augmentation; this flow is to be increased by 4.5 $m^3\ s^{-1}$ (2000) and 4.08 $m^3\ s^{-1}$ (2015) respectively due to waste water effluents.

[z]) the flows are increased by 0.72 $m^3\ s^{-1}$ by waste water effluents.

therefore, reduced, which must be regarded as the main impact of WRS on recreation.

The main energies of water quality management in the General Water Plan were focused on organic waste. In the Odra River basin, however, the salt concentration in some waste waters is enormous, due to industrial waste water and water pumped from mines. In WRS design this problem is treated preferentially by calculation the low flow augmentation for waste water dilution. Therefore, the possibility of regulating the flow of the Odra River by means of reservoirs, in the town of Ostrava, was investigated as a secondary goal, so as not to interfere with the primary goals – river flow regulation at other control points of the WRS. Water quality enhancement in the lower part of the Odra River can provide for future withdrawals of water for cooling purposes in industry.

The water *requirements* in the WRS have been collected, processed and grouped at control points of the WRS in the form of system inputs. The grid of control sites in the simulation model was determined in such a way as to specify the points of the main impact and main requirements and the points where it was possible to sum up these impacts. Some reservoir sites are also control points, if water is diverted from reservoirs or if water is withdrawn immediately downstream of the reservoir. The

Table 12.5 The stream gauging stations on the WRS of the Odra River basin

Station	River (Stream)	Station	River (Stream)
Spálov	Odra	Opava, Polní street	Opava
Studénka	Odra	Kružberk	Moravice
Petřvald	Lubina	Branka	Moravice
Svinov	Odra	Děhylov	Opava
N. Heřmínovy	Opava	Šance	Ostravice
Morávka, reservoir site	Morávka	Jablunkov	Olše
Sviadnov	Ostravice	D. Lomná	Lomná
Olešná, reservoir site	Olešná	Ropice	Olše
Žermanice, reservoir site	Lučina	Těrlicko	Stonávka
Ostrava l.	Ostravice	Věřňovice	Olše
Bohumín	Odra	Mikulovice	Bělá

water supply requirements were classified as withdrawals without consumption $W_{m,j}$, consumption $C_{m,j}$ and acceptable minimum flows MI_j in month m and site j (see section 6.3.6). Alternative values are listed in Table 12.4.

The basic objective of the WRS in the Odra River basin was the necessary reliability of drafts and river flow regulation for water supply. The deterministic simulation model was used for operation without water shortages with the observed input time series in the period 1931 – 1970. The reliability of this model was estimated at 98.7% (based on Chegodayev's method). This value was verified by a chance-constrained model of the sub-system, including the main Slezská Harta and Kružberk reservoirs and for the corresponding alternatives the reliability was approximately

Table 12.6 Reservoirs in the WRS of the Odra River basin

Reservoir, River (Stream)	Storage $[10^6\ m^3]$		Minimum reservoir release $[m^3\ s^{-1}]$
	Total	Active	
1 *Existing reservoirs*			
Kružberk, Moravice R.	35.5	24.0	0.74
Šance, Ostravice R.	54.2	45.8	0.30
Morávka, Morávka R.	10.1	4.4	0.11
Žermanice, Lučina R.	20.1	18.5	0.07
Těrlicko, Stonávka S.	24.4	22.0	0.10
2 *Reservoirs planned by 2000*			
H. Lomná, Lomná S.	16.1	15.6	0.07
Slezská Harta, Moravice R.	158.0	150.0	0.67
Vlčovice, Lubina S.	21.0	15.0	0.11
3 *Reservoirs after 2000*			
N. Heřmínovy, Opava R.	100.5	95.0	0.57
Spálov, Odra R.	121.0	119.6	0.13

97%. This value was acceptable for the objectives of the General Water Plan. The monthly time series in streamflow gauging stations were the basic hydrological input data. These data were corrected, adjusted and completed and in some cases reduced to uncontrolled flows using the values of water level variation, withdrawals and effluents. The time series with a shorter observation period than 1931 – 1970 were completed to this period. The models for the analysis of the WRS in the Odra River basin used the monthly flows at points listed in Table 12.5.

The uncontrolled monthly flows at control points of the WRS were defined as a linear function of the nearest gauging station flows, using the ratio of the mean annual flows.

The hydrological data for flood control design were focused on the sub-system of the Ostravice River basin, where the systems approach was necessary. The observed and standard design floods in the station Šance (Ostravice R.), Morávka (Morávka R.), and Sviadnov (Ostravice R.) were used as input data.

In the basin, 19 reservoir sites were examined for the purposes of the WRS. Five reservoirs have been constructed, three have been designed and are to be built by the year 2000 and two reservoirs after the year 2000 (see Table 12.6). The diversion channels in WRS are the Rašovický channel from the Rašovice weir to the Žermanice reservoir (see Fig. 12.3), the diversion channel from the Těrlicko reservoir to the industrial centre of Třinec and after the year 2000, the transbasin diversion channel from the Teplice reservoir to the Odra River. The function of further channels was incorporated by summing up the requirements at the control points of the WRS.

In the design two groups were considered, i.e., satisfying the demands on the WRS by: a) the water resources of the Odra River basin using the reservoirs and the diversion channels inside the basin, and b) by the co-operation of these reservoirs with the transbasin diversion from the Teplice reservoir in the Water Resources and Transportation System Danube – Odra – Labe.

Conclusions

According to the problematic balance between the demands and water resources, the optimization of water resources of the WRS in the Odra River basin was based on such methods of reservoir operation that would prove the feasibility of meeting the increase of requirements until the year 2000 and 2015.

The reservoirs on some rivers have been built and their hydrological potential is used and almost exhausted. The possibility of enlarging the system by further reservoir storage is limited, and by the year 2015, the utilization of all the suitable reservoir sites is assumed with the exception of the Spálov reservoir, the construction of which depends on the Water Resource and Transportation System Danube – Odra – Labe and not on the demands in the basin of the Odra River alone.

The optimization in the simulation models was focused on two groups of choices, i.e.:

a) groups with a water supply objective and other requirements satisfied by the reservoirs in the Odra River basin, and

b) groups with objectives satisfied by a combination of the reservoirs operating in the Odra River basin and the Water Resources and Transportation System Danube – Odra – Labe, or its first stage, the transbasin diversion from the Teplice reservoir on the Bečva River.

The optimization of the reservoir operation was based on the issues of the municipal water supply of the waterworks system "Northern Moravia" (that is to include the present sub-system of the Ostrava Area Water Works) and the sub-system of the

Kružberk and Slezská Harta reservoirs and their co-ordination with other WRS requirements.

Having found and verified such an operation we interrupted the optimization process as the feasibility proof of such operation was sufficient for the discriminating level of the General Water Plan and the accuracy of the input parameters.

The economic comparison of these two groups: a) and b) proved the greater effectiveness of the second group, b), which has better economic indices and a higher development flexibility surpassing the time horizon of the year 2015 and making the WRS more adaptable for further demands. However, this group of choices is related to the implementation of the Water Resources and Transportation System Danube – Odra – Labe, which will in future cover the whole territory of Czechoslovakia and form an international inland waterways system.

The examples given may seem to overestimate the importance of the input and output data and to underestimate the role of the inner structure of the simulation model and the computer process. This attitude conforms to the methods used before the computer era with better insight into computation, and an objection might be raised that in present simulation models the calculation procedure is not explicitly described in the application to the investigated WRS. However, the number of relationships and constraints that have to be considered has increased in such a way that a cybernetic and systems approach with simulation modelling is now necessary. The simulation procedures provide, in the mutual relationship and dependence, the basic water resources calculation (see Section 6.3.6) in the system, and they form the basis for the simulation SIM-WRS programming language (or other languages). Using such programs or simulation languages, the designer or planner of the WRS (with the exception of programmers that try to improve the inner structure of computation) are not interested in the inner structure and use the cybernetic "black box" approach to study the system response to the set of system inputs by the simulation model.

An explicit (step-by-step) description of computation, including the inner structure, is only possible for simple examples (see Chapter 4). Such a description is not possible for large systems (e.g., with three or more reservoirs and say 10 sites) and in a large scale WRS, and the purpose of such description would be questionable as it would be retrogressive compared to the systems approach method used for WRS analysis on different discriminating levels.

13 PROSPECTS OF WATER RESOURCE SYSTEMS ANALYSIS AND SYNTHESIS

13.1 GENERAL ASSUMPTIONS OF THE SYSTEMS APPROACH IN WATER SCIENCE

The consequence of the economic, social and cultural development of mankind is the growth of requirements for the quantity and quality of water resources, their more intensive and comprehensive use. The unequal distribution of water resources capacity and water demands in time and place will require co-ordination of water management over ever increasing territories with a growing number of relationships between resources and users. The conflicts between different objectives of water resource development are likely to sharpen with the growing imbalance in supply and demand integration. The increasing dynamic rates of technological and economic changes will accelerate the dynamics of water management. The impacts of anthropological factors on the hydrological cycle, on quantitative and qualitative movement of water in nature, i.e., the feedback relationship between man and water resources will grow stronger. All these phenomena have a probabilistic nature, and their future behaviour cannot be predicted accurately.

The future development of water resource management, as outlined, shows that treatment of all its problems will require a dialectic approach that is, and will be, performed by the systems approach with the use of stochastic, heuristic and prognostic methods.

WRS design and operation will become a more comprehensive and more frequently used tool of water resources development studies. In spite of progress in science and technology and their working methods, it cannot be assumed that the WRS problems will be treated in all their complexity as they occur in nature and society. A purposeful simplifying abstraction will be necessary in the identification of the systems and their modelling in order to make them mathematically tractable and to offer results reflecting the real problems in such a way as to aid decision-making, design, construction and the operation of WRS.

Direct experiment with WRS, as compared with other systems, cannot be undertaken in view of high costs, operational and environmental consequences, the long-term function of its components, etc. These circumstances influence the prospects of the application of optimizing techniques of system modelling of this kind. Therefore, modelling of the optimal operation of WRS tends towards the application of mathematical models. The development of mathematical models of WRS is the main direction of scientific research of WRS.

The systems identified with complex and large-scale WRS are open systems – they have both inputs and outputs.

Hydrological and water quality data are among the most important input values. Therefore, systems hydrology should be developed to offer input variables for WRS, i.e., data with probability description and invariant characteristics in the future, which can be used to reflect the anthropological factors impact on uncontrolled hydrological conditions. Hydrology, too, is to be orientated prognostically rather than statistically as it has been up to now. Therefore, more attention should be given to the basic relationships in the mass balance equation, to the relationships between surface and ground water, evapotranspiration and changes in the human impact on the landscape and impact on the generic elements of the hydrological regime. The co-operation of hydrologists in research and design teams then becomes essential.

We believe that water quality requirements on WRS will grow faster than those of water quantity. The change of water quality, particularly in recent years, is a more dynamic element in WRS than its quantity. Despite measures for water pollution prevention the optimistic predictions have a relatively low reliability.

During recent years, sources of pollution that had been neglected grow in importance. It is difficult to control the point source of pollution and the sources of non-point pollution grow exponentially. The prospect of future development and search for effective technical and administration measures for water quality management are very important activities for the solution of WRS issues. In these considerations the problem of reliability cannot be overlooked; at present potential sources of pollution of enormous importance are emerging. Therefore, the point is not only operational reliability, but also reliability related to the risk of failure of WRS functions concerning water quality.

The boundary problems between the water quantity and water quality are formed by the quantity – quality relationships in water resources facilities. A start has been made in dealing with these relationships. Their importance is essential for WRS, and, therefore, research in the laboratory and under field conditions has to be accelerated.

All the problems discussed require an enormous research capacity, which cannot be carried by one country alone. International co-operation is necessary, and this is also the case for other WRS issues. This co-operation has been developed in the CMEA countries and also on a broader scale (IIASA, IWRA, etc.).

There are a number of methodological questions that have not yet been answered effectively. These questions concern the decision theory, the application of which is to be used for scientifically approved optimal decisions. The currently used optimization methods are often too "weak" for decision-making in complex WRS.

A further development of the theory of games and heuristic methods and their application will be necessary.

An important aspect of decision analysis is the reliability of systems. In spite of the fact that reliability is a classical notion in water management, its importance in application to WRS has often been reduced and simplified. A theory of risk and reliability, taking into account many aspects of WRS, has to be developed.

We are persuaded that a single algorithmic activity in WRS is often cumbersome and cannot hit the target. In this situation it is sometimes necessary to deviate from the algorithmic experience and use a shorter relationship between intuition and decision. It will be necessary to analyse methodologically the fact that man deals with most of his problems heuristically, i.e., he uses a method based on the development of inner models that are orientated towards a single clear purpose without superfluous complexity. Some indisputable advantages of heuristic methods, distinguished by creativity, are a sufficient impetus for more attention despite resistance on the part of certain "experienced" experts, who are not able to change their views due to accumulated, but one-sided experience. Correct, true and extensive is a component of qualification and it is indispensable in the verification of modern ideas.

A surviving phenomenon of skilled, but conservative activity, based on past experience, is the methodological approach based entirely on the past and frightened of uncertainty in the future. However, in a society which builds up its future on a scientifically elaborated plan, the necessity to create an image of the future world is essential, and prognostic rather than static aspects need to penetrate into all theories.

In WRS the prediction of the future state of water resources is an integral component of their design and operation. The extrapolation methods are very useful for prediction and often are, and will be, applied in water management, whatever other methods follow, which might reduce the danger of false trends and sources of errors inherent in it.

Further, in water management prediction it is often necessary to abandon mathematical and exact procedures, which do not entail the anticipation of a new mathematically unexpected reality – a discovery. Subjective methods involving imagination and brainstorming of properly chosen experts should be applied here.

WRS analysis and design is creative work. The interdisciplinary and large-scale character and novelty of the topic requires investigation and research by broad teams. Scientific knowledge should be used to set up creative teams, their style of work, with the right to use new approaches, including the risk of failure, the psychology of research teams, and the right to overcome administrative obstacles that hinder team work in WRS.

The most difficult prediction problems include the prediction of the future demands of WRS, as water management mostly serves other branches of the national economy and the development of society. Flood control and the municipal water supply are the most reliable components of this prediction. The water management administration has to intervene in all components of its system environment to gain relatively reliable predictions of their requirements on WRS dozens of years ahead of time.

A necessary corollary is planning of budgetary and construction capacities to meet these requirements in place and time.

All these aspects of water management development have been concentrated in the General Water Plan of Czechoslovakia. Subsequent planning in economic and construction studies should incorporate the relatively fixed time schedules of WRS development. Unless this condition is met, WRS development will not be rational and the best WRS design and operation will not be able to attain the maximum possible benefits.

High initial costs and a long physical and economic life are a special feature of water management which has serious economic, environmental, and other impacts. The effort to express certain environmental effects and intangibles in economic terms can add homogeneity to comparisons of studies and alternative designs. The lack of reliable indices of economic effectivity in water management and, on the other hand, poor knowledge of loss functions due to water shortages are the weakest points of optimization computations. The organization of comprehensive collection and processing of various data concerning WRS operation is an integral part and assumption of a reliable economic basis of WRS and the whole water management sector.

Modelling WRS we are interested, in the first place, in the behaviour of the identified system; in the cybernetic models the structure of the system need not be the central point of our interest, bearing in mind the fact that WRS are predominantly dynamic, continuous, non-linear, multi-dimensional and stochastic systems.

In model design two different groups can be chosen:

a) methods of operation research,

b) cybernetic methods.

The difference between them lies in the complexity and probability character of the modelled system. Up until now, we have not been able to cross the borders between operation research methods and cybernetics, especially in practical applications.

Optimization of WRS functions, in theoretical studies and practical case studies, used, at best, methods of operation research with varying success and mathematical tractability. Some methods have failed even in dealing with relatively simple and not extended WRS with single (or a dominating) objective, as the model was not mathematically tractable.

Recently, some attempts have succeeded in using quite a new approach on a cybernetic basis. It can be stated that solution of some theoretical problems prepared prerequisites for the further development of these methods.

A promising optimizing method of operation research, suitable for less complex, but a stochastic systems, that can be described, e.g., statistically, is dynamic programming with some of its modifications, e.g., DDDP or incremental dynamic programming. All models of WRS using an optimization of this kind are, however, simpler than the actual WRS.

In principle, there are three possibilities of evolving a suitable method:

– simplification of the actual system in the model,

– decomposition of the system into several subsystems and their separate treatment,

– application of cybernetic methods.

In using the first two possibilities there is often a danger of oversimplification. The application of cybernetic methods seems to be effective if the system identified with the WRS is, in principle, cybernetic. The cybernetic adaptive control system can maintain its behaviour in the given acceptable limits and adapt to conditions that will occur in future.

13.2 CHARACTERISTIC PROPERTIES OF CYBERNETIC ADAPTIVE SYSTEMS

The processes in the operation of WRS have an adaptive character. Their properties are not Markovian, i.e., their state at a given moment does not depend on the states in a limited and constant number of previous time points, but on the whole history of the process. In the present methods of operation research applied in the reservoir systems operation and in the attempts of their use for dealing with the problems of WRS, the Markovian character of the process has mostly been assumed (stochastic methods of Markovian optimization). The most effective self-organizing adaptive control model is an ultrastable model, i.e., its organization tends to return, from each state and under all conditions, to the stable state. Its ultrastable feature is in contradiction to the Markovian character of the process.

The adaptivity of a system is given by maintaining the important variables within the required limits. The mechanism of this maintenance is automatic, and the system has an ability to adapt to any changes in its environment that might be the consequence of the impact of certain factors.

In principle, the variables and their feasible states, given within a certain range, must be determined in advance. In multi-purpose WRS these variables may include technological and economic values.

A precise definition of the boundary between the system and its environment, with stochastic impacts, is desirable.

The system is often decomposed into subsystems using the functional aspects, in fact, system behaviour aspects. We are primarily interested in information relationships, which may have an impact on the system variety reduction. Therefore, in the multi-purpose WRS, e.g., the geographical standpoint used for the identification of the system structure, must be regarded as less important.

The identified subsystems have to be connected by information relationships so that the output of one subsystem should be the input of a second one and vice-versa, developing in this way a feedback relationship. In such a manner the boundary subsystems and the system environment are interconnected.

The adaptive process takes place separately in the individual subsystems. Their states change as long as the stable state is reached when the principal variables are within the required limits. Each subsystem has its stability objectives, which can be mutually exclusive. The stabilization of one subsystem in a stable state can cause deviations of outputs of some other subsystems, which are influenced in their inputs. Therefore each subsystem can ignore the decision of any other subsystem, which would otherwise reach its stable state. It is necessary to solve this rather complex problem if an ultrastable system is to be reached, the system that searches automatically for ultrastability and tends, as a complex system, to a stable state for all stimuli that can enter on its input. The aim is an acceptable combination of given values (or ranges of values) of principal variables in the individual subsystems.

In Markovian processes, the current state depends only on the previous state or on several previous states and in controlled processes on the decision that was made immediately before the process transition to the current state, i.e.

$$S_k = S_k(S_{k-1}; S_{k-2}; \ldots, S_{k-s}; R_k) \tag{13.1}$$

where $S_k, S_{k-1}, \ldots, S_{k-s}$ are the states of the process in discrete times, R_k is the control that occured between defined instants or in a discrete instant k.

Unlike Markovian processes, in the adaptive process the state S_k depends on all possible information before the instant k, which is called historical information H_k.

The system, the behaviour of which is identified by this adaptive process, records in its "memory" all the couples of output responses and corresponding input impulses (i.e., all historical information). The effectiveness of these responses is evaluated, and the results of this evaluation are also recorded in its "memory". Such system "knows" which set of responses $\bar{R}$ corresponds best to the set of stimuli $\bar{P}$ as the set $\bar{R}$ has been evaluated as the most successful under given conditions of the process. This learning process of the adaptive control system is generally stochastic and non-Markovian. In modelling of WRS the transitive probabilities of moving the system to any feasible subsequent state can usually be determined beforehand in relation to the sequence of all previous stages, satisfying in this way the assumption of practical implementation of the model of such a system. The principle of development of the adaptive control model are, in general, very simple.

The way of learning in cybernetic adaptive systems with a non-Markovian character is a progressive form of learning as compared with the evolutionary process that generally occurs in nature and that is Markovian. The latter process is often denoted as a cumbersome learning process with relatively slow adaptation. An adaptive system identified with WRS has to control relatively fast changes and has to learn quickly from the historical information to identify and specify the purposeful behaviour, i.e., homeostatic ultrastability. It is not the process of survival typical for certain biological species. The evolutionary process is controlled only by the feedback loops (the successful mutations are strengthened by a positive feedback, unsuccessful ones are suppressed). The changes of environment which form the stimuli for the state change, often random, occur very slowly. Therefore this learning process is cumbersome and gradual.

Variety of the system is one characteristic of its complexity. The multi-purpose WRS are complex systems with a great variety of disturbing elements that affect them. In managing the system we try to control these disturbing effects and decrease their variety; decreasing variety is one possible way of system control.

In investigating the system functions we try to determine some relationships occurring in the set of disturbing effects, to attain the invariants of this set and in this way to reduce its actual variety; this can be achieved by adding further information.

The system, whose environment is thus investigated, cannot thus be simplified, but the conditions for better control are formed in this way as each step in the determination of relationships in the set of disturbing elements helps in forecasting the behaviour of the system.

It is necessary to distinguish the cybernetic adaptive systems and the pure stochastic systems (Aoki, 1971) with unique probability distribution functions of random parameters (or at least some statistical moments). Sometimes they are called systems with maximum incomplete, i.e., stochastic information. In the cybernetic adaptive systems a part of the basic stochastic information is missing and the probability distribution functions of some stochastic parameters depend on certain additional unknown random variables.

Modelling of WRS as a cybernetic adaptive control system will proceed along the following steps:

- definition of the cybernetic adaptive control system,
- description of the behaviour of the cybernetic adaptive control system,
- choice of the model with an adaptive character.

A definition of a system of this kind for WRS is not difficult in principle. It is only an unusual information standpoint in an investigation into relationships.

A detailed mathematical description of such system is not possible nor necessary in the investigation of its behaviour. In addition, if the cybernetic adaptive control system and its model are properly realized with ultrastability, they are able to give precision to their description during their functioning. A precise initial description is, therefore, not necessary.

The model is a mathematical and logical abstraction and its behaviour can be investigated so that it is realized:

– on a single-purpose facility, which is created for one task or for a group of tasks,

- on an analog computer (continuous modelling),
- on a digital computer (discrete modelling),
- on a hybrid computer system.

The application of the analog computers seems to be the most natural way for the first experiments with modelling of such cybernetic adaptive systems, although only certain problems with relative low accuracy can be dealt with in this manner.

In the preliminary stage their application is cheapest and computation is most effective. Digital computers seem to be the basic computer tool for the concrete tasks of modelling, while hybrid computer systems put the system into perspective.

The development of adaptive models of WRS can be performed in principle. Cybernetic adaptive models have been successfully developed in other technical branches.

13.3 CONCLUSION

Taking the basic assumption of dealing successfully with water resources problems, i.e., in the analysis of technological, economic and social impacts, much has been accomplished that facilitates the formulation of adequately simplified tasks and finding a form that is mathematically tractable under present conditions.

The development of new methodological approaches with corresponding means (technology) are so dynamic that there is a real hope for quick improvement in dealing with ever more complex task formulations, i.e., closer to reality. A mere extrapolation of the present achievements cannot meet our requirements. Futurological studies are hypotheses that certainly cannot be proved now. Without a doubt all over the world water management, water conservation, and the prevention of water resources pollution are the necessary prerequisites for the further development of mankind.

It is necessary to persuade the political and administrative decision-makers that water is irreplaceable in many functions and that there is a great danger arising from the limitation and pollution of water resources.

Not only in the minds of water resources engineers and planners, but public opinion also has to made aware of the idea of water as a renewable but exhaustible resource. Only in this way can the application of the results of scientific and technological progress in water resources management and engineering and in water resources systems be achieved.

REFERENCES AND BIBLIOGRAPHY [1])

Abramowitz, M. (ed.): Handbook of Mathematical Functions. Nat. Bureau of Standards, 1964.

Ackoff, L. L.: General Systems Theory and Systems Research – Contrasting Conceptions of Systems Science. *General Systems*, **VIII**, 1963, pp. 117–121.

Akmanoglu, N. O. and Akin, U.: A Simulation Study for the Maximization of Hydroelectric Energy in the Keban Reservoir. Proc. IAHR, 4, 1973, pp. 221–226.

Alekseyev, G. A.: Graphical and Analytical Methods of Determination of Sample and Population Parameters of Distribution Functions. Trudy GGI, 73, Gidrometizdat, Leningrad, 1960 (in Russian).

Alexy, J.: Operations Analysis in Business Practice. SVTL, Bratislava, 1967, p. 221 (in Czech and Slovak).

Amstadter, B. L.: Reliability Mathematics: Fundamentals; Practices; Procedures. McGraw-Hill, New York, 1971.

Anděl, J. and Balek, J.: Mathematical-Statistical Analysis of Hydrologic Series Generation. *Vodohosp. čas.*, 1, 1970, pp. 3–28 (in Czech).

Aoki, M.: Optimization of Stochastic Systems. Nauka, Moscow, 1950, p. 424 (in Russian).

Ashby, W. R.: An Introduction to Cybernetics. Chapman and Hall, London, 1956.

Babič, I.: On Modelling of Water Management System for Long-Term Predictions. *Vodní hospodářství*, A, 12, 1972, pp. 305–308 (in Czech).

Babič, I.: An Attempt to Define the System of Water Resources. Proc. WRS, Vol. II, Karlovy Vary, 1972, pp. 42–48 (in Czech).

Babič, I.: Use of Mathematical Models for Water Management Planning and Control. *Vodní hospodářství*, A, 6, 1975, pp. 146–150 (in Czech).

Babič, I.: Long-Range Planning Methods of Development and Protection of Water Resources. Proc. IWRA, Vol. II, 1975, pp. 231–236.

Bailey, N. T. J.: On Queueing Processes with Bulk Service. *J. Roy. Statist. Soc.*, B 16, 1954, p. 8.

Bajard, Y. G.: A Description of a Conceptual Model for a Quantitative Simulation of Water Behaviour in a Region (Runoff, Surface Storage, Ground Water and Water Use). Proc. IWRA, Vol. IV, 1975, pp. 97–110.

Bakayev, A. A., Kostina L. I., and Yarovicky, N. V.: Automated Models of Economic Systems. Naukova Dumka, Kiev, 1970, p. 190 (in Russian).

Balek, J. *et al.*: Mathematical Modelling of Regional Flow Genesis and Hydrological and Meteorological Stochastic Processes. Final Rep. II. 7-3/13, Czechoslovak Academy of Sciences, Inst. for Hydrodynamics, Prague, 1975, p. 57 (in Czech).

Balkov, A. M.: Generalization of the Method of Form-Transforming Filters to Non-Stationary Random Processes. *Avtomatika i telemechanika*, **XX**, 8, 1959 (in Russian).

Banayoun, R. and Tergny, J.: Mathematical Programming with Multiobjective Functions: A Solution by P.O.P. *Metra*, 2, 1970.

---

[1]) In the references and bibliography the following abbreviations were used:

WRS – Water Resource System or Water Resource Systems

Proc. IWRA, 1975 – Proceedings of the Second World Congress on Resources "Water for Human Needs", IWRA, New Delhi 1975

Proc. IAHR, 4, 1973 – Proceedings – Research and Practice in the Water Environments, 15th Congress of the IAHR, Vol. 4, Istanbul, 1973

Proc. WRS, Karlovy Vary, 1972 – Proceedings of the 1st Symposium "Water Resources Systems", Karlovy Vary, 1972

Proc. WRS, Hradec Králové, 1976 – Proceedings of the 2nd Symposium "Water Resources Systems", Hradec Králové, 1976

Barbour, E.: Multiobjective Water Resources Planning. Proc. IWRA, Vol. II, 1975, pp. 203–214.

Bašta, B.: A Study of the Comprehensive Development of the Sava River Basin in Yugoslavia. Proc. WRS, Vol. II, Karlovy Vary, 1972, p. 49 (in Czech).

Bašta, B.: Planning Decision Processes and their System. Academia, Prague, 1977, p. 228 (in Czech).

Beard, L. R.: Models for Optimizing the Multipurpose Operation of a Reservoir System. Symposium and Workshops on the Application of Mathematical Models in Hydrology and Water Resources Systems, Bratislava, 1975, p. 9.

Beard, L. R.: Status of Water Resources Analysis. *Journ. of the Hydraulic Div.*, Proc. ASCE 99, HY 4, 1973, pp. 559–565.

Becker, A., Krippendorf, H., and Sosnowski, P.: Real Time and Long-Term Aspects in Operating Multipurpose Reservoirs, Workshop IIASA/IMGW. Operation of Multiple Reservoir Systems, Poland, Jodlowy Dwor, 1979, p. 21.

Becker, L. and Yeh, W. W.: Optimization of Real Time Operation of Multiple-Reservoir System, *Water Resources Research*, 6, 1974, pp. 1107–1112.

Bečvář, V.: Water Management System Comprehensive Evaluation. Water Research Institute, SZN, Prague, 1986, p. 110 (in Czech).

Belayev, L. S. and Churkveidse, S. S.: On Optimization Methods of Flow Regulation by Hydroelectric Stations Cascades Working in Power and Water Resources Systems. Proc. IAHR, 4, 1973, pp. 181–188.

Belenson, S. M. and Kapur, K. C.: An Algorithm for Solving Multicriterion Linear Programming Problems. *Operational Research Quarterly*, 1, 1973.

Bellman, R.: Adaptive Control Processes: A Guided Tour. Princeton University Press, Princeton, New Yersey, 1961, p. 343.

Bellman, R.: Dynamic Programming. Princeton University Press, Princeton, New Yersey, 1957.

Bellman, R.: Mathematical Optimization Techniques. Princeton University Press, Princeton, New Yersey, 1963, p. 291.

Bellman, R. and Dreyfus, S. E.: Applied Dynamic Programming. Princeton University Press, Princeton, New Yersey, 1962, p. 423.

Bellman, R. and Kalaba, R.: Dynamic Programming and Modern Control Theory. Academic Press, New York – London, 1965, p. 111.

Belonogov, G. G. and Bogatyrev, V. I.: Automated Information Systems. Sovetskoye radio, Moscow, 1973, p. 328 (in Russian).

Beneš, J.: Statistical Dynamics of Control Circuits. SNTL, Prague, 1961, p. 336 (in Czech).

Beneš, J.: Cybernetic Systems with Automatic Organization. Academia, Prague, 1966, p. 184 (in Czech).

Beneš, J.: Perspectives of a New Branch: Theory of Large-Scale Systems Control. *Acta Polytechnica*, Technical University, Prague, **III**, 2, 1967, p. 5 (in Czech).

Beneš, J.: Stochastic Methods of Control Theory. Technical University, Prague, 1970, p. 200 (in Czech).

Beneš, J.: Systems Theory (Complex Control). Academia, Prague, 1974, p. 200 (in Czech).

Bennis, W. G.: Towards a True Scientific Management: The Concept of Organization Health. *General Systems*, **VII**, 1962, pp. 269–282.

Bertalanffy, L. von: Modern Theories of Development. An Introduction to Theoretical Biology. Oxford University Press, Oxford, 1934, Harper Torchbooks, New York, 1962.

Bertalanffy, L. von: The Theory of Open Systems in Physics and Biology, *Science*, 3, 1950.

Bertalanffy, L. von: Biophysik des Fliessgleichgewichts (Einführung in die Physik offener Systeme und ihre Anwendung in der Biologie). Vieweg, Braunschweig, 1953.

Bertalanffy, L. von: General Systems Theory. *General Systems*, **I**, 1956, pp. 1–10.

Bertalanffy, L. von: General Systems Theory: Foundations, Development, Applications. George Braziller, New York, 1968.

Berztiss, A. T.: Data Structure, Theory and Practice. Univ. of Pittsburg, Academia Press, New York, 1971.

Bestchinsky, A. A. and Reznikovsky, A. Ch.: Control and Management of Multipurpose Utilization of

Water Resources and Development of Water Economy and Hydropower. Proc. IWRA, Vol. IV, 1975, pp. 261 – 264.

Biswas, A. K.: Systems Analysis for Water Management in Developing Countries. Proc. IWRA, Vol. IV, 1975, pp. 39 – 45.

Blackwell, D. and Girshick, M. A.: Theory of Games and Statistical Decisions, Wiley, New York, 1954.

Blažej, Z.: Quantification of the Time Factor in the Branch of Water Management. Proc. WRS, Vol. II, Karlovy Vary, 1972, pp. 162 – 165 (in Czech).

Blažek, V.: Water Utilization Model in the Sáva River Basin. Proc. WRS, Vol. II, Karlovy Vary, 1972, pp. 59 – 66 (in Czech).

Blažek, V.: Economic Data for the Solution of Multi-Purpose Water Resources Systems. Proc. WRS, Vol. II, Karlovy Vary, 1972, pp. 166 – 167 (in Czech).

Blažek, V.: System Approach to Water Resources Systems. *Vodní hospodářství*, A, 5, 1975, pp. 119 – 123 (in Czech).

Blažek, V., Vergner, Z., and Záruba, L.: Presumption Analysis of the Construction in Stages of the Channel Danube – Odra – Labe. *Vodní hospodářství*, A, 6, 1972, pp. 135 – 142 (in Czech).

Bogárdi, I.: Uncertainty in Water Decision-Making. Proc. UN Seminar: River Basin and Interbasin Development, Vol. I, Budapest, 1975, New York – Budapest, pp. 188 – 197.

Bogárdi, I. and Dávid, L.: Hydroeconomic Analysis of a Multipurpose Water Management System. Proc. WRS, Vol. II, Hradec Králové, 1976, pp. 40 – 56.

Boháč, M.: New Programme of Water Management Development in Czechoslovakia. Bulletin of Water Management, 3, CMEA, Moscow, 1968 (in Russian).

Boháč, M. and Krajčí, J.: First Draft of the General Water Plan of ČSR and SSR. *Vodní hospodářství*, A, 1, 1973, pp. 1 – 3 (in Czech).

Bohun, V. and Vischer, D. L.: Risk Analysis – Its Use for Evaluation of Marginal Projects and Sensitivity of Results to Different Probability Distributions of Input Data. Proc. IAHR, 4, 1973, pp. 241 – 248.

Boulding, K.: General System Theory – The Skeleton of Science. *General Systems*, **I**, 1956, pp. 11 – 17.

Bovbel, E. I., Daneyko, I. I., and Izok, V. V.: Elements of Information Theory. BGU, Minsk, 1974, p. 112 (in Russian).

Buras, N.: A Three-Dimensional Optimization Problem in Water Resources Engineering. *Operations Research Quarterly*, **16**, 4, 1965, p. 10.

Buras, N. : Scientific Allocation of Water Resources. Am. Elsevier Publ. Co., New York, 1972, p. 208.

Busacker, R. G. and Saaty, T. L.: Finite Graphs and Networks. McGraw-Hill, New York, 1968.

Buslenko, N. P. and Shreyder, Yu. A.: Statistical Computation Methods (Monte Carlo) and their Use on Digital Computers. Fizmatiz, Moscow, 1961 (in Russian).

Calabro, S. R.: Reliability Principles and Practices. McGraw-Hill, New York, 1962.

Cannon, W. B.: Organization for Physiological Homeostasis. *Physiol. Rev.*, 9, 1929.

Cavadias, G. S.: River Flow as a Stochastic Process. Statistical Methods in Hydrology. Gidrometizdat, Leningrad, 1970, pp. 214 – 240 (in Russian).

Cembrowicz, R. G.: Determination of Design Periods for Water Supply Systems. Proc. IAHR, 4, 1973, pp. 249 – 257.

Central Data Base of Technical Data System. UVTEI, Prague, 1974 (in Czech).

Chaemsaithong, K., Duckstein, L., and Kisiel, C.: Hydrologic and Social Inputs into a Cost-Effective Design of a Water Resources System. Proc. IAHR, 4, 1973, pp. 233 – 240.

Chandor, A., Graham, J., and Wiliamson, R.: Practical Systems Analysis. R. Hart-Davis. Educ. Publ. Ltd., London, 1972.

Chang, T. P. and Toebes, T. H.: Iterative Simulation Algorithm in Reservoir Systems Operation. Proc. IAHR, 4, 1973, pp. 97 – 106.

Chaturvedi, M. C.: Planning Methodology and First Stage Model of a Large River Basin. Proc. IAHR, 4, 1973, pp. 165 – 174.

Chaturvedi, M. C. and Rogers, P.: Large-Scale Water Resources Systems Planning with Reference to Ganges Basin. Proc. IWRA, Vol. II, 1975, pp. 283–297.

Cheland, D. J. and King, W. R.: System Analysis and Project Management. McGraw-Hill, New York, 1968.

Chernoff, H. and Moses, L. E.: Elementary Decision Theory. John Wiley, New York, 1959, p. 364.

Chervony, A. A., Lukyashenko, V. I., and Kotin, L. V.: Reliability of Complex Systems. Mashinostroyeniye, Moscow, 1976, p. 286 (in Russian).

Chilingaryan, L. A. and Mkhitaryan, S. A.: Application of the Mathematical Modelling to a Complex Water Economy System. Proc. IAHR, 4, 1973, p. 1–8.

Chin-Liu and Tedrow, A. C.: Multilake River System Operation Rules. *Journ. of the Hydr. Div.*, Proc. ASCE, HY 9, 1973, pp. 1369–1381.

Chorafas, D. N.: Statistical Processes and Reliability Engineering. D. van Nostrand Co., Princeton, 1960, p. 438.

Chorafas, D. N.: Systems and Simulation. Academic Press, New York, 1965.

Chow, Ven-Te *et al.*: Handbook of Applied Hydrology. McGraw-Hill, New York, London, 1964, p. 1418.

Chow, Ven-Te, Maidment, D. R., and Tauxe, G. W.: Computer Time and Memory Requirements for DP and DDDP in Water Resources Systems Analysis. *Water Resources Research*, **11**, 5, 1975, pp. 621–628.

Chowadze, F. I. *et al.*: Analysis of Factors Causing Failing and Catastrophies of Dams – Evaluation of Dam Reliability Indices. *Gidrot. stroyitelstvo*, 7, 1980, pp. 34–38 (in Russian).

Chrtek, M. and Paul, J.: Inland Waterway in the Middle Reaches of the Labe River. *Vodní hospodářství*, A, 5, 1973, pp. 115–120 (in Czech).

Churchman, C. W.: Prediction and Optimal Decision: Philosophical Issues of Science of Values. Prentice Hall Inc., New Yersey, 1961.

Churchman, C. W.: An Approach to General Systems Theory. Views on General Systems Theory. Proc. of the Second Systems Symposium at Case Inst. of Technology 1963, Wiley, New York, 1964, pp. 173–175.

Churchman, C. W.: The Systems Approach. Delacorte Press, New York, 1968.

Churchman, C. W., Ackoff, R. L., and Arnoff, E. L.: Introduction to Operations Research. Wiley, New York, 1957.

Chvátal, V.: Hypergraphs and Ramseyian Theorems. Proc. Amer. Math. Soc., **27**, 1971, pp. 434–440.

Chvátal, V., Erdös, P., and Hedrlín, Z.: A Lower Bound for the Capacity of a Graph. *J. Comb. Th. B.*, **13**, 1973, pp. 200–202.

Chvátal, V. and Hammer, P. L.: Aggregation of Inequalities in Integer Programming. *Ann. Diser. Math.*, **1**, 1977, pp. 145–162.

Cicchetti, C. J. *et al.*: Recreation Benefits Estimation and Forecasting: Implication of the Identification Problem. *Water Resources Research*, **8**, 4, 1972, pp. 840–850.

Cigánik, M.: Information Systems in Science, Technology and Economics. Alfa, Bratislava, 1973, p. 442 (in Slovak).

Cohon, J. L.: Applications of Multiple Objectives to Water Resources Problems, XXIInd International Meeting of the Institute of Management Sciences, Kyoto, July 1975.

Cohon, J. L. and Marks, D. H.: Multiobjective Screening Models and Water Resource Investment. *Water Resources Research*, **9**, 4, 1973.

Cohon, J. L. and Marks, D. H.: A Review and Evaluation of Multiobjective Programming Techniques. *Water Resources Research*, 2, 1975, pp. 208–220.

Comprehensive Water Resources Operation – 1st Stage. VRV, Prague, 1972, p. 103 (in Czech).

Coutagne, A.: Méthodes pour déterminer débit de crue maximum qu'il est possible de prévoir pour un barrage et pour lequel barrage doit être établi. Fourth Congress on Large Dams, Vol. II, New Delhi, 1951, pp. 687–712.

Croley, T. E.: Sequential Stochastic Optimization for Reservoir Systems. *Journ. of the Hydr. Div.*, Proc. ASCE, 1, 1974, p. 201 – 219.

Cvetkov, E. V.: Probability Methods for Optimal Policy Determination for Power Systems with Water Power Plants and Carry-Over Reservoirs. Trudy Vsesoyuz. Nauch.-iss. inst. elektroenergetiki, **13**, Gosenergoizdat, 1961, p. 41 (in Russian).

Cvetkov, E. V., Reznikovsky, A. Sh., Korobova, D. N., and Ismailov, G. H.: The Methods of Complex Reservoirs Cascades Control and the Experience of their Use in the USSR. Workshop IIASA/IMGW, Operation of Multiple Reservoir Systems, Poland, Jodlowy Dwor, 1979, p. 23.

Curry, G. L., Helm, J. C., and Clark, R. A.: Chance-Constrained Model of Systems of Reservoirs. *Journ. of the Hydr. Div.*, Proc. ASCE, **99**, HY 12, 1973, pp. 2353 – 2356.

Čábelka, J.: Water Resources and Transportation System Danube – Odra – Labe and its Relation to the Ostrava Region. *Vodní hospodářství*, 5, 1969, pp. 139 – 144 (in Czech).

Čábelka, J.: The Inland Waterways and Water Transport in Czechoslovakia – The Present State and Development Programme. *Vodní hospodářství*, A, 5, 1975, pp. 107 – 114 (in Czech).

Čermák, L., Geier, J., Hep, L., and Kvasnička, J.: Computation of the Problems of the Ostrava – Karviná Region in Water Management by Computers. *Vodní hospodářství*, A, 11, 1971, pp. 294 – 296 (in Czech).

ČSN – Czechoslovak National Standards: 01 1001 Mathematical Symbols, 1961; 36 9001 Terminology of Digital and Analog Computers, 1974; 36 9030 Outlines of Flowcharts for Information and Data Processing Systems, 1973 (in Czech).

Danskin, J. M.: The Theory of Max-Min. Springer, New York, 1967, p. 127.

Dantzig, G. B.: Linear Programming and Extensions. Princeton University Press, Princeton, New Yersey, 1963, p. 478.

Dávid, L. and Duckstein, L.: Long Range Planning of Water Resources: A Multiobjective Approach. Proc. UN Seminar: River Basin and Interbasin Development, Vol. I, Budapest, 1975, New York – Budapest, 1976, pp. 160 – 173.

Davis, D. R.: Decision Making Under Uncertainty in Systems Hydrology. Tech. Rep. 2, Dept. of Hydrol. and Water Resources, Univ. of Arizona, Tuscon, 1971, pp. 1 – 179.

Davis, D. R., Kisiel, Ch. C., and Duckstein, L.: Bayesian Decision Applied to Design in Hydrology. *Water Resources Research*, 1, 1972, pp. 33 – 41.

Dee, N., Baker, J., Drobny, N., and Duke, K.: An Environmental Evaluation System for Water Resources Planning. *Water Resources Research*, 3, 1973, p. 525 – 535.

Delleur, J. W. and Rao, R. A.: Linear Systems Analysis in Hydrology – The Transform Approach. The Kernel Oscillations and the Effect of Noise. Systems Approach to Hydrology. Proc. of the First Bilateral U.S. – Japan Seminar in Hydrology, W.R.P. Fort Collins, Colorado, 1971, pp. 116 – 142.

Demenchuk, V. M., Razumkin, B. S., and Dunin-Barkovsky, L. V.: Application of the Graph Theory to the Allocation of Water Resources. *Vodnyye Resursy*, 2, 1976, pp. 73 – 80 (in Russian).

Dimitrov, P., Ijjas, I., and Winter, J.: Out-of-Kilter Algorithm for a Land Drainage System. Proc. WRS, Vol. I, Hradec Královė, 1976, pp. 59 – 75.

Ditkin, V. A. and Kuznyecov, P. I.: Handbook of Calculus of Operations. Theoretical Basis, Tables of Operators. Academia, Prague, 1954, p. 338 (in Czech).

Djordjević, B. and Opricović, S.: Optimal Design of a Water Supply Reservoirs System, Proc. WRS, Vol. I, Karlovy Vary, 1972, pp. 208 – 219.

Djordjević, D., Opricović, S., and Simonović, S.: Optimal Long-Term Control of a Cascade of Multipurpose Reservoirs. Proc. WRS, Hradec Královė, 1976, p. 76 – 89.

Doležal, M.: Water Resources Systems Analysis in the General Water Plan. Proc. WRS, Karlovy Vary, 1972, p. 48 (in Czech).

Doležal, M.: Water Reservoirs Development as the Main Element of Water Resources Systems. Proc. WRS, Hradec Královė, 1976, pp. 90 – 104 (in Czech).

Dooge, J. C. I.: Mathematical Models of Hydrologic Systems. Int. Symp. on Modelling Techniques in Water Res. Systems, Proc., Vol. 1, Ottawa, 1972, pp. 171 – 189.

Dorfman, R.: Formal Models in the Design of Water Resources Systems. *Water Resources Research*, **1**, 3, 1965, pp. 329 – 336.

Downton, F.: A Note on Moran's Theory of Dams. *Quart. J. Math.*, Oxford, Ser. 2, **8**, 1957, p. 5.

Dráb, Z.: Basic Notions of Systems Engineering. *Slaboproudý obzor*, 7, 1964 (in Czech).

Dráb, Z.: What is System Engineering. *Hospodářské noviny*, 25, 1967, *Výběr informací*, 1969 (in Czech).

Dráb, Z.: Some Problems of Systems Engineering, 1 – 4, Technocentrum, Prague, 1969, p. 119 (in Czech).

Dráb, Z.: Systems Engineering – Problems and Possibilities. Proc. of SI 69, Vol. I, SAK, Prague, 1970 (in Czech).

Dráb, Z.: Introduction to Systems Engineering. SNTL, Prague, 1973, p. 316 (in Czech).

Dráb, Z. and Klímek, A.: Systems Engineering. Ústř. pro rozvoj automatizace a výp. techniky, Prague, 1969, p. 47 (in Czech).

Draškovič, D.: The Problems of Future Water Supplies for Domestic, Industrial and Agricultural Needs in the Morava River Watershed. Proc. IWRA, Vol. III, 1975, pp. 105 – 114.

Dub, O. and Němec, J.: Hydrology. Tech. Handbook 34, SNTL, Prague, 1969, p. 380 (in Czech).

Duckstein, L.: Decision-Making and Planning for River Basin Development: General Report. Proc. UN Seminar: River Basin and Interbasin Development, Vol. I, Budapest, 1975, New York – Budapest, 1976, pp. 111 – 122.

Dudley, N. J. and Burt, O. R.: Stochastic Reservoir Management and System Design for Irrigation. *Water Resources Research*, **9**, 3, 1973, pp. 507 – 522.

Dupač, V.: On Markov Chains. Sov. věda, matematika-fyzika, V, Prague, 1955, pp. 321 – 336 (in Czech).

Dupač, V. and Dupačová, J.: Markov Processes. SPN, Prague, 1975, p. 134 (in Czech).

Dupač, V. and Hájek, J.: Probability in Science and Technology, NČSAV, Prague, 1962 (in Czech), 1970 (in English).

Dupačová, J. and Kos, Z.: Chance-Constrained and Simulation Models of Water Resources Systems. *Ekonomicko-matematický obzor*, **15**, 2, 1979, pp. 178 – 191.

Dupačová, J., Gaivoronski, A., Kos Z., and Szántai, T.: Stochastic Programming in Water Resources System Planning. IIASA, WP-86-40, Laxenburg, 1986, p. 39.

Dyckman, T. R. and McAdams, A. K. *et al.*: Management Decision-Making under Uncertainty. MacMillan, New York, 1969, p. 704.

Eastman, J. and ReVelle, Ch.: Linear Decision Rule in Reservoir Management and Design 3. Direct Capacity Determination and Interseasonal Constraint. *Water Resources Research*, **9**, 1, 1973, pp. 29 – 42.

Economic Study of Water Management Project in South Moravia. SOGREAH, Grenoble, 1970 (in Czech).

Eisel, L. M.: Watershed Management: A Systems Approach. *Water Resources Research*, 2, 1972, pp. 326 – 338.

Erdélyi, A., Magnus, W., Oberhettinger, F., and Tricomi, F.: Tables of Integral Transform. McGraw-Hill, New York, 1954.

Evans, T. G.: Algorithm 85 Jacobi. *Communication of the ACM*, **5**, 4, 1962.

Fandel, G.: Optimale Entscheidung bei mehrfacher Zielsetzung. Springer, Berlin, 1972.

Fleming, G.: "Real-Time" Operation of a Water Resource System. Proc. IAHR, 4, 1973, pp. 15 – 22.

Ford, L. R. and Fulkerson, D. R.: Flows in Networks. Princeton University Press, Princeton, New Yersey, 1962, p. 194.

Foster, F. G.: On a Stochastic Matrices Associated with Certain Queueing Processes. *Ann. Math. Statist.*, 24, 1953, p. 6.

Freeman, A. M. and Haveman, R. H.: Benefit-Cost Analysis and Multiple Objectives: Current Issues in Water Resources Planning. *Water Resources Research*, 6, 1972, pp. 1533 – 1539.

Fukuda, K.: On Fluctuation of Water Storage in Drought Years in Reservoirs Used for Irrigation. Proc. IWRA, Vol. IV, 1975, pp. 163 – 170.

Gardner, U. F. and Barnes, J. L.: Transient in Linear Systems I. John Wiley, New York, 1950.

Gass, S. I.: Linear Programming – Methods and Applications. McGraw-Hill, New York, 1969.

General Water Plan (GWP), Ministry of Forestry and Water Management, Prague, 1976 (in Czech).

Glückaufová, D. and Trčka, V.: Use of Computer for Mixed-Integer Bivalued Linear Programming. *Ekonomicko-mat. obzor*, 3, 1972 (in Czech).

Glückaufová, D. and Sýkora, M.: Model Analysis of Decision Making in Development of Water Resources System. Conf. Systems of Automatic Control of Water Resources Systems, Brno, Czechoslovakia, Sept. 1980 (in Czech).

Glushkov, V. M.: Introduction to Cybernetics. Izd. AN SSSR, Kiev, 1964 (in Russian).

Gnedenko, B. V. and Kovalenko, I. N.: Introduction to Queueing Theory. Nauka, Moscow, 1966, p. 291 (in Russian).

Goicoechea, A., Duckstein, L., and Fogel, M. M.: Multiobjective Programming in Watershed Management: A Study of the Charleston Watershed. *Water Resources Research*, 6, 1976, pp. 1085–1092.

Goicoechea, A., Duckstein, L., and Fogel, M. M.: Multiple Objectives Under Uncertainty: An Illustrative Application of Protrade. *Water Resources Research*, **15**, 2, 1979, pp. 203–210.

Golubyev, B. N. and Christoforov, A. V.: On Some Issues of Water Resources Control. *Vodnyye Resursy*, 1, 1978, p. 83–88 (in Russian).

Goode, H. H. and Mackol, R. E.: System Engineering. An Introduction to the Design of Large-Scale McGraw-Hill, New York, 1957, p. 551.

Gorokhov, V. G.: Scientific Analysis of the Systems Technique. Systems Research. Sistemmniye Issledovaniya. Nauka, Moscow, 1974, p. 44–56 (in Russian).

Graef, M.,Greiller, R., and Hechtová, G.: Real-Time Data Processing. Alfa, Bratislava, 1974, p. 233 (in Slovak).

Gregor, J.: Terminology and Conception of Information Systems. SNTL, Prague, 1974 (in Czech).

Gundelach, J. and ReVelle, Ch.: Linear Decision Rule in Reservoir Management and Design 5. A General Algorithm. *Water Resources Research*, 2, 1975, p. 204–207.

Gutiérrez Atencio, F. J.: Control of Floods with Multipurpose Hydropower Schemes. Proc. IWRA, Vol. III, 1975, pp. 357–367.

Gvishiani, D. M. and Lishichkin, V. A.: Prognostics. Moscow, 1968 (in Russian).

Gvozdjak *et al.*: Digital Computer Programming. Alfa, Bratislava, 1974, p. 213 (in Slovak).

Habr, J.: Systems Programming. Statistika a demografie, V, Academia, Prague, 1965 (in Czech).

Habr, J.: Simple Optimization Methods for Practical Economics. SNTL, Prague, 1st ed. 1966, 2nd ed. 1970 (in Czech).

Habr, J.: System Aspects of Model Building. Ek. mat. lab. – EÚ ČSAV, Prague, 1970 (in Czech).

Habr, J. and Vepřek, J.: Systems Analysis and Synthesis. SNTL, Prague, 1972, p. 272, 2nd ed. 1986 (in Czech).

Hadley, G. and Within, T. M.: Analysis of Inventory Systems. Prentice Hall, Englewood Cliffs, New York, 1963, p. 427.

Haimes, Y. Y.: Hierarchical Analyses of Water Resources Systems. Modelling and Optimization of Large-Scale Systems. McGraw-Hill, New York, 1977, p. 478.

Haimes, Y. Y. and Hall, W. A.: Multiobjectives in Water Resources Systems Analysis: The Surrogate Worth Trade-Off Method. *Water Resources Research*, **10**, 4, 1974.

Haimes, Y. Y., Hall, W. A., and Freedman, H. T.: Multiobjective Optimization in Water Resources Systems: The Surrogate Worth Trade-Off Method. Elsevier, Amsterdam, 1975, p. 200.

Hall, A. D.: A Methodology for Systems Engineering. Van Nostrand, Princeton, 1962.

Hall, W. A. and Dracup, J. A.: Water Resources Systems Engineering. McGraw-Hill, New York, 1970, p. 372.

Hájek, J. and Anděl, J.: Stationary Processes. SPN, Prague, 1969, p. 147 (in Czech).

Hájek, V.: Systems Engineering. Selected Parts. Technical University, Prague, 1970, p. 40 (in Czech).

Halada, M.: Introduction to Systems Engineering. Nadas, Prague, 1970 (in Czech).

Hall. W. A. and Shephard, R. W.: Optimum Operations for Planning of Complex Water Resources Systems. Contribution 122, University of California, Water Resources Center, Los Angeles, 1967, p. 112.

Handbook of Cybernetics, Vol. 2., Kiev, 1975, p. 620 (in Russian).

Hanssmann, J.: Operations Research in Production and Inventory Control. New York, 1962, p. 366.

Hanzl, A. and Bureš, K.: Multipurpose Water Resources Systems Analysis. *Vodní hospodářství*, A, 7, 1975, pp. 187 – 195 (in Czech).

Hare, W. C. I.: Systems Analysis: A Diagnostic Approach. Hartcourt, Brace et Wold, New York, 1967.

Harrary, F.: Graph Theory. Addison and Wesley, Reading, 1969.

Havel, I.: Algorithms and Automated Programming. Práce, Prague, 1968, p. 127 (in Czech).

Heidari, M., Chow, V. T., Kokotovic, P. V., and Meredith, D. D.: Discrete Differential Dynamic Programming Approach to Water Resources Systems Optimization. *Water Resources Research*, **7**, 2, 1971, pp. 273 – 282.

Heilbrunn, L. V.: The Dynamics of Living Protoplasm. New York, 1956.

Hep, L.: Water Management and Systems Engineering. *Vodní hospodářství*, A, 11, 1970, pp. 289 – 293 (in Czech).

Hep, L.: On Water Management Control. *Vodní hospodářství*, A, 9, 1971, pp. 233 – 238 (in Czech).

Hep, L.: System Approach in Water Resources Systems. Proc. WRS, Karlovy Vary, 1972, p. 67 (in Czech).

Hep, L.: History of Water Resources Systems in ČSR. Proc. WRS, Karlovy Vary, 1972, pp. 475 – 481 (in Czech).

Hino, M.: On-Line Prediction of Hydrologic System. Proc. IAHR, 4, 1973, pp. 121 – 130.

Hipel, K. W., Ragade, R. K., and Unny, T. E.: Metagame Analysis of Water Resources Conflicts. *Journ. of the Hydr. Div.* ASCE, HY 10, 1974, pp. 1437 – 1455.

Hipel, K. W., Ragade, R. K., and Unny, T. E.: Metagame Theory and its Applications to Water Resources. *Water Resources Research*, 3, 1976, pp. 331 – 339.

Hirsch, R. M., Cohon, J. L., and ReVelle, Ch. S.: Gains from Joint Operation of Multiple Reservoir Systems. *Water Resources Research*, 2, 1977, pp. 239 – 255.

Hitch, C. J.: An Appreciation of Systems Analysis. RAND, Santa Monica, 1955.

Hoag, M. W.: An Introduction to Systems Analysis. RAND, Santa Monica, 1956.

Holý, M.: Social Aspects of Different Use of Water in Biosphere. Proc.: The Man and Hydrological Cycle, Prague, 1973 (in Czech).

Holý, M.: Irrigation Structures. Central Board of Irrigation and Power, 135, New Delhi, p. 323.

Holý, M.: Erosion and Environment. Pergamon, London, p. 390.

Holý, M. and Kos, Z.: Water Resources Planning for Irrigation Systems. *Water Supply & Management*, **5**, 1981, pp. 195 – 200.

Holý, M. and Kos, Z.: New Methods of Storage Control. State-of-the-Art. Irrigation, Drainage and Flood Control. ICID, New Delhi, 1982, pp. 1 – 22.

Holý, M. and Říha, J.: Environment and Water Resources Systems and Water Management Development. *Vodní hospodářství*, A, 1, 2, 1972, pp. 1 – 4 and 21 – 30 (in Czech).

Holý, M., Říha, J., and Sládek, J.: Society and Environment, Svoboda, Prague, 1975, p. 171.

Holý, V.: Control Computers in ČSSR. Výběr informací z org. a výp. tech., 4, 1972 (in Czech).

Howard, Ch. D. D.: Methodology for Multiobjective Allocation of a Scarce Water Resource. Proc. IWRA, Vol. IV, 1975, pp. 111 – 117.

Howard, R. A.: Dynamic Programming and Markov Processes. The Technology Press of the Massachussets Institute of Technology, New York – London, 1960, p. 182.

Hromada, Š.: Problems of Power and Economic Evaluation of Water Power Plants and Pumping Plants in Water Resources Systems Design. Proc. WRS, Karlovy Vary, 1972, pp. 410 – 415 (in Czech).

Hufschmidt, M. M. and Fiering, M. B.: Simulation Techniques for Design of Water Resource Systems. Harvard University Press, Cambridge, Massachusetts, 1966, p. 212.

Huthmann, G.: Short-Term Forecasting of Stream Flows with the Aid of Multiple Frequency Response Function. Symp. and Workshops on the Appl. of Math. Model in Hydrology and WRS, Bratislava, 1975, p. 9.

Hýbl, J., Adamec, S., and Škabrada, J.: Computer Programming. SNTL – Alfa, Prague – Bratislava, 1974, p. 218 (in Czech).

Hýbl, J. and Škabrada, J.: Cybernetics (with Respect to Economic Management). SUPRO, Prague, 1968, p. 115 (in Czech).

Hydrotechnic Department, Faculty of Civil Engineering, Technological University of Bratislava: The History of the First Water Resources Systems in Slovakia. Proc. WRS, Karlovy Vary, 1972, pp. 485 – 492 (in Slovak).

Ignatyev, A. A.: Models of Science in Science Research. Sistemniye Issledovaniya, Nauka, Moscow, 1974, pp. 19 – 35 (in Russian).

IIASA (International Institute of Applied Systems Analysis), CP-76-3: Evaluating Tisza River Basin Development Plans Using Multiattribute Utility Theory. IIASA, Laxenburg, March 1976, p. 24.

IIASA, CP-76-5: Workshop on the Vistula and Tisza River Basins. 11 – 13 February, 1975, Laxenburg, 1976, p. 136.

IIASA, Expect the Unexpected, Exec. Rep. 1, 1979, Laxenburg.

Indelicato, S. and Rossi, G.: Methodological Aspects of Water Resources Planning by Simulation Techniques. Proc. IWRA, Vol. II, 1975, pp. 263 – 277.

INIS: Thesaurus. International Atomic Energy Agency, Vienna, 1972, p. 498.

Institute of Management: International Institute for Applied Systems Analysis. Institut řízení, Prague, R 01, 1974, p. 290 (in Czech).

International Symposium on Mathematical Models in Hydrology. Proceedings, IAHS-AISH, Warsaw, 1971.

Ireson, W. G. (ed.): Reliability Handbook. McGraw-Hill, New York, 1966.

Jacoby, H. D. and Loucks, D. P.: Combined Use of Optimization and Simulation Models in River Basin Planning. *Water Resource Research*, **8**, 6, 1972, pp. 1401 – 1414.

Jaglom, A. J.: General Theory of Stationary Random Functions. Sov. věda, mat.-fyz.-astron., V, Prague, 1955, pp. 108 – 138 (in Czech).

James, L. D. and Lee, R. R.: Economics of Water Resources Planning. McGraw-Hill, New York, 1971, p. 615.

Jamieson, D. G.: A Hierarchical Approach to the Analysis of Water Resource Systems. Proc. of the UN Interregional Seminar: River Basin and Interbasin Development, Budapest, Sept. 1975.

Jamshidi, M. and Mohseni, M.: On the Optimization of Water Resources Systems with Statistical Inputs. Proc. of the IFIP Working Conf. on Biosystems Simulation in Water Resources and Water Problems, North-Holland P.C., 1976, pp. 393 – 408.

Janáč, H. and Korsuň, S.: Some Possibilities of Linear Programming Application in Irrigation. *Zemědělská ekonomika*, 1970, pp. 11 – 12, 751 – 759 (in Czech).

Jermář, M.: Step-by-Step Comparison of the Resources Development and Potentials as Decision-Making Tool. Proc. WRS, Hradec Králové, 1976, pp. 148 – 161 (in Czech).

Joeres, E. F.: The Use of Chance-Constrained Programming in the Design and Management of Water Resources Systems. Proc. WRS, Vol. II, 1972, Karlovy Vary, pp. 93 – 107.

Joeres, E. F., Liebman, J. C., and ReVelle, Ch.: Operating Rules for Joint Operation of Raw Water Resources. *Water Resources Research*, **7**, 2, 1971, pp. 225 – 235.

Kabele, J. and Plecháč, V.: Problems of Determination of Effectivity of Multipurpose Water Resources System Investment. *Vodní hospodářství*, 4, 1968, pp. 180 – 184 (in Czech).

Kabele, J. and Plecháč, V.: Effectivity of the System Danube – Odra – Labe: Problems. *Vodní hospodářství*, 5, 1968, pp. 234 – 238. Development. *Vodní hospodářství*, A, 10, 1971, pp. 269 – 277. Evaluation. *Vodní hospodářství*, A, 4, 1972, pp. 81 – 86 (in Czech).

Kaczmarek, Z., Krajewski, K., Kornatowski, T., Filipkowski, A., Kindler, J., and Kebler, D.: The Multi-Step Method for Simulation and Optimization of Vistula River Planning Alternatives. Proc. of the Warsaw Symp. on Math. Models in Hydrology, 1971, pp. 1072–1077.

Kapur, K. C. and Lamberson, L. R.: Reliability in Engineering Design. John Wiley, New York, 1977, p. 586.

Karlin, S.: Mathematical Methods and Theory in Games. Programming and Economics, Vol. I, II, Pergamon Press, London, 1959, pp. 433 and 386.

Károlyi, A., Károlyi, C., Radványi, R., Salamin, A., and Selényi, E.: The Zagyva – Tarna Complex Water Management System. VITUKI, Budapest, 1974, p. 71.

Kartvelishivili, N. A.: Optimal Operation Policy of Reservoirs of Water Power Plants. Gos. en. izd., Moscow – Leningrad, 1963, p. 88 (in Russian).

Kartvelishvili, N. A.: River Flow Regulation. Gidrometizd., Leningrad, 1970, p. 218 (in Russian).

Kartvelishvili, N. A.: Theory of Probability Processes in Hydrology and River Flow Regulation. Gidrometizd., Leningrad, 1967, p. 291 (in Russian).

Kaufmann, A. and Cruon, R.: Les phénomenes d'attente. Dunod, Paris, 1961, p. 303.

Kedrov, B. M.: Principles of History and their Application in Systems Analysis of Development of Science. Sistemniye Issledovaniya, Nauka, Moscow, 1974, pp. 5–18 (in Russian).

Keeney, R. L. and Wood, E. F.: An Illustrative Example of the Use of Multiattribute Utility Theory for Water Resources Planning. *Water Resources Research*, **13**, 4, 1977, pp. 705–712.

Keeney, R. L. and Raiffa, H.: Decisions with Multiple Objectives. John Wiley, New York, 1976.

Kendall, D. G.: Some Problems in the Theory of Dams. *J. Roy. Statist. Soc.*, B 19, 1957, p. 6.

Kendall, M. G. and Stuart, A.: The Advanced Theory of Statistics. Ch. Griffin and Co. Ltd., London,1958, 1966, 1970, 1973.

Kilasoniya, A. N. and Grigoliya, G. L.: Preparation of Hydrologic Record for the Computation of Water Resources Systems. Proc. WRS, Hradec Králové, 1976, pp. 162–173 (in Russian).

Kindler, J.: Some Criteria for Optimization of the Long-Term Water Resources Investment Programs. Proc. WRS, Karlovy Vary, 1972, pp. 169–179.

Kindler, J.: The Monte Carlo Approach to Optimization of the Operation Rules for a System of Storage Reservoirs. Symp. and Workshops on the Appl. of the Math. Models in Hydrology and WRS, Bratislava, 1975, p. 21.

Kindler, J.: The Out-of-Kilter Algorithm and Some its Applications in Water Resources. Proc. UN Seminar: River Basin and Interbasin Development, Vol. I, Budapest, 1975, New York – Budapest, 1976, pp. 198–210.

Kiping, E. S.: Nonparametric Methods in Statistics. Statist. Methods in Hydrology. Gidrometizd., Leningrad, 1970, pp. 124–161 (in Russian).

Kisiel, C.: Time Series Analysis of Hydrologic Data. Advances in Hydroscience, Vol. 5, 1969.

Kitson, T.: The Operation Systems in Great Britain. Workshop IIASA/IMGW, Operation of Multiple Reservoir Systems, Poland, Jodlowy Dwor, 1979, p. 13.

Klimov, G. P.: Stochastic Queueing Systems. Nauka, Moscow, 1966, (in Russian).

Klír, J. and Valach, M.: Cybernetic Modelling. New York, 1965.

Kolev, N. and Nikolova, N.: On the Optimum Management of Multipurpose Reservoir. Proc. IAHR, 4, 1973, pp. 189–196.

Kolev, N. D. and Jancheva, S. D.: Optimal Operational Distribution of Water Resources for Water Resources Systems Control in People's Republic of Bulgaria. Proc. WRS, Hradec Králové, 1976, pp. 174–189 (in Russian).

Kolmogorov, A. N.: Theory of Information Communication. Moscow, 1956 (in Russian).

Konrád, K.: Dynamic Programming and its Use in Economic Water Management Computation. *Vodní hospodářství*, 12, 1967, pp. 552–555 (in Czech).

Kopylov, V. A.: Development of Automated Information Retrieval Systems. Energiya, Moscow, 1974, p. 145 (in Russian).

Korda, B.: The Textbook of Linear Programming. SNTL, Prague, 1962, p. 201 (in Czech).

Korda, B. *et al.*: Mathematical Methods in Economics. SNTL, Prague, 1967, p. 601 (in Czech).

Koris, K. and Nagy, B.: Regulation Model of the Lake Velence Reservoir System. Proc. WRS, Hradec Králové, 1976, pp. 190–207.

Korsching, P. F. and Nowak, P. J.: Soil Erosion Awareness and Use of Conservation Tillage for Water Quality Control. *Water Resources Bulletin*, American Water Resources Association, **19**, 3, June 1983, pp. 459–462.

Korsuň, S.: Optimization of Principal Design of the System with a Reservoir and Irrigation of Agricultural Lands. *Meliorace*, 1, 1972 (in Czech).

Korsuň, S.: Optimization of Complex Irrigation Systems Planning. *Vodní hospodářství*, 9, 1972, pp. 239–242 (in Czech).

Korsuň, S.: Optimization of Water Resources and Irrigation Systems in the Kladénka and Olšava Streams Basins. *Vodní hospodářství*, 6, 1976, pp. 159–165 (in Czech).

Korsuň, S.: Irrigation – The Subsystem of the Water Resources System. Proc. WRS, Vol. II, Karlovy Vary, 1972, pp. 108–110 (in Czech).

Korsuň, S.: Optimizations of Complex of Water – Agricultural Systems Conceptions. Proc. WRS, Vol. III, Karlovy Vary, 1972, pp. 13–26.

Korsuň, S.: Optimization Solution of the Water Resources System of the River Svratka. Proc. WRS, Vol. I, Hradec Králové, 1976, pp. 208–223 (in Czech).

Korsuň, S. and Janáč, J.: Optimal Design of an Irrigation System. ČSVTS Agroplan, Brno, 1975 (in Czech).

Kos, Z.: Irrigation Construction Planning and Water Resources Reliability. *Vodní hospodářství*, 3, 1964 (in Czech).

Kos, Z.: Simulation of Run-Off for Design of Water Resources Systems. Proc. Int. Hydrology Symp. Fort Collins, Col., USA, 1967.

Kos, Z.: Economic Level of Irrigation Withdrawals in Water Resource Systems Design. PhD Thesis, Technical University, Prague, 1968, p. 235 (in Czech).

Kos, Z.: Irrigation Withdrawals in Water Resource Systems. *Vodní hospodářství*, 7, 1969, pp. 198–200 (in Czech).

Kos, Z.: Linear Regression Model and its Application in Hydrology. Works and Studies, 6, Vodní toky, Prague, 1969 (in Czech).

Kos, Z.: Simulation Models of Water Supply for Irrigation in Water Resource System. ICID Bull., 1970, pp. 47–52.

Kos, Z.: Optimization Methods in Simulation Models of Water Resources Systems. Proc. WRS, Vol. III, Karlovy Vary, 1972, pp. 27–38.

Kos, Z.: Chance-Constrained Model of Water Resources Systems. Symp. and Workshops on the Application of Mathematical Models in Hydrology and Water Resources Systems, Bratislava, 1975, pp. 143–148.

Kos, Z.: Stochastic Hydrology and its Application for Water Resources Systems Analysis. *Vodohospodársky časopis*, 4–5, 1975, pp. 424–434 (in Czech).

Kos, Z.: Purposeful Transformation of River Flow Conditions by Reservoirs in Water Resources Systems. Proc. WRS, Vol. I, Hradec Králové, 1976, pp. 224–238 (in Czech).

Kos, Z.: Stochastic Models of Water Resources Systems. Works and Studies 146, Water Research Institute, SZN, Prague, Vol. I, 1978, Vol. II, 1979, p. 468.

Kos, Z.: Operation of Water Resources Systems in Czechoslovakia. Workshop IIASA/IMGW, Operation of Multiple Reservoir Systems, Poland, Jodlowy Dwor, 1979, p. 17.

Kos, Z.: Multimodelling of Water Resources Systems. 12th European Regional Conf. ICID, Dubrovnik, Yugoslavia, 1979.

Kos, Z.: Stochastic Water Requirements for Supplementary Irrigation in Water Resource Systems. IIASA (Int. Inst. for Applied Systems Analysis), Laxenburg, Austria, RR-82-34, 1982, p. 61.

Kos, Z. and Novák, D.: Optimization of Water Resources Utilization in Water Resource Systems in ČSSR. Symp. ECE to Problems of Water Resources Planning. Proc. WATER/SEM, 4/R3/CCM.4, Varna, Bulgaria, 1976.

Kos, Z., Pometlo, J., and Zeman, V.: Analysis of WRS in the Hornád River Basin. Water Research Institute, Prague, 1974 (in Czech).

Kos, Z. and Zeman, V.: Water Resource Systems Design in the General Water Plan. SZN, Prague, 1976, p. 271 (in Czech, English summary).

Kos, Z. and Zeman, V.: Odra River Water Resources System (A Case Study). Workshop IIASA/IMGW, Operation of Multiple Reservoirs Systems, Poland, Jodlowy Dwor, 1979.

Kos, Z. *et al.*: Model of Comprehensive Water Management with Application of Computer Hardware and Software, Vodohospodářský rozvoj a výstavba, Prague, 1967 – 1970, p. 890 (in Czech).

Kos, Z. *et al.*: Problems of Multipurpose Water Resources Systems: Principles and Methods and their Verification. Final Research Report, VRV, Prague, 1971 (in Czech).

Kos, Z. *et al.*: Study of the Important Water Resource System in the Upper and Middle Reaches of the Labe River Basin. Preliminary Investigation of the General Water Plan. VRV, Prague, 1973.

Kos, Z. *et al.*: Study of the Important Water Resource System in the River Odra Basin. Preliminary Investigation of the General Water Plan, VRV, Prague, 1974.

Kovalenko, I. N.: Research of Analysis of Complex Systems Reliability. Naukova Dumka, Kiev, 1975, p. 210 (in Russian).

Kozák, J. and Seger, J.: Simple Statistical Methods in Prognostics. SNTL, Prague, 1975, p. 280 (in Czech).

Krajčí, J.: Importance of Construction of Large-Scale Water Resource Systems. *Vodní hospodářství*, A, 9, 1972, pp. 217 – 219 (in Slovak).

Kravets, A. A.: Probability and Systems. Izd. Voronezh. univ., Voronezh, 1970, p. 192 (in Russian).

Kreuz, Z.: Systematic River Training of the Odra River and its Tributaries in the Northern Moravia Region. *Vodní hospodářství*, A, 5, 1974, pp. 115 – 119 (in Czech).

Krishna Murthy, V. R., Liu, R. S., and Crow, L. I.: Priority Allocation of Available Surface Water Resources in a River Basin. Proc. IWRA, Vol. IV, 1975, pp. 47 – 59.

Kritsky, S. N. and Menkel, M. F.: Computation of the River Flow Regulation with Respect to Carry-Over and Correlation Relation Among the Consecutive Years. Problemy regulirovaniya verchovnogo stoka, 8, Izd. AN SSSR, Moscow, 1959 (in Russian).

Kritsky, S. N. and Menkel, M. F.: Computation in Water Management. Gidrometizd., Moscow – Leningrad, 1952 (in Russian).

Krutilla, J. V. and Eckstein, O.: Multiple Purpose River Development. Studies in Applied Economic Analysis. J. Hopkins Press, Baltimore, 1958.

Kubec, J. and Pavlík, Z.: Utilisation of Inland Waterways for Transbasin Diversion of Water to Regions with Water Deficits. *Vodní hospodářství*, A, 7, 1973, pp. 167 – 171 (in Czech).

Kubíček, J.: Water Resources Systems for Municipal Water Supply of the Liberec, Jablonec and Frýdlant Regions (Thesis). Dept. of Hydrotechnics, Technical University, Prague, 1972 (in Czech).

Kubík, S. *et al.*: Optimal Systems of Automated Control. SNTL, Prague, 1972, p. 500 (in Czech).

Kučera, S. and Maroušek, J.: System Programming II. Stage. Representation of the Project Development by a Network. Kovoprojekta, VIII, 8, 1966, Prague, p. 26 (in Czech).

Kudláček, V.: Notes to the Theory of Systems and their Modelling. University of Technology, Brno, 1970.

Kurbakov, I. I.: Information Coding and Retrieval in Automated Dictionary. Izd. Sov. Radio, Moscow, 1968, p. 247 (in Russian).

Kutta, E. and Soukup, N. *et al.*: Management and Control in the Period of Scientific and Technological Revolution. Svoboda, Prague, 1973, p. 340 (in Czech).

Kuuvisto, E.: Operation of Multiple Reservoir Systems in Finland. Workshop IIASA/IMGW, Operation of Multiple Reservoir Systems, Poland, Jodlowy Dwor, 1979, p. 3.

Lancaster, W. F.: Information Retrieval Systems: Characteristics, Testing Evaluation. John Wiley, New York, 1968, p. 223.

Lank, B.: Water Resources Systems Network Model. Water Research Institute, Prague, 1986, p. 60.

Lapa, V. G.: Mathematical Principles of Cybernetics. Izd. Vishcha shkola, Kiev, 1971, p. 418 (in Russian).

Laudát, J.: System Co-operation of Power Systems with Water Resources Systems and Development Modelling. Proc. WRS, Vol. II, Karlovy Vary, 1972, p. 111 (in Czech).

Lazarov, C.: Sur une méthode simulation de rangs de probabilité hydrologiques pour la recherche de systems hydrauliques et économiques. Proc. IAHR, 4, 1973, pp. 62–70.

Lebedyev, V. L.: On Spectral Analysis of Non-Stationary Random Processes. *Radiotechnika i elektronika*, 2, 1959 (in Russian).

Lebedyev, B. D., Podinovsky, V. V., and Styrikovich, R. S.: Optimization with an Ordered Set of Criteria. *Ekonomika i matem. metody*, 4, 1971 (in Russian).

Lee, K., Day, H. J., and Cheney, P. B.: Progress in Simulation of Integrated Flood Plain Analysis and Management. Proc. IWRA, Vol. IV, 1975, pp. 265–275.

Lee, W.: Decision Theory and Human Behaviour. John Wiley, New York, 1971, p. 352.

Leipold, T. and Spiegel, R. P.: Stochastic Modelling of Multipurpose Reservoir under Consideration of the On-Line Operation. Workshop IIASA/IMGW, Operation of Multiple Reservoir Systems, Poland, Jodlowy Dwor, 1979, p. 15.

Lektorsky, V. A. and Sadovsky, V. N.: On the Principles of Systems Research (Related to Bertalanffy's General Systems Theory). *General Systems*, V, 1960, pp. 171–179.

Leonov, Yu. P.: On Problems of Filtering of Non-Stationary Random Functions. *Avtomatika i telemekhanika*, XVII, 2, 1956 (in Russian).

Levin, B. R.: Theory of Random Processes and its Application in Radio-Engineering. SNTL, Prague, 1965, p. 570 (in Czech).

Lindley, D. V.: The Theory of Queues with a Single Server. Proc. Cambridge Phil. Soc., 1952, p. 62.

Linhart, J.: Activity and Recognition. Academia, Prague, 1976, p. 574 (in Czech).

Linhart, J. and Hovorka, K.: On the Character of Heuristic Processes. *Ekonomika a řízení věd. techn. rozvoje*, 4, 1973, pp. 132–153 (in Czech).

Liška, M.: Problems of Economic Evaluation of Multipurpose Water Resources Systems. Proc. WRS, Karlovy Vary, 1972, pp. 121–129 (in Slovak).

Liška, M.: Recognitions from the Cost-Benefit Analysis of Existing and Prepared Multipurpose Water Projects and Systems in ČSSR. Proc. WRS, Hradec Králové, 1976, pp. 274–293 (in Slovak).

Little, J. D. C.: The Use of Storage Water in a Hydroelectric System. *Journal of the Operations Research Society of America*, 3, 1955, p. 11.

Liapunov, A. M.: On Logical Schematic Representation of Programmes. Znaniye, Moscow, 1962 (in Russian).

Llamas, J.: Mathematical Approach for Water Resources Development in Yamaska River Basin. Proc. IWRA, Vol. III, 1975, pp. 119–124.

Lotti, C. and Pandolfi, C.: System Approach to Water Resources Planning. Proc. IWRA, Vol. II, 1975, pp. 279–281.

Loucks, D. P., Sedinger, J. R., and Haith, D. A.: Water Resource Systems Planning and Analysis. Cornell University, Prentice Hall, Inc. Englewood Cliffs, New Yersey 07632, 1981, p. 545.

Lozansky, V. R. and Kuzin, A. K.: Forecasting, Planning and Control of Water Quality in the USSR. Proc. IWRA, Vol. IV, 1975, pp. 231–238.

Luce, R. D. and Raiffa, H.: Games and Decisions. John Wiley, New York, 1957, p. 507.

Maass, A. *et al.*: Design of Water Resources Systems. Harvard University Press, Cambridge, Massachusetts, 1962, p. 644.

Macháček, L.: Simulation Model of Water Resources System. Water Research Institute, Prague, 1986, p. 82.

Macháček, V.: Further Preparation of the Channel Danube – Odra – Labe. *Vodní hospodářství*, 3, 1967, pp. 119 – 121 (in Czech).

Machol, R. E., Tanner, W. P. J., and Alexandr, X. N.: System Engineering Handbook. McGraw-Hill, New York, 1965.

Maidment, D. R., Chow, V. T., and Tanec, G. W.: Computer Memory Requirement for DP and DDDP in Water Resources Analysis. Proc. 55th Annual AGU Conf., Washington, 1974.

Maňas, M.: Theory of Games and Optimal Decision. SNTL, Prague, 1974, p. 256 (in Czech).

Maňas, M., Vejmola, S., Walter, J., and Zelinka, J.: Operations Research. SPN, Prague, 1971, p. 362 (in Czech).

Mandl, P.: A Mathematical Model of a Storage. *Kybernetika*, **17**, 2, 1971.

Maniak, U. and Trau, W.: Determination of Optimal Operation Rules for Reservoirs on the Basis of a Stochastic Model. Proc. IAHR, 4, 1973, pp. 89 – 96.

Maré de, L.: Cost-Benefit Analysis – Some Notes on the System Objectives and Valuation of Environmental Benefits. Proc. IWRA, Vol. II, 1975, pp. 225 – 230.

Markov, A. A.: Theory of Algorithms. Trudy Math. Inst. AN SSSR, Moscow, 1954.

Matalas, N. C.: Mathematical Assessment of Synthetic Hydrology. *Water Resources Research*, **3**, 4, 1967, pp. 937 – 945.

Matsuura, E. and Matsui, K.: Water Space in Urban Environment. Proc. IWRA, Vol. III, 1975, pp. 349 – 356.

Mawer, P. A. and Thorn, D.: Improved Dynamic Programming Procedures and their Practical Application to Water Resources Systems. *Water Resources Research*, 6, 1969, pp. 1174 – 1178.

McBean, E. A., Hipel, K. W., and Unny, T. E. (ed.): 1. Reliability in Water Resources Management. 2. Inputs for Risk Analysis in Water Systems. Water Resources Publication, Littleton, 1979.

McBean, E. A. and Schaake, J. C.: Estimation of Response Surface Gradients in Multiobjective Water Resources Planning. *Water Resources Research*, 4, 1976, pp. 592 – 598.

McKerchar, A. I.: Optimizing the Monthly Operation of Two New Zealand Hydropower Storage. Proc. IAHR, 4, 1973, pp. 215 – 220.

McKinsey, J. C. G.: Introduction to the Theory of Games. McGraw-Hill, New York, 1952.

Meadow, Ch. T.: The Analysis of Information Systems. A Programmer's Introduction to Informational Retrieval. John Wiley, New York, 1967.

Medla, A. *et al.*: Comprehensive Water Resources Systems: Methodology, 1st Stage, VRV, Prague – Brno, 1969, (in Czech).

Mejia, J. M., Egli, P., and Leclerc, A.: Evaluating Multireservoir Operating Rules. *Water Resources Research*, 6, 1974, pp. 1090 – 1098.

Menkyna, V.: Extension of Multipurpose Water Power Plant Utilization – Forced Aeration of Streams in Hydro-Turbines and Connected Problems. Proc. WRS, Vol. III, Karlovy Vary, 1972, pp. 39 – 50 (in Slovak).

Mesarović, M. D.: Views on General Systems Theory. Proc. of the Second Systems Symposium, Case Inst. of Technology, 1963, John Wiley, New York, 1964.

Mesarović, M. D., Macko, D., and Takahara, Y.: Theory of Hierarchical Multilevel Systems. Academic Press, New York, 1970.

META Systems Inc.: Systems Analysis in Water Resources Planning. Water Information Center, Inc., Port Washington, New York, 1975, p. 393.

Meyer-Zurwelle, J.: Optimum Release Strategies for Systems of Flood Protection Reservoirs. Proc. IAHR, 4, 1973, pp. 205 – 214.

Mihál, J.: Multipurpose Water Reservoirs as an Economic Category. VÚV, Bratislava, 1969 (in Slovak).

Mihoc, G. and Nadejde, I.: Parametric Nonlinear and Stochastic Programming. Alfa, Bratislava, 1974, p. 784 (in Slovak).

Mikoláš, J. and Bárta, V.: Algorithmization of Water Treatment Processes – A Subsystem of Water Resources Systems. Proc. WRS, Karlovy Vary, 1972, p. 320–329 (in Czech).

Miková, L. and Říha, L.: Economic Optimization of Water Resources in Industry. *Vodní hospodářství*, A, 10, 1969, pp. 272–275 (in Czech).

Mikulec, S.: Water Resources Problems in Yugoslavia. Proc. WRS, Vol. II, Karlovy Vary, 1972, pp. 71–75 (in Czech).

Mirtskhulava, C. E.: Reliability of Irrigation and Drainage Systems. Kolos, Moscow, 1974, p. 274 (in Russian).

Molnár, Z.: Heuristic Convergent Models for Multipurpose Optimization. *Ekonomicko-matematický obzor*, 2, 1976, pp. 145–164 (in Czech).

Monarchi, D. G., Kisiel, C. C., and Duckstein, L.: Interactive Multiobjective Programming in Water Resources. *Water Resources Research*, **9**, 4, 1973.

Mookerjea, D.: Management of Water Resources. Proc. IWRA, Vol. IV, 1975, pp. 33–38.

Moran, P. A. P.: The Theory of Storage. Methuen and Co. Ltd. – Wiley and Sons Inc., London – New York, 1959, p. 95.

Morávek, S.: Municipal Water Resources and Water Supply System of Towns of Liberec and Jablonec. *Vodní hospodářství*, A, 7, 1974, pp. 171–172 (in Czech).

Morse, P. M. and Kimball, G. E.: Methods of Operations Research. M.I.T., Cambridge, 1951.

Mosný, M.: Systems Engineering, B. Mathematical and Logical Modelling in Technology. VÚSTE, Prague, 1970 (in Czech).

Mosonyi, E. and Buck, W.: Some Aspects of Risk for Improved Water Resources Planning. Proc. IWRA, Vol. II, 1975, pp. 221–224.

Mrázek, K.: Water Resources Systems and Water Quality. Proc. WRS, Vol. II, Karlovy Vary, 1972, pp. 135–148 (in Czech).

Müller, B.: Water Management in the Ohře River Basin. *Vodní hospodářství*, A, 1, 1973, pp. 5–9 (in Czech).

Můra, D.: Systems Engineering, Vol. 3. Ústav výpočetní techniky, Ostrava, 1970, p. 200 (in Czech).

Murray, D. and Yakowitz, S.: Constrained Differential Dynamic Programming and its Application to Multireservoir Control. *Water Resources Research*, **15**, 4, 1979, pp. 1017–1027.

Mužík, V. and Vostrý, Z.: Real-Time Adaptive Prediction Methods of River Flows. Symposium IIASA: Recent Development in Real-Time Forecasting and Control of Water Resource Systems, Laxenburg, 1976.

Nacházel, K.: Application of Monte Carlo Methods in Reservoirs Cooperation. *Vodní hospodářství*, A, 7, 1970, pp. 176–183 (in Czech).

Nacházel, K.: Water Resources Systems Analysis with Sample Realizations of Random Flow Series. *Vodní hospodářství*, A, **26**, 9, 1976, pp. 229–239 (in Czech).

Nacházel, K. *et al.*: Stochastic Models of Within-Year Flow Variation and their Implementation for Rational Use of Water Resources. (Research Report II-7-2/15) Dept. of Hydrotechnics, Faculty of Civil Engineering, Technical University, Prague, 1975, p. 87 (in Czech).

Nacházel, K. *et al.*: Dynamics of Long-Term Time Patterns of Flows and its Influence on Rational Use of Water Resources. Dept. of Hydrotechnics, Technical University, Prague, 1975, p. 288 (in Czech).

Nacházel, K. and Bureš, P.: Application of Monte Carlo Method for Computation of Reservoirs with Carry-Over. *Vodohospodársky čas.*, 1, 1973, pp. 3–32 (in Czech).

Nacházel, K. and Patera, A.: Filtering of Hydrological Time Series. *Vodohospodársky čas.*, 6, 1974, pp. 569–592 (in Czech).

Naur, P.: Certification of Algorithm 85 Jacobi. *Communication of the ACM*, **6**, 8, 1963.

Nayac, S. C. and Arora, S. R.: Optimal Capacities for a Multireservoir System Using the Linear Decision Rule. *Water Resources Research*, **7**, 3, 1971, pp. 485–498.

Nayac, S. C. and Arora, S. R.: Linear Decision Rule – A Note on Control Volume Being Constant. *Water Resources Research*, **10**, 4, 1974, pp. 637–642.

Neumann, J. von and Morgenstern, O.: Theory of Games and Economic Behaviour. Princeton University Press, Princeton, 1944.

Newell, A.: Heuristic Programming: Ill-Structured Problems. Progress in Operation Research. Aranovski (ed.). Vol. 3, 1969, pp. 361–414.

Newel, A., Shaw, J. C., and Simon, H. A.: Report on a General Problem Solving Program. Proc. Inter. Conf. Inform. Processing, Paris, 1959.

Newell, A., Shaw, J. C., and Simon, H. A.: The Process of Creative Thinking. Univ. Colorado Symp. on Cognition. Rand Corp. Rep., P 1320 (b).

Nopmongcol, P. and Askew, A. J.: Multilevel Incremental Dynamic Programming. *Water Resources Research*, 6, 1976, pp. 1291–1297.

Novák, V., Gabriel, P., and Kazda, I.: Application of Analog Computers in Hydrotechnic Computation. *Vodní hospodářství* (Ref. No.), 1969 (in Czech).

Novotný, S.: Problems of Optimization of Multipurpose Water Resources Systems. Proc. WRS, Karlovy Vary, 1972, pp. 439–445 (in Czech).

Novotný, S. and Polenka, E.: Selection of an Optimal Alternative of a Single-Purpose Water Resource System by Decision Analysis Method. *Vodní hospodářství*, A, 6, 1973, pp. 135–139 (in Czech).

Ogurtsov, A. P.: Stages in System Interpretation of Scientific Knowledge (Ancient and New Times). Sistemniye Issledovaniya, Nauka, Moscow, 1974, pp. 154–186 (in Russian).

Opricović, S. and Djordjević, B.: Analysis and Synthesis of Water Resources Systems by System Engineering. Proc. IAHR, 4, 1973, pp. 9–14.

Optimal Allocation of Water Resources to System Elements and Determination of Water Supply Reliability. Proc. Energetichesky Institut F. M. Krzhizhanovskogo, Moscow, 1969 (in Russian).

Ore, O.: Theory of Graphs. Amer. Math. Soc. Coll. Publ., 1962.

O'Riordan, J.: An Approach to Evaluation in Multiple Objective River Basin Planning. Canada Dept. of Environment, Vancouver, B. C., 1973.

Ostrom, A. R.: A Review of Conflict Resolution Model in Water Resources Management, IIASA, CP-76-5, Laxenburg, 1976, pp. 95–105.

Ošlejšek, J.: Application of Network Analysis Methods for Optimal Design of Water Mains. Czech Scientific and Technical Society, Brno, 1975 (in Czech).

Özis, Ü.: The Maximization Approach in Water Resources Planning. Proc. IWRA, Vol. II, 1975, pp. 255–261.

Palla, I. and Partl, Q.: Program for Optimization of a Reservoir System with Water Power Plants. Comp. Centre of the Prague School of Economics, Prague, 1970, p. 34 (in Czech).

Palla, I. and Partl, Q.: The Stochastic Approach to the Optimization of the Operating Regime of a Hydroelectric Power Plant System. Symposium ECE, Dubrovnik, 1970, p. 5.

Palla, I. and Partl, Q.: The Usage of Dynamic Programming for the Optimum Design and Operating Regime of Hydro-Electric Power Plants. Symposium ECE. Varna, 1970, p. 7.

Palla, I. and Váňová, H.: Design of a Method of Complex Water Resources Systems Analysis. Proc. WRS, Karlovy Vary, 1972, pp. 332–340 (in Czech).

Parikh, S. C.: Linear Dynamic Decomposition Programming of Optimal Long-Range Operation of a Multiple Multipurpose Reservoir System. Rep. ORC 66-28, Operations Research Center, University of California, Berkeley, 1966, p. 141.

Partl, Q.: Study of Technical and Economic Comparison of Reservoirs Vestřev and Trotina. ŘVT, Prague, 1967, p. 17 (in Czech).

Partl, Q.: Optimal Reservoir Release Policy. Works and Studies 3, ŘVT, Prague, 1968, p. 107 (in Czech).

Partl, Q.: Optimal Co-operation of Reservoirs in the Ohře River Basin. VRV, Prague, 1969, p. 74 (in Czech).

Partl, Q.: Design of a Cybernetic Adaptive Model of Multipurpose Water Resources System. VRV, Prague, 1970–1971 (in Czech).

Partl, Q.: Paper on the Methods for Control of Water Resources Systems. Proc. WRS, Vol. III, Karlovy Vary, 1972, pp. 63–70.

Partl, Q.: Water Resources Systems – A New Scientific Branch. *Vodní hospodářství*, A, 11, 1974, pp. 292–294 (in Czech).

Partl, Q. and Bečvář, V.: Optimal Methods of Multipurpose Water Resources Systems Analysis (2nd Stage) GWP, VRV, Prague, 1970 (in Czech).

Patera, A.: On Adaptive Models of Reservoirs and Water Resources Systems. *Vodohospodársky čas.* **26**, 2, 1978, pp. 144–153 (in Czech).

Patera, A.: Adaptation of Reservoir Control with Short Realizations of Hydrological Processes. *Vodohospodársky čas.* **26**, 3, 1978, pp. 228–244 (in Czech).

Patera, A.: Reliability and Adaptation of Long-Term Operation of Reservoirs. *Vodohospodársky čas.* **27**, 1, 1979, pp. 3–13 (in Czech).

Patera, A.: Adaptive Models of Reservoirs and Water Resources Systems with Respect to their Future Operation. *Vodní hospodářství*, 8, 1978, pp. 207–211 (in Czech).

Pawinger, E. and Hanzl, A.: Flood Control in the Sava River Basin. Proc. WRS, Vol. II, Karlovy Vary, 1972, pp. 76–84 (in Czech).

Pearce, S. F.: Computer-Based Trend Projection: Data Processing. *Org. and Control*, 4, 1973.

Peter, P. and Lukáč, M.: The Štiavnica System of Water Reservoirs in the Past and Today. Proc. WRS, Karlovy Vary, 1972, pp. 193–201 (in Slovak).

Peter, P. and Lukáč, M.: The Štiavnica System of Water Reservoirs – The Present Requirements on their Operation. *Vodní hospodářství*, A, 8, 1972, pp. 203–207 (in Slovak).

Peter, P. and Lukáč, M.: Solution of the Co-operation of Two Water Supply Reservoirs. Proc. WRS, Hradec Králové, 1976, pp. 306–327 (in Slovak).

Petrova, T. M.: Mathematical Models in Scientific Research. Sistemniye Issledovaniya, Nauka, Moscow, 1974, pp. 57–74 (in Russian).

Pírko, Z.: Principles of Calculus of Operations. SPN, Prague, 1951 (in Czech).

Pivoda, B.: Economic Operation of Drinking Water Systems in a Non-Stationary Regime. Proc. WRS, Karlovy Vary, 1972, pp. 180–188 (in Czech).

Plander, I.: Mathematical Methods and Programming of Analog Computers. SAV, Bratislava, 1969, p. 629 (in Slovak).

Planning Comprehensive Development of the Vistula River System. Final Report, UNDP, New York, Hydroprojekt, Warsaw, 1972, p. 132.

Plecháč, V.: On Economic Effectiveness of the Danube–Odra–Labe System. *Vodní hospodářství*, A, 6, 1972, pp. 238–240 (in Czech).

Plecháč, V.: General Water Plan and Water Management in River Basins. *Vodní hospodářství*, A, 12, 1972, pp. 309–312 (in Czech).

Plecháč, V.: Second Edition of the General Water Plan. *Vodní hospodářství*, A, 6, 1975, pp. 141–145 (in Czech).

Plecháč, V.: Long-Term Planning in Water Management. SZN, Prague, 1978, p. 451 (in Czech).

Posekaný, T.: System of Ponds in South Bohemia. *Vodní hospodářství*, 9, 1968, pp. 431–437 (in Czech).

Poston, R. and Steward, I. N.: Catastrophe Theory and its Applications. Pitman, London, 1979, p. 512.

Prabhu, N. U.: Some Solutions for the Finite Dam. *Ann. Math. Stat.* 29, 1958, p. 10.

Prabhu, N. U.: Queues and Inventories. John Wiley, New York, 1965, p. 355.

Prakash, A.: Environmental Impact of Water Resources Utilization. Proc. IWRA, Vol. V, 1975, pp. 405–411.

Preul, H. C.: Urban Runoff Management. Proc. IWRA, Vol. IV, 1975, pp. 209–214.

Prigogine, I.: Introduction to the Thermodynamics of Irreversible Processes. Springfield, Illinois, 1955 Liège, 1947 (in French).

Proceedings: Research and Practice in the Water Environment. 15th Congress of the IAHR, Vol. 4, Istanbul, 1973, p. 257.

Programme in FORTRAN for Optimization of Utilization of Water Resources in a Large-Scale Water Resources System – Subset 2 Topic CMEA – 1A – 3.01. Warsaw. Inst. met. i vod. choz. Poland, 1975, p. 110 (in Russian).

Pryazhinskaya, V. G. and Shabanov, V. V.: Optimization of Stochastic Process of Water Rate Control of Agricultural Crops. Proc. IAHR, 4, 1973, pp. 175–180.

Pugachev, V. S.: Theory of Random Functions and its Use for Problems of Automatic Control. Gostechizdat, Moscow, 1957 (in Russian).

Pushkin, V. N., Pospelov, D. A., and Sadovsky, V. N.: The Problems of Heuristics. Inst. Psychol. AN SSSR, Moscow, 1969 (in Russian).

Pushkin, V. N., Pospelov, D. A., and Yefimov, E. I.: Psychological Theory of Thinking and Some Way of Development of Cybernetics. *Voprosy psichologii*, 2, 1971, pp. 66–79 (in Russian).

Quimpo, G. R.: Stochastic Identification of Nonlinear Hydrologic Systems. Symp. and Workshops on the Appl. of Math. Models in Hydrology and WRS, Bratislava, 1975, p. 8.

Ragade, R. K., Hipel, K. W., and Unny, T. E.: Metarationality in Benefit-Cost Analyses. *Water Resources Research*, 5, 1976, pp. 1069–1075.

Raiffa, H.: Decision Analysis: Introductory Lectures on Making Decisions Under Uncertainty. Addison-Wesley, Massachusetts, 1968, p. 309.

Raichl, J.: Programming in ALGOL. Academia, Prague, 1967, p. 177 (in Czech).

Raichl, J.: Computer Programming. Academia, Prague, 1972, p. 297 (in Czech).

Ramana Murthy, K. V.: The Role of Direct Analog Computers in the Solution of Water Resources Problems. Proc. IWRA, Vol. V, 1975, pp. 283–290.

Randák, J.: Design of a Reservoir System for Flood Control (Thesis). Dpt. of Hydrotechnics, Technical University, Prague, 1972 (in Czech).

Rao, G. V. V. and Williams, T. T.: Sequential Optimization of Multiple Non-Monetary Objectives in Integrated Operation of Reservoirs Systems. Proc. IWRA, Vol. IV, 1975, pp. 71–81.

Rapoport, A.: Mathematical Aspects of General Systems Analysis. *General Systems*, **XI**, 1966, pp. 3–11.

Rapoport, A.: Remarks on General Systems Theory. *General Systems*, **VIII**, 1963, pp. 123–124.

Rational Water Management in the Odra River Basin. Povodí Odry, VRV, Ostrava – Prague, 1972, p. 225 (in Czech).

Raven, F. H.: Mathematics of Engineering Systems. McGraw-Hill, New York, 1966, p. 524.

Recommendations of the Ministry of Forestry and Water Management ČSR for Use of Principles of Economic Evaluation in Water Management. Prague, 1976.

Rektorys, K. *et al.*: Survey of Applicable Mathematics. SNTL, Prague, 1963, p. 1136 (in Czech). English translation, M.I.T. Press, 1969, p. 1369.

ReVelle, Ch., Joeres, E., and Kirby, W.: The Linear Decision Rule in Reservoir Management, 1. Development of the Stochastic Model. *Water Resources Research*, **5**, 4, 1969, pp. 767–777.

ReVelle, Ch. and Kirby, W.: Linear Decision Rule in Reservoir Management and Design, 2. Performance Optimization. *Water Resources Research*, **6**, 4, 1970, pp. 1033–1044.

Reznikovsky, A. Sh.: Computation of Hydropower in Water Resources Systems Using the Monte Carlo Method. Energia, Moscow, 1969, p. 304 (in Russian).

Reznikovsky, A. Sh.: Generation of Hydrologic Time Series and Computation of Release of Reservoir Cascade with Carry-Over for Water Power Generation. Izd. AN SSSR, Energetika i transport 2, Moscow, 1964 (in Russian).

Roberts, N. H.: Mathematical Methods in Reliability Engineering. McGraw-Hill, New York, 1964, p. 300.

Robiček, M.: Water Resources Systems as an Subsystem of the Integrated Technological Infrastructure. Proc. WRS, Karlovy Vary, 1972, pp. 113–121 (in Czech).

Roefs, T. G. and Guitron, A. R.: Stochastic Reservoir Models: Relative Computational Effort. *Water Resources Research*, 6, 1975, pp. 801–804.

Rohde, F. G. and Langer, U.: The Maximum Principle of Pontryagin Applied to Hydropower Storage Problems. Proc. IWRA, Vol. I, 1975, pp. 165–169.

Romanovicz, R. and Budzianowski, R.: Some New Approach to Large-Scale Hydrological Systems. Proc. WRS, Hradec Králové, 1976, pp. 328–343.

Rossman, L. A.: Reliability-Constrained Dynamic Programming and Randomized Release Rules in Reservoir Management. *Water Resources Research*, 2, 1977, pp. 247–255.

Roy, R.: Problems and Methods with Multiple Objective Functions. 7th Int. Symp. on Mathem. Programming, Hague, 1970.

Rubin, M. L.: Principles of Modern Systems Techniques. Mir, Moscow, 1975, p. 527 (in Russian).

Rubinštein, S. L.: On Thinking and Ways of its Research. SPN, Bratislava, 1960 (a), (in Slovak).

Rubinštein, S. L. (ed.): Process of Thinking and Principles of its Analysis, Synthesis and Generalization. AN SSSR, Moscow, 1960 (b), (in Russian).

Russel, S. O.: Use of Decision Theory in Reservoir Operation. *Journ. of the Hydraulics Div.* ASCE, HY 6, 1974, pp. 809–817.

Rychetník, L., Zelinka, J., and Pelzbauerová, V.: A Collection of Linear Programming Tasks. SNTL – ALFA, Prague – Bratislava, 1968, p. 314 (in Czech).

Rydzewski, J. R.: Multiobjective Planning of Irrigation Projects in Less Developed Countries. Proc. IWRA, Vol. II, 1975, pp. 215–219.

Rydzewski, J. R. and Rashid, H. A.-H.: Optimization of Water Resources for Irrigation in East Jordan. AWRA, *Water Resources Bulletin*, **17**, 3, 1981, pp. 367–372.

Ryšková, A.: Water Resources System as an Subsystem of an Economically Important Region. Proc. WRS, Karlovy Vary, 1972, pp. 454–461 (in Czech).

Říha, J.: Are There Any Limits to the Saturation of Man's Ecosystem? Proc. WRS, Vol. III, Karlovy Vary, 1972, pp. 71–88.

Říha, J.: A Method of Analytical Computation of an Optimal Irrigation Area. *Vodní hospodářství*, A, 4, 1974, pp. 103–106.

Saaty, T. L.: Elements of Queueing Theory with Applications. McGraw-Hill, New York, 1961, p. 510.

Saaty, T. L.: Exploring the Interface between Hierarchies, Multiple Objectives and Fuzzy Sets. *Fuzzy Sets and Systems*, **1**, 1, 1978.

Salton, G.: Automatic Information Organization and Retrieval. McGraw-Hill, New York, 1968.

Scheidegger, A. E.: A Random Graph Pattern of Drainage Basins. IASH, Bern, 1967.

Schavelev, D. S.: The Optimal Criteria in Designing Hydraulic and Water Management Systems. Proc. IAHR, 4, 1973, pp. 227–232.

Schleifer, R.: Computer Programs for Elementary Decision Analysis. Harvard Univ. Press, Mass., Cambridge, 1971, p. 346.

Schmidt, E.: New Prices of Electric Power. *Energetika*, **26**, 1, 1976, pp. 19–21 (in Czech).

Schneeweiss, H.: Decision Criteria with Risk. Springer, Berlin, 1967, p. 215 (in German).

Schultz, G. A.: Operation of Multiple Reservoir Systems. Workshop IIASA/IMGW, Operation of Multiple Reservoir System, Poland, Jodlowy Dwor, 1979, p. 13.

Schweig, Z. and Cole, A.: Optimal Control of Linked Reservoirs. *Water Resources Research*, **4**, 3, 1968, pp. 479–498.

Shannon, C. E.: A Mathematical Theory of Communication, *Bell Syst. Techn. Journ.*, 27, 1948.

Sigvaldason, O. T.: A Simulation Model for Operating a Multipurpose Multireservoir System. *Water Resources Research*, 2, 1976, pp. 263–278.

Sigvaldason, O. T.: Multireservoir Management of the Trent, Severn Riden and Cataraqui System –

A Case Study. Workshop IIASA/IMGW, Operation of Multiple Reservoir Systems, Poland, Jodlowy Dwor, 1979.

Silva de, S. H. C.: Water Resources Ecology. Proc. IWRA, Vol. V, 1975, pp. 361 – 372.

Simeonov, D.: Optimum Utilization of the Water Resources and Their Potential Energy. Proc. IWRA, Vol. II, 1975, pp. 245 – 253.

Sládeček, M.: Creative Teams. *Ekonomika a řízení věd. tech. rozvoje*, 4, 1973, pp. 58 – 67.

Sládek, J.: Control Theory. Technical University, Prague, 1974, p. 260 (in Czech).

Smith, W. L.: On the Distribution of Queueing Times. Proc. Cam. Phil. Soc., 49, 1953, p. 13.

Solomon, S.: Statistical Relations among Hydrological Variables. Statis. met. v gidrologii. Gidrometizd., Leningrad, 1970, pp. 18 – 73 (in Russian).

Solomoniya, O. G.: Problems of Design of Optimal Irrigation Systems with Reservoirs. Proc. Sbornik Vsesoyuznoy konf., Novosibirsk, 1968 (in Russian).

Solomoniya, O. G., Gershkovich, M. I., and Ediberidze, M. G.: Optimum Policy of Water Resources Systems Control. Proc. WRS, Hradec Králové, 1976, pp. 357 – 362 (in Russian).

Staš, L. *et al.*: Water Management and Operation of Reservoir in the Odra River Basin. *Vodní hospodářství*, A, 9, 1973, pp. 229 – 234 (in Czech).

Stoerner, H.: Mathematical Theory of Reliability. Introduction and Application. Oldenbourg, München, 1970, p. 329 (in German).

Stone, P. J. and Lenton, R. L.: A Decentralized Allocation Analysis to Aid Negotiations for International River Basin Planning. Proc. WRS, Vol. I, Hradec Králové, 1976, pp. 363 – 377.

Strupczewski, W. G. *et al.*: Stochastic Properties of the Process Transformed in the Linear Hydrologic Systems. Symp. and Workshops on the Appl. of Math. Models in Hydrology and WRS, Bratislava, 1975, p. 10.

Strupczewski, W. G., Budzianowski, R. J., and Kundzewicz, Z. W.: Optimal Control of a Storage Reservoir During the Summer Flood Season. Proc. IWRA, Vol. IV, 1975, pp. 171 – 178.

Study for the Regulation and Management of the Sava River in Yugoslavia. Final Report, Hydroprojekt – C. Lotti, Prague – Rome, 1972, p. 2200.

Sundar, A. and Butcher, W. S.: Optimal Operation of Multi-Reservoir Water Resources Systems. Proc. IWRA, Vol. IV, 1975, pp. 89 – 95.

Svanidze, G. G.: Principles of Computation of River Flow Regulation by the Monte Carlo Method. Izd. Mecniereba, Tbilisi, 1964, p. 272 (in Russian).

Svanidze, G. G. and Reznikovsky, A. S.: Stochastic Methods of Modelling for Designing Water Resource Systems. Proc. IAHR, 4, 1973, pp. 53 – 61.

Sýkora, M.: Sensitivity Analysis in Water Resources Systems Planning. Proc. IVth Conf. SI, Vol. 2, Dům techniky ČSVTS, Prague, 1977 (in Czech).

Sýkora, M.: Decision Analysis Method for Water Resources Systems. Conference: Systems of Automatic Control of WRS, Brno, Sept. 1980 (in Czech).

System Engineering Department: Introduction in Systems Engineering. Vol. I, Mining University, Ostrava, 1973, p. 192 (in Czech).

Szidarowski, F., Bogárdi, I., Duckstein, L., and David, D.: Economic Uncertainties in Water Resources Project Design. *Water Resources Research*, 4, 1976, pp. 573 – 580.

Szölösi-Nagy, A.: On the Short-Term Control of Multiobjective Reservoir Systems: A Case Study for the Kapos Basin, Hungary.

Szölösi-Nagy, A. and Wood, E. F.: Bayesian Strategies for Dynamic Water Resources Control Systems. Symp. IIASA: Recent Developments in Real-Time Forecasting and Control of Water Resources Systems, Laxenburg, 1976.

Šalomon, M.: Mathematics in Regulation and Automation. SNTL, Prague, 1957, p. 205 (in Czech).

Šembera, J.: Control System in Water Management. SZN, Prague, 1979, p. 216 (in Czech).

Šerek, M.: Computer Use for Drinking Water Supply Systems. *Vodní hospodářství*, 1, 1967, pp. 14–17 (in Czech).

Šerek, M.: Program TM-38 for Discharge Analysis in Systems of Drinking Water Mains. *Vodní hospodářství*, 12, 1967, pp. 555–558 (in Czech).

Šerek, M.: Discharge Analysis in Waterworks Networks by Matrix Algebra. *Vodohospodársky čas.*, **XVI**, 4, 1968 (in Czech).

Šerek, M.: Matrix Analysis of Water Resources Systems. Proc. WRS, Karlovy Vary, 1972, pp. 376–389.

Škabrada, J., Hýbl, J., and Adamec, S.: Computers in Economic Management. SNTL – Alfa, Prague – Bratislava, 1974, p. 280 (in Czech).

Šor, J. B.: Statistical Method of Analysis of Quality and Reliability Control. SNTL, Prague, 1965, p. 456 (in Czech).

Štichová, V.: A Network Model of Watercourses in Water Management. Thesis, VŠE, Prague, 1972, p. 41 (in Czech).

Švec, J. and Kotek, Z. *et al.*: Theory of Automatic Control. SNTL, Prague, 1969, p. 464 (in Czech).

Švec, Z. and Vitha, O.: Open Problems of Water Resources Systems Design. Proc. WRS, Hradec Králové, 1976, pp. 378–395.

Takeuchi, K. and Moreau, D. H.: Optimal Control of Multiunit Interbasin Water Resource System. *Water Resources Research*, 3, 1974, p. 407–414.

Tapiero, Ch. S.: Managerial Planning: An Optimum and Stochastic Approach Control, Vol. 1 and 2, Gordon and Breach, London, 1977.

Taylor, L. E.: A Decade of Water Resources Planning in England and Wales. Proc. IWRA, Vol. II, 1975, pp. 361–371.

Technical Committee: Water Resources Planning, Social Goals and Indicators, Methodological Development and Empirical Tests. Water Res. Lab. Utah, State Univ. Logan, 1974.

Tekel, L.: Problems of Water Power Utilization of Multipurpose Reservoirs in Water Resources Systems. Proc. WRS, Karlovy Vary, 1972, pp. 137–145 (in Slovak).

Ter-Manuelianc, A.: Dynamic Programming in Economic Practical Use. *Ek.-mat. obzor*, 3, 1966, p. 28 (in Czech).

Ter-Manuelianc, A.: Mathematical Models of Inventory Control. *Ek.-mat. obzor*, 4, 1968, pp. 1–2 (in Czech).

Ter-Manuelianc, A.: Possibilities of Use of Scientific Methods in Practical Management. Horizont, Prague, 1971, p. 44 (in Czech).

Tersine, R. J.: Materials Management and Inventory Systems. North Holland, New York – Amsterdam, 1976.

Tinbergen, J.: Introduction to Econometrics. Panst. wydaw. naukowe, Warsaw, 1957, p. 282 (in Polish).

Toebes, G. H.: Technology Transfer in Reservoir Systems Operation. Workshop IIASA/IMGW, Operation of Multiple Reservoir Systems, Poland, Jodlowy Dwor, 1979, p. 10.

Tolstov, G. P.: Fourier Series. Gostechizdat, Moscow, 1951 (in Russian).

Toman, P.: JSEP (Joint System of Electronic Processors). SPN, Prague, p. 112 (in Czech).

Trakhtengerts, E. A.: Software for Automated Control Systems. Statistika, Moscow, 1974, p. 288 (in Russian).

Trott, W. J. and Yeh, W. G.: Optimization of Multiple Reservoir System. *Journal of the Hydraulic Div.* ASCE, HY 10, 1973, pp. 1865–1884.

Urban, J. and Stránský, J.: River Flow Regulation by a Reservoir System. *Vodní hospodářství*, A, 12, 1975, pp. 294–298 (in Czech).

Ushakov, I. A. and Fishbeyn, F. I.: Bibliography on Reliability. Nadezhnost i kontrol kachestva, Moscow, 6–12, 1978, 1–3, 1979 (in Russian).

Vedula, S. and Rogers, P. P.: Multiobjective Analysis of Irrigation Planning in River Basin Development. *Water Resources Research*, **17**, 5, 1981, pp. 1304–1310.

Velev, D., Radkov, M., and Nedkova, N.: Determination and Selection of an Optimal Alternative of Water Resource Systems. Proc. WRS, Hradec Králové, 1976, pp. 396–408.

Velikanov, A. L. and Korobova, D. N.: Dynamic Programming as Applied to Substantion of a Reliable Water Supply in Integrated Water Management Systems. Proc. IWRA, Vol. V, 1975, pp. 177–182.

Velikanov, A. L. and Pryzner, V. I.: Use of Statistical Methods for Water Resources Systems Parameters Optimization. *Vodniye Resursy*, 4, 1976, pp. 58–64 (in Russian).

Vemuri, V.: Multiple-Objective Optimization in Water Resources Systems. *Water Resources Research*, 1, 1974, pp. 44–48.

Ventcel, E. S.: Probability Theory. Gos. izd. fiz. mat. lit., Moscow, 1962, p. 564 (in Russian).

Ventcel, E. S.: Elements of Dynamic Programming. Nauka, Moscow, 1964, p. 175 (in Russian).

Ventcel, E. S.: Introduction in Operations Research. Mir, Moscow, 1966, p. 287 (in Russian).

Vepřek, J.: Operations Analysis in Economic Practical Use. SNTL, Prague, 1970, p. 138 (in Czech).

Vepřek, J.: Basic Notions of Operations Research, Large-Scale Systems Research, and Control Systems. Conf.: Control Systems, Inorga, UVTA, Karlovy Vary, 1970.

Verghaeghe, R. J.: Overview Paper on Operation of the Largest Reservoir System in the Netherlands: The IJssel Lake. Workshop IIASA/IMGW, Operation of Multiple Reservoir Systems, Poland, Jodlowy Dwor, 1979, p. 10.

Veselý, J.: A Survey of Basic System Terms. Centrální databanka FSÚ, Prague, 1975, p. 40 (in Czech).

Vicens, G., Rodrigues-Iturbe, I., and Schaake, J. C.: A Bayesian Framework for the Use of Regional Information in Hydrology. *Water Resources Research*, 3, 1975, pp. 405–414.

Vicens, J. A.: Application of Statistical Methods for Operation of Systems of Water Power Plants. Gidrometizd., Moscow, 1963, p. 88 (in Russian).

Vischer, D. L. and Spreafico, M.: The Rapid Emptying of Water Reservoir System. Proc. IWRA, Vol. V, 1975, pp. 161–170.

Vitha, O.: Design of Water Resources Systems: The Role of Man and Computer. Proc. WRS, Karlovy Vary, 1972, p. 158–166 (in Czech).

Vitha, O. and Doležal, M.: Design of Water Resources Systems. Works and Studies 141, VÚV, Prague, 1975, p. 185 (in Czech).

Vitha, O., Doležal, M., and Morava, S.: Recommendations and Data for Design and Economic Evaluation of Multipurpose Water Resources Systems. SRVH, VÚV, Prague, 1972 (in Czech).

Vito-Don: Annotated Bibliography on Systems Cost Analysis. RAND, Santa Monica, 1967.

Vlček, J.: Systems Analysis. *Ekon.-mat. obzor*, 4, 1968 (in Czech).

Vlček, J.: Systems Analysis and Systems Approach. *Ekon.-mat. obzor*, 4, 1969 (in Czech).

Vlček, J.: Computers and Control (The Efficiency of their Use). SNTL, Prague, 1971, p. 132 (in Czech).

Vlček, J.: Computer System Data Processing, SNTL, Prague, 1972, p. 216 (in Czech).

Vlček, J.: Computer Hardware and Software in CMEA Countries: ČSSR. SNTL, Prague, 1975, p. 80 (in Czech).

Vlček, R. *et al.*: Analysis of Value. SNTL, Prague, 1973, p. 220 (in Czech).

Vogel, J. and Kazda, I.: Use of Computers in Hydrotechnics. *Vodní hospodářství* (Ref. No.), 1969 (in Czech).

Volf, F.: Value Analysis in Civil Engineering (Thesis). Technical University, Faculty of Civil Engineering, Prague, 1977, p. 501 (in Czech).

Volkov, G.: Birth of Science. Mladá fronta, Prague, 1975, p. 144 (in Czech).

Votruba, L.: Water Resources Systems and their Construction. Conf.: Systémové inženýrství ve stavebnictví. ČVTS, Mar. Lázně, 1974 (in Czech).

Votruba, L.: Application of Systems Theories and Methods on Problems of Water Management. Proc. WRS, Hradec Králové, 1976 (in Czech).

Votruba, L. and Broža, V.: Reservoirs Water Management. SNTL, Prague, 1966, p. 324, 2nd rev. ed. 1980, p. 443 (in Czech).

Votruba, L. and Nacházel, K.: Principles of Theory of Stochastic Processes and their Application in Water Management. Technical University, Prague, 1969, 1971, p. 183 (in Czech).

Votruba, L., Nacházel, K., and Patera, A.: Systems Engineering in Water Management. Technical University, Prague, 1974, p. 252 (in Czech).

Wagner, H. M.: Statistical Management of Inventory Systems. John Wiley, New York, 1962, p. 292.

Wagner, H. M.: Principles of Operations Research with Applications to Managerial Decisions. Prentice Hall Int. Ed., London, 1975, p. 1039.

Walter, J.: Stochastic Models in Economics. SNTL – ALFA, Prague – Bratislava, 1970, p. 195 (in Czech).

Walter, J. *et al.*: Operation Research. SNTL – ALFA, Prague – Bratislava, 1973, p. 190 (in Czech).

Walter, J. and Lauber, J.: Simulation Models of Economic Processes. SNTL – ALFA, Prague – Bratislava, 1975, p. 213 (in Czech).

Water for Human Needs. Proc. of the Second World Congress on Water Resources. IWRA, Vol. I – V, New Delhi, Summaries of Congress Papers, 1975.

Water Resources Systems Determination in SSR. Vodorozvoj, Bratislava, 1974 (in Slovak).

Waziruddin, S. and Altinbilek, H. D.: Mathematical Modelling and Optimization of the Ceyhan Aslantas Project in Turkey. Proc. IAHR, 4, 1973, pp. 23 – 32.

Weis, L.: Statistical Decision Theory. McGraw-Hill, New York, 1961, p. 195.

Weissenberger, I.: Analysis of Information Systems. Svoboda, Prague, 1972, p. 276 (in Czech).

White, D. J.: Decision Theory. Aldine Publ. Co., Chicago, 1969, p. 194.

Whithin, T. M.: The Theory of Inventory Management. Princeton University Press, Princeton, New Yersey, 1957, p. 302.

Wiener, N.: Cybernetics: Communication and Control in the Animal and the Machine. M.I.T., Cambridge, 1948.

Wilkinson, J. C.: River Dee Research Program. 2. A Long-Term Control Strategy for a Multipurpose Reservoir. *Water Resources Research*, 4, 1972, pp. 904 – 910.

Wilson, I. G. and Wilson, M. E.: Information. Computers and System Design. John Wiley, New York, 1965.

Won, T. S.: The Objects and Syllabi in Water Resources Education. Proc. IWRA, Vol. IV, 1975, pp. 419 – 422.

Wunderlich, W. O.: Overview of Water Management Methods for the TVA – Operated Reservoir System. Workshop IIASA/IMGW, Operation of Multiple Reservoir Systems, Poland, Jodlowy Dwor, 1979, p. 23.

Yakowitz, S.: Dynamic Programming Applications in Water Resources. *Water Resources Research*, **18**, 4, 1982, pp. 673 – 696.

Yakowitz, W. and Rutherford, B.: Contribution to Discrete-Time Differential Dynamic Programming, Submitted to Optimal Control, Water Resources Systems, 1981.

Yeh, W. W. G. and Becker, L.: Optimization of Real-Time Daily Operation of a Multiple Reservoir System. Proc. WRS, Hradec Králové, 1976, pp. 409 – 423.

Yermolyev, Yu. M. and Melnik, J. M.: Extremal Problems on Graphs. Naukova Dumka, Kiev, 1968, p. 176 (in Russian).

Yevjevich, V.: Stochastic Processes in Hydrology. Water Resources Publications, Fort Collins, 1976, p. 276.

Yevjevich, V.: Overview of Research on Operation of Multiple Reservoir Systems (Colorado State University Activities), Workshops IIASA/IMGW, Operation of Multiple Reservoir System, Poland, Jodlowy Dwor, 1979, p. 13.

Young, G. K., Moseley, J. C., and Evenson, D. E.: Time Sequencing Element Construction in a Multi-Reservoir System. *Water Resour. Bull.*, 4, 1970, pp. 528 – 541.

Zahradník, J.: Linear Programming in Economic Management. SNTL, Prague, 1969, p. 233 (in Czech).

Zavřel, F.: Principles of Automatic Control Theory. ALFA, Bratislava, 1970, p. 230 (in Czech).

Zeman, V.: On the Use of Economic Objective Function. Proc. WRS, Karlovy Vary, 1972, p. 189–197 (in Czech).

Zeman, V.: Computer Programming in Analysis of Water Resources Systems by Simulation Technique. *Vodní hospodářství*, A, 7, 1974, pp. 181–183 (in Czech).

Zeman, V.: Water Resources Systems Planning Tasks. Water Research Institute, Prague, 1986, p. 96 (in Czech).

Zeman, V. and Sýkora, M.: Interrogative Computer System for Water Resources Systems. Conf.: ASŘ vodního hospodářství, Brno, 1980 (in Czech).

Zítek, F.: Lost Time (Elements of Queueing Theory). Academia, Prague, 1969, p. 180 (in Czech).

Zubek, L.: Conditions of Water Management Development in the Odra River Basin. *Vodní hospodářství*, A, 5, 1974, pp. 119–121 (in Czech).

INDEX